DISASTER BY DESIGN: THE ARAL SEA AND ITS LESSONS FOR SUSTAINABILITY

RESEARCH IN SOCIAL PROBLEMS AND PUBLIC POLICY

Series Editor: Ted I. K. Youn

Recent Volumes:

Volume 10:	The Environmental State Under Pressure – Edited by Arthur P.J. Mol and Frederick H. Buttel, 2002
Volume 11:	Terrorism and Disaster: New Threats, New Ideas – Edited by Lee Clarke, 2003
Volume 12:	The Organizational Response to Persons with Mental Illness Involved with the Criminal Justice System – Edited by Stephanie W. Hartwell, 2005
Volume 13:	Long-Term Management of Contaminated Sites – Edited by Thomas M. Leschine, 2006
Volume 14:	Cultures of Contamination: Legacies of Pollution in Russia and the U.S. – Edited by Michael R. Edelstein, Maria Tysiachniouk and Lyudmila V. Smirnova, 2006
Volume 15:	Equality and the Environment – Edited by Robert Wilkinson and William R. Freudenburg, 2008
Volume 16:	Integrating the Sciences and Society: Challenges, Practices, and Potentials – Edited by Harriet Hartman, 2008
Volume 17:	New Approaches to Social Problems Treatment – Edited by Mark Peyrot and Stacy Lee Burns, 2010
Volume 18:	Environment and Social Justice: An International Perspective – Edited by Dorceta E. Taylor, 2010
Volume 19:	Government Secrecy – Edited by Susan Maret, 2011

RESEARCH IN SOCIAL PROBLEMS AND PUBLIC POLICY
VOLUME 20

DISASTER BY DESIGN: THE ARAL SEA AND ITS LESSONS FOR SUSTAINABILITY

EDITED BY

MICHAEL R. EDELSTEIN
Ramapo College of New Jersey, USA

ASTRID CERNY
Ramapo College of New Jersey, USA

ABROR GADAEV
Samarkand State Architectural and Civil Engineering Institute, Uzbekistan

United Kingdom – North America – Japan
India – Malaysia – China

Emerald Group Publishing Limited
Howard House, Wagon Lane, Bingley BD16 1WA, UK

First edition 2012

British Library Cataloguing in Publication Data
A catalogue record for this book is available from the British Library

ISBN: 978-1-78190-375-9
ISSN: 0196-1152 (Series)

CONTENTS

LIST OF CONTRIBUTORS

Bakhtiyor Abdullayev	Department of Soil Science, Samarkand Agricultural Institute, Uzbekistan
Mirza Nomman Ahmed	Institute of Agricultural Policy and Market Research, Justus Liebig University, Giessen, Germany
Farhod Ahrorov	Agricultural Economics and Management Department, Samarkand Agricultural Institute, Uzbekistan
Flavia Alaya	Professor Emerita of Cultural History, Ramapo College of New Jersey, Mahwah, NJ, USA
Laylo L. Almatova	Uzbekistan Health and Medical Statistics Institute, Uzbekistan
Eshkuvat U. Arzikulov	Department of Physics, Samarkand State University, Uzbekistan
Rakhman Ulashevich Arzikulov	Medical doctor, Samarkand, Uzbekistan
Murad Askarov	Representative of the Republic of Uzbekistan to the United Nations in New York, NY, USA
Gulmira Boboeva	Water Supply and Water Resources Protection and Rational Use Department, Samarkand State Architectural and Civil Engineering Institute, Uzbekistan
Astrid Cerny	Environmental Studies Program, Ramapo College of New Jersey, Mahwah, NJ, USA
Utkur Djanibekov	Center for Development Research (ZEF), Bonn University, Germany

Gavkhar Salaevna Durdieva	Mamun Academy, Khiva, Uzbekistan
Michael R. Edelstein	Environmental Studies Program (BA) and Sustainability Studies Program (MA), Ramapo College of New Jersey, Mahwah, NJ, USA
Abror Gadaev	Water Supply and Water Resources Protection and Rational Use Department, Samarkand State Architectural and Civil Engineering Institute, Uzbekistan
Ibragim Ganiev	Department of Agricultural Economics and Management, Samarkand Agricultural University, Samarkand, Uzbekistan
Debjani Ghatak	Institute of Marine and Coastal Sciences, Rutgers University, New Brunswick, NJ, USA
Michael H. Glantz	Consortium for Capacity Building (CCB), Institute for Arctic and Alpine Research (INSTAAR), University of Colorado-Boulder, Boulder, CO, USA
Lola Gulyamova	Department of Geography, National University of Uzbekistan, Uzbekistan
Shavkat Hasanov	Department of Agricultural Economics and Management, Samarkand Agricultural University, Uzbekistan
Howard Horowitz	Environmental Studies Program, Ramapo College of New Jersey, Mahwah, NJ, USA
Andrew Jones	Independent consultant, Baja California Sur, Mexico
Yusup S. Kamalov	Union for Defense of the Aral Sea and Amudarya (UDASA) from Nukus, Karakalpakstan, Uzbekistan

Ashraf B. Khodjaev	Permanent Mission of the Republic of Uzbekistan to the United Nations in New York, New York, NY, USA
Ramona Lall	New York City Department of Health, New York, NY, USA
John P.A. Lamers	Center for Development Research (ZEF), Bonn University, Germany
William J. Makofske	Professor Emeritus of Physics, Ramapo College of New Jersey, Mahwah, NJ, USA
James R. Miller	Department of Marine and Coastal Sciences, Rutgers University, New Brunswick, NJ, USA
Aziz Murtazaev	Samsung Electronics, Republic of Korea, Suwon, Korea
Olim Murtazaev	Agricultural Economics Department, Samarkand Agricultural University, Uzbekistan
Ilkhomjon Niyazov	Agricultural Economics Department, Samarkand Agricultural Institute, Uzbekistan
Kudrat Nurmetov	Slovak University of Agriculture, Nitra, Slovakia
Imtiaz Rangwala	Department of Marine and Coastal Sciences, Rutgers University, New Brunswick, NJ, USA
Leslie Raucher	New Jersey Higher Education Partnership for Sustainability, Ramapo College of New Jersey, Mahwah, NJ, USA
Inna Rudenko	ZEF/UNESCO Khorezm project, Urgench, Uzbekistan
Saodat Safarova	Samarkand Medicine Institute, Uzbekistan

Qamariddin N. Srojev	Department of Physics, Samarkand State University, Uzbekistan
Mukhtor Turaev	Ecology department, Bukhara State University, Uzbekistan
Michael Wilson	Environmental Studies Program, Ramapo College of New Jersey, Mahwah, NJ, USA
Zikrilla Yasakov	Water Supply and Water Resources Management Department, Samarkand State Architectural and Civil Engineering Institute, Uzbekistan

ACKNOWLEDGMENTS

This volume and the project it grew out of occurred with the help of many friends. We begin by acknowledging the support of the Trust for Mutual Understanding who graciously underwrote the project "Exchanging Lessons of the Aral Sea Disaster." The Trust for Mutual Understanding has been a wonderful supporter of exchange projects between the United States and the former Soviet Union states and has come through again for this project. Without their diligent pursuit of this worthwhile mission, this volume would be impossible.

Ramapo College of New Jersey and its Institute for Environmental Studies have supported this project since its inception, sponsoring Abror Gadaev as a visiting Fulbright Scholar and allowing the exchange project to be proposed and administered. We benefited from their sponsorship for the October 2011 full day conference on the Aral Sea. Our appreciation to Dr. Peter Mercer, College President, for his encouragement.

Dr. Ronald Kase, Associate Provost and friend, has helped in every way to inspire and provide administrative support. Without his help, and the help of Claudia Esker, and the staff of the Ramapo College Foundation, this project would have been impossible. It is a pleasure to work in the supportive environment you provide.

Ted Youn, Editor of the Journal of Research in Social Problems and Public Policy, has been an excellent colleague in the past and remains so, affording an important public and academic voice to topics such as this.

Emma Bruun, Editor at Emerald Publishers, has been helpful, supportive, and very patient in bringing this project to publication. We are grateful, Emma.

To our American team, both those who participated in the Uzbek trip and those who assisted on their reverse visit and our Aral Sea conference. Thank you Ramona Lall, Mike Wilson, Flavia Alaya, Leslie Raucher, Bill Makofske, Jim Miller, Sister Miriam MacGillis, and Trent Schroyer. A special thanks to Howie Horowitz for preparing a brilliant field trip to study the American analogs he writes about here. And to Mickey Glantz, Andrew Jones, and Tobias Siegfried for their Skype participation at the conference.

For our Bridgeton field trip, our thanks go to Flavia Alaya and her husband Sandy for hosting us in their lovely, historic Bridgeton home and organizing scenic tours and meaningful exchanges. Our thanks to Michael De Luca, Salvatore Mangiafio, and Lisa Calvo of Rutgers Cooperative Extension for meeting with us and sharing with the Uzbeks how agricultural extension provides benefits over the long-term in rural areas.

Particular thanks go also to Mike Wilson and his wife Pam who offered to host us for a Saturday night party with music and food at their home, prepared it for days and then we had to cancel because of the unprecedented snowstorm on October 29. Not to be deterred, Pam packed up her pot roast and scalloped potatoes and put out a delicious farewell meal for our Uzbek guests the next day at the college residence before we left for the airport.

A special thank you to Mickey Glantz, whose work first inspired us and who then personally mentored and guided with generosity of time and ideas.

A very warm thanks or "kata rakhmat" to His Excellency, Ambassador Murad Askarov for supporting and contributing to this project and for openness to discussing projects that address the problems affecting the Aral region. And to Deputy Secretary Ashraf Khodjaev for your ongoing assistance.

In Uzbekistan: In Samarkand, to the administration, faculty and students at Samarkand Architecture and Engineering Institute, for their support and hospitality during our stays in Samarkand and to the Uzbek team – Eshkuvat Arzikulov, Sobir Boboev, Erkin Isakov, Farhod Ahrorov, Shavkat Hasanov, Shaymon Khushmurodov, Ilkhomjon Niyazov, Shovqiddin Qarshiev, their friends and family, and to our gracious hosts along our travels, and, of course, to Jamal, our inexhaustible driver. Thank you to Dr. Rakhmon Arzikulov for attempting to heal us with acupuncture needles and cotton oil, as well as educate us. A special thanks to Mansur Abduraimov, Chair of the NGO Zarafshan River Protection for your informative support and friendship.

In Khiva: To Shokat Kadirov, Director, Mamun Academy, Khiva for hosting and joining the two-day seminar there. And with particular gratitude to Gavhar Durdieva for her unforgettable tour of Ichan Kala and for being such a gracious host, with her husband Odilbek Zargarov, in town and at the relaxing and lovely lake restaurant. Abdullah, Eshkuvat's classmate, graciously invited us for dinner at his house, providing important cultural insights on Central Asian hospitality.

In Nukus: To Ubbiniyaz A. Ashirbekov, Director, Executive Committee, Nukus Branch, International Fund for Saving the Aral Sea, for taking extensive time to describe the work of the agency. To the Minister of Construction of Karakalpakstan, Sultan Atabaev, for being a gracious host, arranging our tour of Muynak and introducing us to the Amu Darya delta.

And Marinika Babanazarova, Director, Karakalpak Museum of Arts, for opening your museum's doors to us for extended viewing time of the magnificent Savitsky collection.

In the Syr Darya region, to Oblakul Kasimov, Senior Scientist, architecture and construction; Gulistan State University, Syr Darya Region. Also his colleagues Ermamat Kurbanov and Khabibulla Abdukarimov, for their important tour of changes in the Syr Darya.

In Bukhara, to Mukhtar Turaev, Yakub Kholov, Sadriddin Bakaev who so well briefed us on salinization issues and wildlife impacts in a lovely garden restaurant. Bukhara master puppet maker Islandar Khakimov's creations Abdullah and Baba Yaga joined the expedition and entertained the teams as they traveled by bus over long stretches in the desert, at rest stops, even in full restaurants, allowing all an opportunity to explore our relationships through humor.

Boris Smirnov and Ludmila Smirnova for creating the website for the project, available at http://www.aralseadisaster.org.

DEDICATIONS

Dedicated to the memory of William Freudenberg, friend and colleague and founding coeditor of this journal. We think you would have enjoyed this issue, Bill. Here's thinking about you.

To the *aka,* or elder brothers, we have gotten to know who fought hard to avoid this disaster and to spur recovery, with special recognition of the lifetime work of Yusup Kamalov.

Dedicated also to Aziza Kaipbergenova for translating and editing the work of her grandfather, the late Tulepbergen Kaipbergenov and to the entire Kamalot Youth Movement in Uzbekistan. This book is about the difficult legacy faced by the youth of your region, and indeed the world, for the errors of sustainability made before your time. It is an unfair inheritance which your grandfather and others of his generation fought hard to avert. Your energy and activism is the hope.

Last but not least, we thank our families for their support, patience and forbearance. From Mike to his wife Ludmila, from Astrid to her mother Christel, and from Abror to his wife Farigat, and children Ilkhom, Anwar, Roxana and Akbar.

In Goshen, N.Y., Michael R. Edelstein
In Ramsey, N.J., Astrid Cerny
In Samarkand, Uz., Abror Gadaev

MAPS OF UZBEKISTAN AND THE GREATER ARAL SEA REGION

Map 1. Map of Uzbekistan and Surrounding Area.

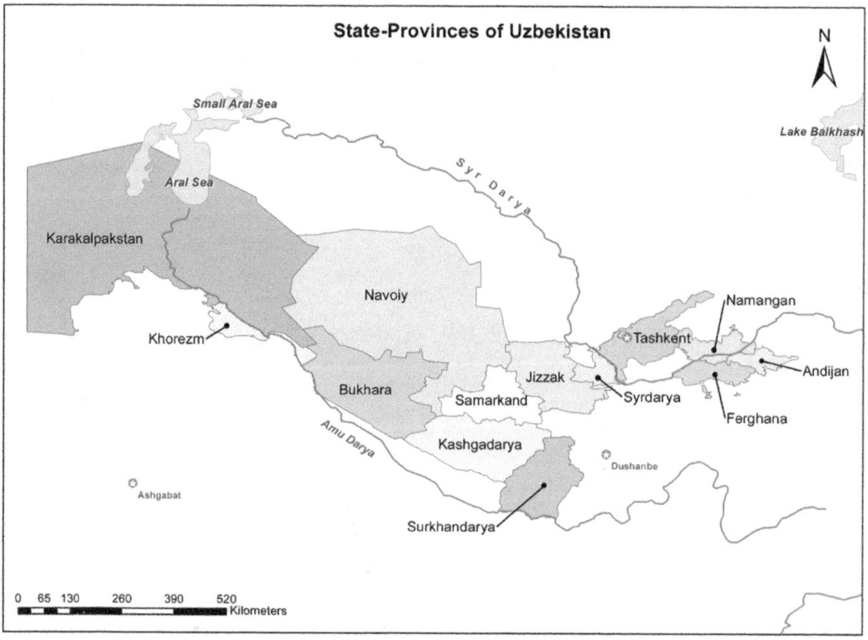

Map 2. Map of Uzbekistan and Fourteen Administrative Regions.

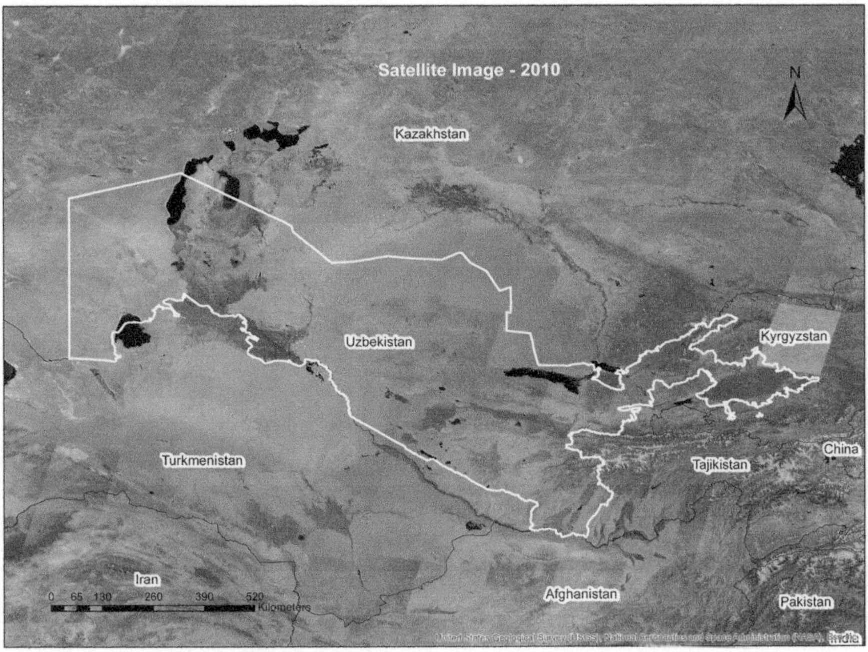

Map 3. Satellite Image of Uzbekistan and Surrounding Area (2010).

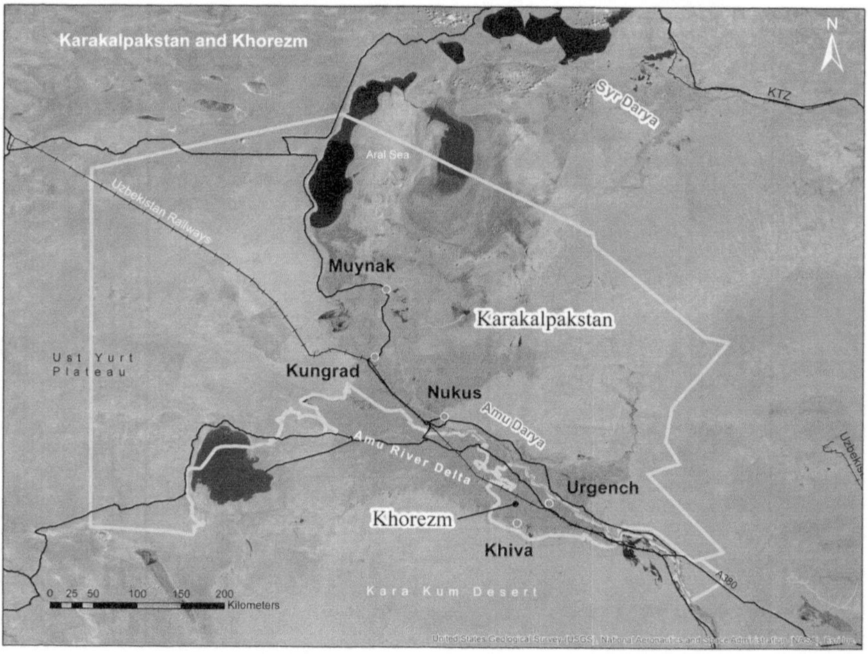

Map 4. Satellite Image of Karakalpakstan and Khorezm (2010).

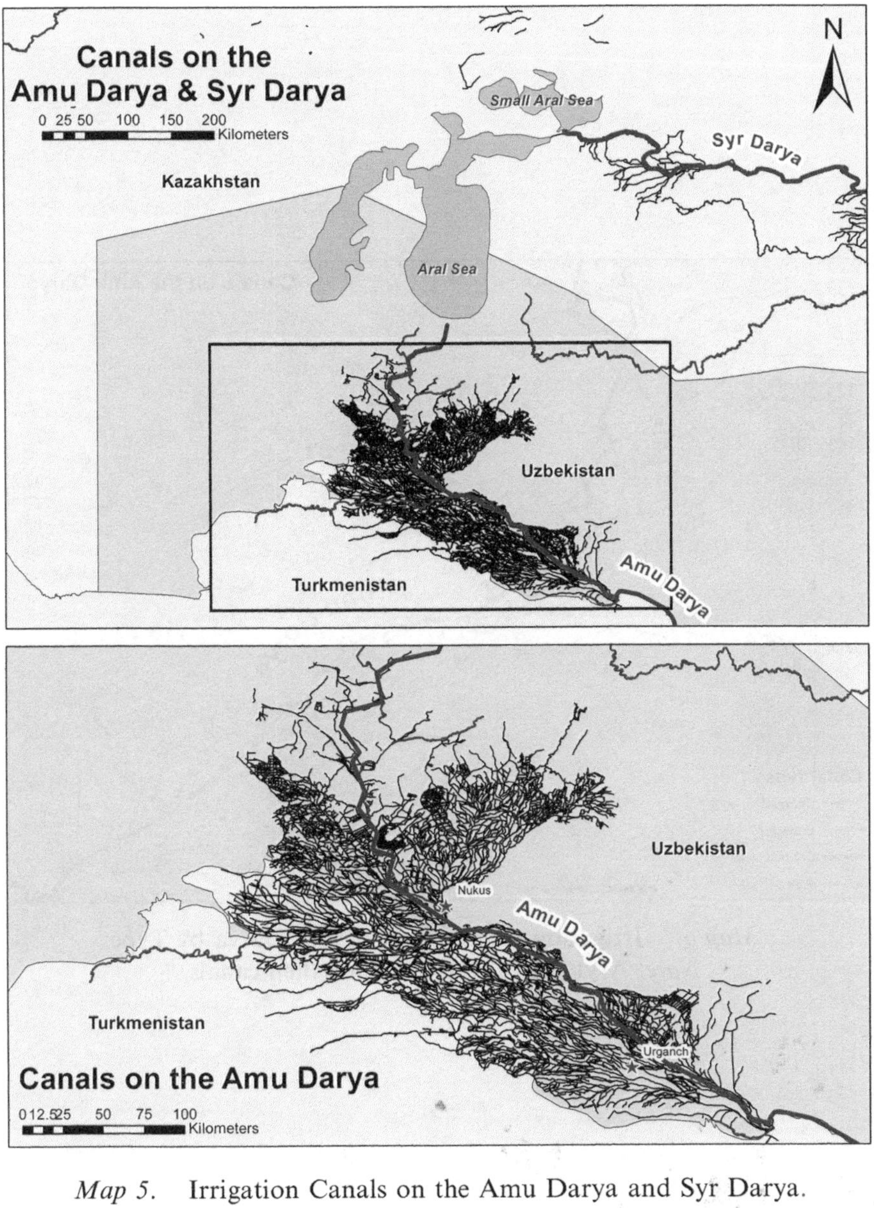

Map 5. Irrigation Canals on the Amu Darya and Syr Darya.

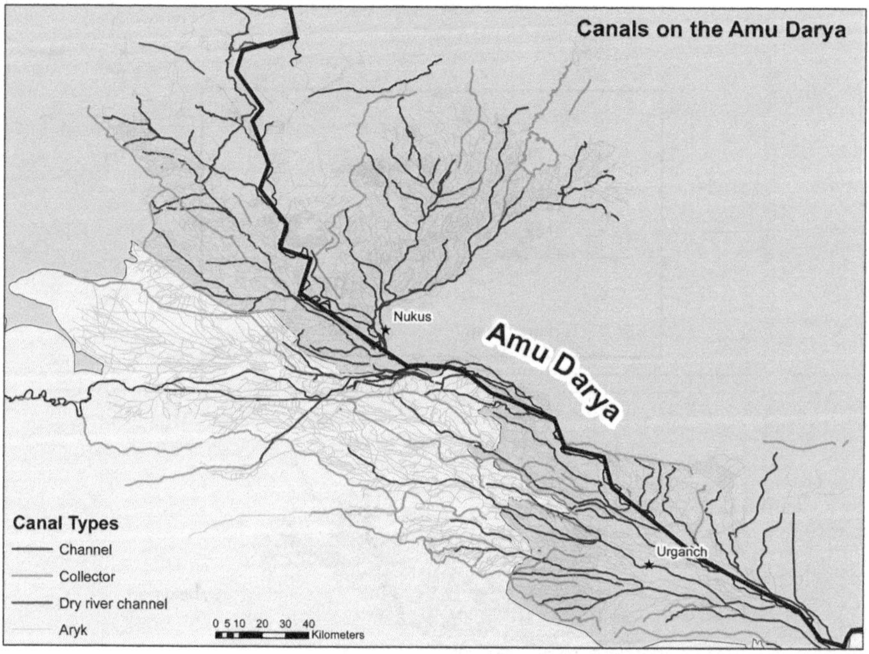

Map 6. Irrigation Canals on the Amu Darya by Type.
Note: Aryk refers to unlined, earthen canals.

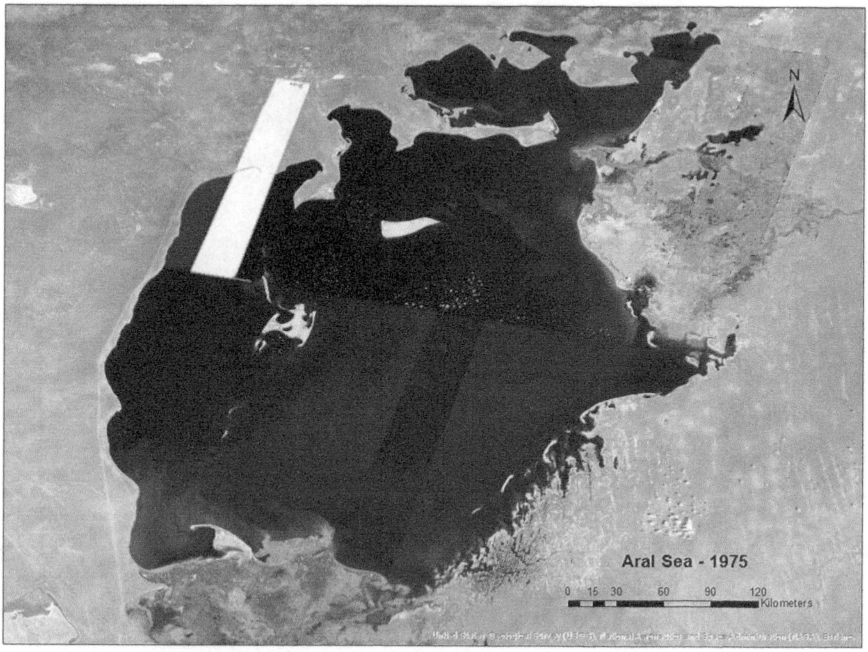

Map 7. The shrinking of the Aral Sea (time lapse series 1975–2010).

Map 8. Satellite Image 1990.

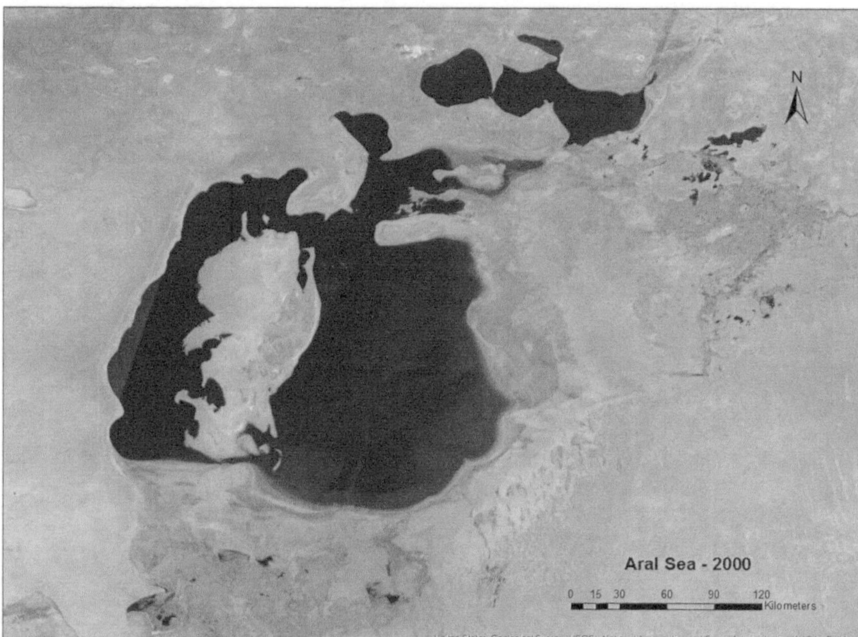

Map 9. Satellite Image 2000.

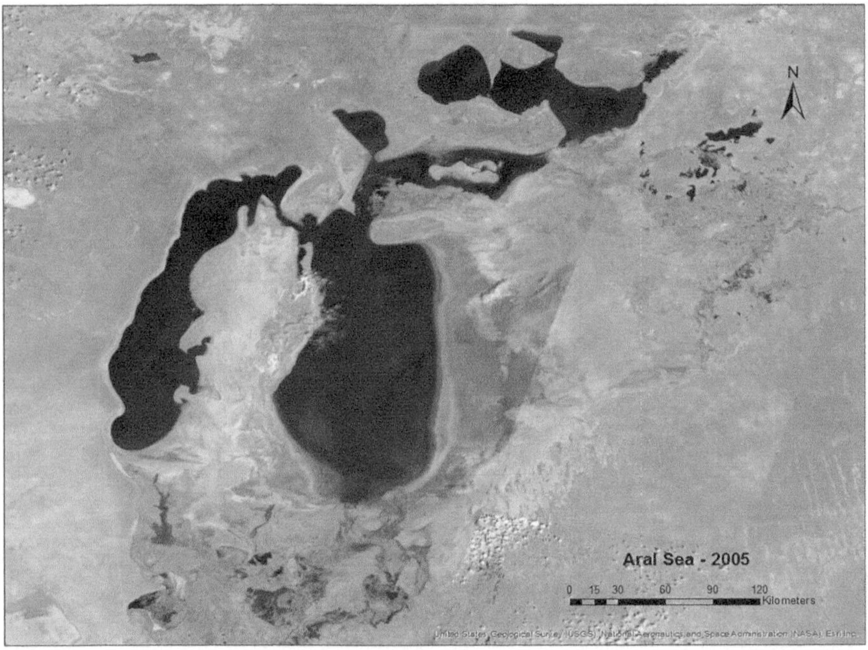

Map 10. Satellite Image 2005.

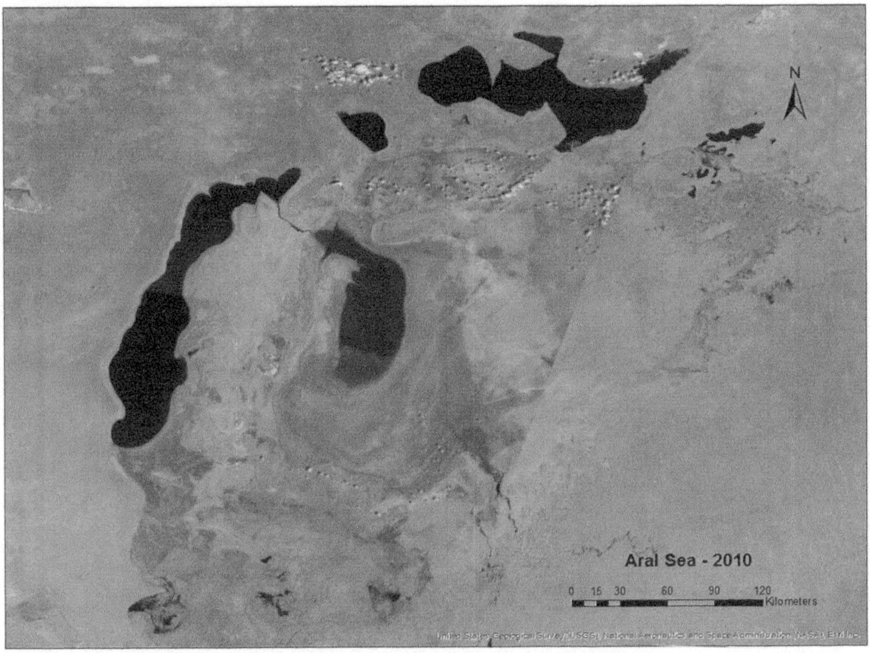

Map 11. Satellite Image 2010.

LIST OF TABLES AND FIGURES

TABLES

FIGURES

LIST OF MAPS

Leslie Raucher

ACKNOWLEGMENTS

We would like to thank the Global Land Cover Facility for the library of images that made the creation of our maps possible. More about GLCF can be found at: http://www.landcover.org and http://glcf.umiacs.umd.edu/data/gls/.

REFERENCES

United States Geological Survey (USGS), National Aeronautics and Space Administration (NASA), Esri Inc. 1975, Global Land Survey, 1975, Landsat MSS, 30m scene GLS1975_Enhanced, USGS, Sioux Falls, South Dakota.

United States Geological Survey (USGS), National Aeronautics and Space Administration (NASA), Esri Inc. 1990, Global Land Survey, 1990, Landsat TM, 30m scene GLS1990_Enhanced, USGS, Sioux Falls, South Dakota.

United States Geological Survey (USGS), National Aeronautics and Space Administration (NASA), Esri Inc. 2000, Global Land Survey, 2000, Landsat ETM+, 30m scene GLS2000_Enhanced, USGS, Sioux Falls, South Dakota.

United States Geological Survey (USGS), National Aeronautics and Space Administration (NASA), Esri Inc. 2005, Global Land Survey, 2005, Landsat ETM+, 30m scene GLS2005_Enhanced, USGS, Sioux Falls, South Dakota.

United States Geological Survey (USGS), National Aeronautics and Space Administration (NASA), Esri Inc. 2010, Global Land Survey, 2010, Landsat ETM+, 30m scene GLS2010_Enhanced, USGS, Sioux Falls, South Dakota.

PREFACE

"Trouble in the Land of Oz." That is what started it all.

The kind of title whose wizardry can transport a team of Americans to a far off land in quest of insight into what many call the world's worst single environmental disaster. An invitation to adventure and learning.

When I invited a visiting Uzbek scholar to give a talk at Ramapo College six years ago, that was the title of his talk. The trouble in the land of Oz (i.e., Uzbekistan) made reference to what was, is, and will be the Aral Sea, its rapid and startling demise and the continuing consequences for the people and environment of Central Asia. Abror Gadaev not only animated the dead aspects of the Aral issue, but he brought to life the living issues, placing them in a complex context of place and Uzbek culture, history and future.

I have a habit of being hooked by the right mix of problems and potentials to travel to far corners of the world and, so, found myself writing a successful Fulbright visiting scholars grant for Abror to spent a year in residence at Ramapo College of New Jersey doing research, teaching and always promoting the culture of his nation. Abror is a great salesman, his charm a combination of charisma, assertiveness, deference, perseverance, and clarity of purpose.

As the year moved toward the end, I was itching to see firsthand the troubled lands we had spent hours discussing. I turned to the Trust for Mutual Understanding, a foundation dedicated to building mutual learning relationships between United States citizens and residents of the Former Soviet Union. I have benefited from their support for a string of wonderful exchange projects that provided significant opportunities to build relationships that are authentic and lasting. The new proposal I sent to the Trust based upon Abror's and my converging interests was entitled "Exchanging Lessons of the Aral Sea Disaster."

The proposal followed the Trust formula of mutual exchange between two teams roughly matched for interest and expertise. In Uzbekistan, we would travel from Tashkent to the Aral region, using Samarkand as our base. In the United States we would focus on creating events that explore applicable solutions and encourage inspiration to help the situation. Our exchanges culminated in a major conference on the Aral demise.

The funding was granted, and I found myself in May 2011 in a travel lounge on Aeroflot's mandatory 10-hour stopover at Moscow's Sheremetyevo airport on the way to Tashkent. With me was a handpicked team that included two of my favorite and most accomplished former students, a beloved former colleague, an amazingly talented new colleague and one of my most cherished current graduate students. We used the long layover to become a tight working group.

When we disembarked the next morning, it was difficult to tell which was brighter, the rising sun or the smile on Abror Gadaev's face. We commenced our trip at the U.S. Embassy, where we were given a rather grim briefing about regional problems and an even grimmer warning about the hazards we might encounter in our travels. We dined that night with my former student Halil Jalolov, and had a wonderful lesson in Uzbek hospitality. A short time later, we were on the road by speeding taxi, with a stop for a beautiful lunch served on the banks of the great Syr Darya River, our hosts, professors from the Syr Darya region. The only thing missing was the Syr Darya itself, which we learned was dramatically lower than its historic levels. In fact, we stood on a dock that used to abut the river and could barely see it on the horizon.

After a stop at Tamerlane's gate, on the mountain divide opening to the Zerafshan valley and Samarkand, we arrived in Samarkand, where over the next several days, we met the Uzbek team and friends, held a small conference on the Aral Sea where we made our first exchange of papers, and were introduced to one of the oldest and most architecturally rich cities in the world. A few days later, both American and Uzbek teams clambered onto the small bus that would be our home and transportation for the next weeks. Off we headed in the hands of our able and affable driver, Jamal. The road to Muynak would take us over some 3,000 kilometers, through Navoiy to Bukhara, then into the Aral region to Khiva in Khorezm, then to Nukus, capital of the Autonomous Republic of Karakalpakstan, with a final hop of four hours to Muynak on the banks of the Aral desert, formally the Aral Sea.

The trip proved a string of captivating experiences, incredible architecture, vast deserts, rare stopping points and a treacherous often one lane road covered over with blowing sand. In Navoiy, we splashed our hands in water inside an ancient *sardoba*, a water storage building designed for deserts to limit evaporation, keep the water cool and keep out animals. We were treated to sumptuous feasts by a series of gracious hosts who invited us into their homes, sometimes with little notice. In Bukhara we lunched with local professors who taught us about the salinization problem, played with charming puppets and admired one of the most beautiful cities on earth.

In the Kizil Kum desert to the west of Bukhara, Astrid, my American colleague, spied a yurt on the horizon. Two children living in the yurt spied us at the same time and waved us to a stop just in time for Astrid to not have to leap from a moving vehicle. The Turkmen woman in the yurt talked to us, as we sat around in a circle on the softly carpeted floor, vibrant colors everywhere and rug hung walls moderating the heat. Her story told us that we were nearing our destination. A nomad whose husband tended sheep, they now found diminished access to water for their flock, forcing their range to progressively shrink to the point that it was barely sufficient to sustain their livelihood. Descendants of people who had earned that same livelihood as shepherds in the desert for thousands of years, they could now see before them an end to their way of life.

Finally we encountered the mighty Amu Darya, or Oxus to the ancient Greeks, only to discover that, as the Syr Darya had already demonstrated, water levels had plunged to the shallow. Everywhere irrigation ditches withdrew water like many straws from a single drink. Scattered farmers toiled in fields in the hot sun or pulled carts or carried heavy loads. We saw no mechanized farming.

Along the Turkmenistan border by the shore of the Amu Darya we passed the tense Uzbek border outpost just as the bus began to stall. Jamal labored mightily to make it far along the stretch of flat, dusty road and out of sight of the border before the shuddering bus stuttered to a halt. We piled out into the heat and gave up our canteens of water to the thirsty radiator. Farmers from a distant field drove up and offered their assistance, as if one more person could fit in the old Soviet era car into which an extended family had been crammed. They stayed with us until the bus had cooled enough for Jamal to get it going. Our reward was a very comfortable hotel next to a park and just steps away from the Ichan Kala, the old city of Khiva. We were tired after a Ten hour drive through the desert and what luckily turned out to be our only bus breakdown.

We spent two days in Khiva. Our first stop was the Mamun Academy, in many ways the home of ancient and modern science, exchanging paper presentations with our talented Uzbek colleagues. Here we learned of the European-funded work of John Lamers and colleagues at Urgench, who had begun to address the agricultural crisis using some good, innovative ideas. We met Gavhar Durdieva who took us on a lengthy tour of the old walled inner city of Khiva, the Ichan Kala, and a living museum of the Silk Road era. Deep inside the city, I was introduced to an elderly man who has headed the archeological restoration of the city for his entire adult life and continues to live in the midst of his life's work. Govhar's husband took us into the house where she was born inside the walled city. The current residents, not related to her, responded as we came to expect from the people of Uzbekistan, by offering fresh naan bread and tea to greet unexpected and foreign visitors with warmth. Outside in the streets, children played soccer up against ancient walls, a hot tandoor oven spit out huge quantities of naan and a pair of musicians whom we met on their way to a performance pulled out their instruments and sang a string of songs to which we could not resist dancing. As Govhar demonstrated, this rich ancient world was in danger of rapid decline because of salt intrusion and other problems requiring active maintenance and repair beyond the means of the caretakers. In Khiva, Eshkuvat's old college classmate, Abdullah, adopted us and we dined with his family at his home in a traditional Uzbek feast of fruits, meats, breads and other delights. Govhar and her husband Ulugbek similarly hosted a dinner at a traditional lakeside fish restaurant where people go to escape the summer heat out on docks built right onto the water.

Two days later, we were off to Nukus. Only a few hours north, as one leaves Khorzm and enters Karakalpakstan, we crossed a cultural divide of language and people as we traveled deeper into the Amu darya delta. We were not exactly prepared to stay at a dilapidated edifice, a classic Soviet era luxury hotel long past its prime, although perhaps the prime had been no better. The overall service experience allowed the younger members of our team a priceless window into what the Soviet past might have been like.

Our conversations in Nukus were no less rewarding than everywhere else in Uzbekistan however, and our learning expanded exponentially. We had a late lunch with the Minister of Construction, Sultan Atabaev, and learned that oil and gas exploration on the floor of the former Aral Sea was the awaited salvation for the region's collapsed economy. It very much changed our understanding of what we thought was going on at the Aral Sea. We gained further insight the same day by speaking to Ubbiniyaz A. Ashirbekov, Director of the Executive Committee, at the Nukus Branch of the International Fund for Saving the Aral Sea. In one day we had spoken to the proponents of the push and pull factors for the future of the Aral Sea.

A remote capital, Nukus is world renowned for its world famous Karakalpakstan Museum of Art which houses the Savitsky collection. We pulled up just at closing time and Marinika Babanazarova, the director, generously offered to remain open for us to make a tour. We then dined at Sultan Atabaev's home, where, as the honored guest, I was graciously offered the head of the lamb slaughtered for the feast and invited to eat the delectable eyeballs. As a total vegetarian and animal rights person, I found myself in one of those interesting cross-cultural quandaries. Fortunately, Abror negotiated a way around, and we proceeded with symbolic gestures but no body parts consumed.

Sultan Aka arranged for our tour of Muynak, and the next morning we found ourselves in the epicenter of the Aral tragedy. Desolate streets, desolate looking people, shuttered fish canning factory and, of course, the rusting ships lined up like tombs on the barren desert sea floor. In the small museum erected inside the former Soviet Palace of Culture, piled relics and photos of a world that no longer existed attested to the fact that there was nowhere but in the past for the population to live. There was certainly no present and perhaps no future. We dined in the only restaurant in town, one small room off of a Korean convenience store filled with strutting chickens and mewing kittens. The only menu item was egg omelet, for which the weary travelers were grateful.

On the way back from Muynak, we paused along the Amu Darya to dine with Sultan Aka one last time at his summer cabana on the river. Here as well, it was evident that the river was well below half its normal depth and width. Our seats would normally have had water lapping just below the floor. Instead, it was like being in a tree house.

We made our way home to Samarkand, where we rested, then spent a day traveling south into the Hissar-Alay Outliers of the Tian Shan Mountains just north of Afghanistan. A second day was spent seeing major World Heritage sites and the huge Siyob food bazaar in Samarkand. And a third day was devoted to a remarkable full day conference held at Hotel Kamila to share our observations with our hosts and Uzbek colleagues. From Samarkand, the speedy taxis returned us through the darkness of night to Tashkent for our early morning flight home.

Our search for the Aral Sea had yielded the Aral desert instead. And our attempt to understand the Aral Disaster had given way to an appreciation of multiple interlocked simultaneous disasters. If the Aral Sea's disappearance alone was recognized as the world's worst ecological catastrophe, then the complex of issues we found was indeed overwhelming.

The next October, a small Uzbek team came to New Jersey. We emphasized touring sites that suggested alternative approaches that might be employed in Uzbekistan, meeting with the cooperative extension, visiting a variety of sustainable farms and landscapes, looking at historical preservation projects and holding a wonderful full day conference on the Aral Sea disaster at Ramapo College. The Uzbek Ambassador to the UN, his Excellency Murad Askarov, greeted the team and took us into the UN during one of the more interesting votes of the year, for the non-standing membership in the Security Council. We all appreciated the special attention.

One of the main tasks of the October visit was to plan this publication. I was fortunate to be able to team with Astrid Cerny, an extremely talented geographer in my program, and with Abror Gadaev, to create this volume. And, as a good indication of the depth of the relationships we developed, we were able to involve most of the people who participated in both legs of our exchange in this venture. They represent diverse interests and unique voices. But together, I think you will agree there is a unique harmony. And in the background, an Uzbek *rubab* playing the irresistible rhythm of Central Asia, one that suggests optimism for transforming the horrible Aral Sea Disaster into a sustainable and happy future for the people affected. That transformation is the goal of this volume.

In Goshen, N.Y.
Michael R. Edelstein
Editor

SECTION ONE
THE MULTIPLE DISASTERS OF THE ARAL SEA

If one were revisiting the list of the Seven Wonders of the World, Uzbekistan and the Silk Road region would offer many candidates. The Registan in Samarkand, the central square in Bukhara, the Ichan Kala in Khiva, all offer architectural marvel after marvel. Our task, sadly, is to consider wonders of another kind. Any list of the worst disasters of the world would also rank Uzbekistan at the top, making clear reference to the massive transformation of western Uzbekistan (and Kazakhstan) from home to the fourth largest inland sea to host of a barren desert. All of the waterborne life force of the Aral Sea has been swept from the region and converted into dry sands and dusty winds. The void that remains is a scar on nature of unprecedented magnitude. In the Aral's collapse, the disaster has sucked into its vacuum the peoples of the region, their lives and livelihood, their health and well being, and certainly their hopes and aspirations. And it has created regional and global issues of great urgency and sensitivity.

As if opening a Russian *matrioshka* doll, one finds here a crisis within a crisis within a crisis. The show-stopping, headlining disappearance of one of the world's major inland bodies of water turns out to be just the outer layer that one beholds, only to find, upon closer inspection, that new layers are hidden within it. A *matrioshka* yields a set of more delightful, yet smaller figures. The Aral Sea disaster has led to a veritable stream of interconnected disasters, many of which encompass a wider area than the sea itself. As we will see, beyond the tragedies for the Aral region, per se, these disasters include spreading deserts, blowing contaminants and heating climates. They include a distributive system of such great inefficiency that more than half of irrigation waters are lost to the surrounding environment. The problem, however, is not merely the loss of clean water. The water reappears, forcing its way up through the soil, transporting salts to the surface to create a dusting of snow-like particles in the fields, robbing the soil of its fertility and the water of its potability. The water intensive approach to agriculture, moreover, not only engenders the inefficiencies of monoculture but has led to massive pesticide application, creating its own legacy of contamination. Human health rivals ecological health in a downward intertwining spiral. And then there are the two great rivers that once fed the Aral Sea like two giant feeding tubes, but have since been diverted extensively to supply irrigation waters. Originating high in the mountains of Central Asia, these course downward across international borders, creating a row of upstream and downstream relationships that invite power plays and conflict. Much as the Aral Sea itself, Uzbekistan is last in line, waiting its turn, hoping that enough water keeps trickling in. The nation's endurance may not, however, match that of the Sea. There are many dolls inside our *matrioshka*, each opening to reveal the next layer of complexity.

The loss of the Aral Sea, and these related layers of disaster, manifested as a loss of nature, but there was little else natural about it. *Disaster by Design* begins with a series of provocative articles that establish the constructed nature of disaster in the Aral region.

In Chapter 1, "An Overview of the Aral Sea Disaster" Abror Gadaev and Zikrilla Yasakov lay out the details and dimensions of the Aral Sea disaster. They set forth the basic outlines of what happened and why. They blame cotton production and poor water management, a lack of critical oversight and cross-agency cooperation and the contrary interests of nations whose fate is tightly bound by water. The authors offer mitigatory options. Gadaev, a water specialist, was the inspiration for our project, headed the Uzbek team and served as the principle host for our trip to Uzbekistan and travels to the Aral region.

In Chapter 2, "Going With the Flow: Economic Impacts from the Overuse of Irrigation," Ilkhomjon Niyazov and Farhod Ahrorov, with Astrid Cerny and Michael Edelstein, lay out the problems of irrigation and their role in the Aral loss. The basic problem is fairly straight forward, in the Aral basin, water use exceeds water supply. And 90% of water use is for irrigation. This process is highly inefficient and polluting. Ironically, salinization associated with farming has resulted in declining cotton production. Thus, the effort to increase cotton production has succeeded, in the long run, in achieving just the opposite.

In the Chapter 3, "Pollution and Salinization: Compounding the Aral Sea Disaster," Farhod Ahrorov, a member of our Uzbek team, collaborating with two colleagues, expands on the entangled story of designed disaster to examine the issues of irrigation, cotton and food crop production, and land use decisions that not only have destroyed the Aral Sea but created other problems for water quality across Uzbekistan and the region.

In clear terms, they describe the crass attitude of Soviet planners who, upon discovering the collapse of the Aral fishery, opted to ship ocean-caught fish by train to keep the region's canneries working, but gave no apparent thought to saving the Aral Sea ecosystem.

Chapter 4 by Michael Edelstein is entitled "Death at Rebirth Island: Secrets in the USSR's Culture of Contamination." Here I tell the story of the secret bio-weapons testing lab on Vozrozdheniye (rebirth) Island in the Aral Sea. The story is horrific. And the lessons are sobering. The Aral Sea disaster was spawned by the same authoritarian system that created these horrors. Rhetoric of national interest and willingness to use all means to achieve questionable ends characterize both situations. The irony is the convergence of the two stories and the long-term legacy that such closed systems create. The only defense against such outcomes is open public discourse, a public empowered to hold governments accountable, make secretive processes transparent and to offer countervailing systems of checks and balances. In short, the antidote for such poisons is to make closed systems open and democratic in form.

The Aral Sea disaster itself demonstrates the key attributes of global climate change. But it is also subject to effects from larger scale alterations of temperature and weather. How much of a wild card is regional climate change for the Aral system? Chapter 5, "Potential Climate and Hydrological Changes in The Aral Sea Region" presents the work of James Miller and his associates. They show that projected climate changes may make for hotter and drier summers in the region, intensifying drought and, thus, creating conditions for desertification. Increased flow in rivers due to enhanced snow melt will be offset by evaporation in the summer, making it harder to preserve the water budget in the two major rivers. They suggest that projected climate changes will make water management in Central Asia an even greater challenge. Miller presented this paper at our October 2011 conference.

Chapter 6, "The Significance of Being Downstream: Uzbek Concerns over the Rogun Dam" is a contribution by His Excellency Murad Askarov, the Uzbek Ambassador to the United Nations (UN). Askarov's paper reads like a passionate speech to the UN. In fact, a key focus of his work at the UN is to avert a new set of disasters for the larger Aral Region connected to the design and construction of the largest dam project in Central Asia, the Rogun hydropower project. Askarov reviews the potential adverse impacts of the project, offers alternatives and places the issue, real time into the contested sphere of the UN. Major international agreements support the rights and concerns raised by Askarov, as does the moral plight of a region already under the sword because of the Aral Sea disaster. Askarov demonstrates that *Disaster by Design* is still being practiced in the Aral Region and threatens to cause consequences far into the future.

In sum, there are indeed multiple disasters of the Aral Sea.

Michael R. Edelstein
Editor

CHAPTER 1

AN OVERVIEW OF THE ARAL SEA DISASTER

Abror Gadaev and Zikrilla Yasakov

ABSTRACT

Central Asian nations all gained their independence from the former Soviet Union in 1991. Unfortunately, with their independence, they also inherited many environmental problems. By far, the most pervasive issue is the shrinking of the Aral Sea. The enormity of the environmental crisis of the Aral Sea has been attested by environmentalists both in Central Asia and in other parts of the world. In this chapter, the broad scope of the disaster is set forth along with its principle causes. An overview of the geography and water supply in the region is given. And the daunting challenge of cross boundary impact and thus need for multinational cooperation is discussed. The challenge in the region today is twofold. First, it must address the Aral Sea disaster. And second, it must create the conditions to sustainably manage transboundary water resources to head off new catastrophe.

INTRODUCTION

In the distant past our nation was known as the center of the Great Silk Road, home to the world's greatest concentration of commerce and trade

Disaster by Design: The Aral Sea and its Lessons for Sustainability
Research in Social Problems and Public Policy, Volume 20, 5–15

ISSN: 0196-1152/doi:10.1108/S0196-1152(2012)0000020009

and known for its wealth and prosperity. Today the Aral Sea and surrounding territories are world-known as an environmental and social disaster zone for ecological catastrophe attributed mainly to human causes and to the devastation that occurs in its wake. The once great sea, fourth largest inland body on earth, is now substantially gone. A combination of factors brought us to this point.

Water resources have been a regional problem in Central Asia since the earliest civilizations. An arid region, agriculture has always been dependent on diversions of water from our rivers. The beginning of irrigated agriculture in the region dates back to the 6th–7th century B.C. (Gadaev & Niyazov, 2010). This time period coincides with a flourishing of ancient civilizations where irrigation was a major decisive factor of historical and socioeconomic development (Gadaev, 2007).

Over the past half century, growing water consumption resulted from intensive cultivation of new irrigated territories, where mainly cotton and rice are grown. Compounded by increasing population and growing agriculture, among other factors, the flow of water to the sea from the two major river systems, the Amu Darya and Syr Darya, virtually stopped. And the Aral Sea, dependent entirely on inflow from these rivers, dried up so precipitously that it seemed almost to vanish (Gadaev, 2006).

In the arid climate of the Aral Sea region, it would have required 60 km^3 per year of inflow to sustain the surface area of the Aral Sea of just over 60,000 km^2 (23,000 square miles). The inflow of water was diverted to other uses. Comparing 1930 to 1960, more than four times the water was withdrawn as the result of a sharp increase in upstream water usage per year (from 25 to 103 km^3). Not only were massive quantities of water wasted, but the quality of water also declined with adverse health impacts.

At the beginning of the 20th century, 7 or 8 million people lived in the Aral region. More than 3.5 million hectares (8½ million acres) was under cultivation using varied levels of irrigation. Agriculture was the foundation of the region's economic base. By comparison, a century later, today's population in Central Asia and the Aral Sea Basin region has increased seven times, exceeding 50 million people. With this increase in population, irrigated lands have doubled to more than 7.5–7.9 million hectares (18.5 million acres). Agriculture not only supports the population with food, but as the source of the region's major exports, it is the backbone of the economy (UNESCO, 2009).

You now have the major parts of the puzzle.

A BROAD GEOGRAPHY OF THE REGION

The map of Central Asia has changed many times over the centuries. Today, Uzbekistan is one of only two double landlocked countries in the world. The nation is divided into 12 *viloyats* (governmental regions) plus one autonomous republic: Karakalpakstan (the Aral Sea region). See the collection of maps at the beginning of this book for political and satellite view of Uzbekistan and Central Asia.

The country can be divided into three ecological zones:

- Desert steppe, and semi-arid regions covering 60% of the country, mainly the central and western parts.
- Fertile valleys (including the Fergana valley) that skirt the Amu Darya and Syr Darya rivers.
- Mountainous areas in the east in the Tian Shan and Gissaro-Alay mountain ranges with peaks of about 4,500 m (14,765 feet) above sea level.

In this volume, our concern is primarily with the steppe and semi-arid regions interspersed with growing deserts that characterize the central and western parts of the Uzbek nation. We refer to this region as the primary Aral impact area. Additionally, the Aral Sea was shared with the neighboring nation of Kazakhstan, and lingering impacts of the disaster are of concern to that nation as well. Two other caveats are required in talking about the overall impact region, as well. First, the entire flow of the two principal rivers unites the mountains, valleys, and steppe regions. The Syr Darya and Amu Darya may have flowed to the Aral Sea, but along the way, between the two, they cross all the nations of the region and meet the water demands that include drinking water, irrigation industrial use, and power generation (IFAS, 2009). Thus, it makes little sense to talk of the Aral Sea disaster without incorporating the accumulation of issues associated with these rivers. Second, wind erosion results in contaminated dusts from the Aral Sea spreading across the region and even the globe. The rapid conversion of the Aral Sea to the Aral desert (loosely named Aralkum) was accomplished by storm and wind erosion. And, just as the waters of the original sea were dispersed widely, so have been its sediments.

SCOPE OF THE DISASTER

It is estimated that the surface area of the Aral Sea is now just 25% of what it was during the Soviet Era. As noted, the Aral Sea has the capacity for just over 60,000 km^2 (23,000 square miles) of water surface, its original surface area in 1960 ("Uzbekistan" World Fact Book). Of course, a body of water has a third dimension, depth, so it makes more sense to talk about the cubic dimensions. While an annual inflow of 60 km^3 per year (the equivalent of 24 million Olympic-size swimming pools) was required in the past to maintain the original surface area, refilling the Aral Sea to its original volume will take 1,000 km^3 of water.

The irony of the Aral Sea disaster is that the water resources exist to have prevented it and to remedy it. However, the necessary water is not making its way to the Aral Sea. One hundred and fifteen cubic kilometer of water is available in the Aral Sea Basin; however in 2002 the water flow to the Aral Sea was less than 1 km^3. Table 1 shows how water flow to the Aral Sea has changed over time to produce the Aral Sea disaster (Gadaev, 2006).

Geopolitical Dimensions and Multinational Responses

It is only recently that Uzbekistan and its neighbors have achieved independence after the fall of the Soviet Union in 1991. Administratively Central Asia is a newly reorganized region consisting of the independent states of Uzbekistan, Kazakhstan, Turkmenistan, Tajikistan, and Kyrgyzstan. Conditions have made it difficult for Central Asia to unite. A key issue is that Central Asian countries including Afghanistan share the same watersheds, aquifers, and reservoirs. Any decision about water usage in this region raises transboundary issues (Gadaev, Capece, & Wood, 2004). Often international organizations need to step in to resolve disagreements that might otherwise become urgent regional conflicts.

The worst environmental crisis inherited from the Soviet Union by far is the shrinking of the Aral Sea. The Aral Sea began to shrink in the 1970s

Table 1. Available Water Flow (in km^3).

	Water Available in the Basin	Natural Water Losses	Upstream Water Use	Inflow into the Aral Sea
1930	115	36	25	54
1960	115	13	48	54
1990	115	7	103	5

Source: Authors.

during the Soviet Era. At this time, little attention was paid to environmental issues; political concerns and hopes of increasing agricultural and industrial yield prevailed. Unfortunately, this neglect of attention to the Aral Sea persisted through the independence of Central Asian countries in 1991. In the early nineties, these new states were preoccupied with political and structural reformations and the Aral Sea disaster worsened. It was not until 1993 that attention was finally paid to the Aral Sea, and efforts are currently in motion to salvage what remains of the once bountiful sea. The challenge is unprecedented.

The Aral crisis prompted creation of an interstate commission, the International Fund for Saving the Aral Sea (IFAS). A key focus is the improper use of water taken from the Aral Sea Basin and its numerous consequences. Unfortunately since the collapse of the USSR, Central Asian economies have not been strong enough to restore the natural productivity of the territory (IFAS, 2009). From the time of their formation, the new Central Asian states have been burdened with the primary water crises, as well as a serious of secondary social responsibilities, which include protecting the population from adverse impacts of desertification, creating new workplaces and job markets, and trying to improve the economic and social conditions by introducing new water-efficient technology. Ecologically, each country must implement new plans for the restoration of flora and fauna diversity and the prevention of any further degradation of the Aral Sea.

There are a large number of contributing factors to the shrinking of the Aral Sea that account for the shift of water resources away from the sea and to upstream water use. Increases in irrigation-intensive agricultural industry and a boost in the population of Central Asia are principle reasons. But there are also significant transboundary factors, as well. In terms of Uzbekistan's ability to manage the situation, much of this increased demand on the Amu Darya and Syr Darya rivers now occurs outside of Uzbekistan, and includes flow control in large hydroelectric generating dams, where considerable water is impounded. This places water into the realm of geopolitical conflict, as noted later in this section by Ambassador Askarov. There is also increasing water pollution in the two rivers involving silting and mineralization of the rivers.

It has been 18 years since the first efforts were made to save the Aral Sea and still conditions are not improving. A variety of factors are responsible, namely the failure of cooperative water management efforts because there is not full participation of all Central Asian countries. It will take the cooperation of many nations to restore the Aral Sea, however barriers to this endeavor may not be easy to overcome (Abdullaev, 2009). Each of the Central Asian and surrounding countries has its own unique conditions which affect its international cooperation. Such factors as population pressure, openness of government, industrial water demand, culture, geography, health, and the history of transboundary tension all influence water-related issues.

Ignoring this issue has produced disastrous results, and it is estimated that the Aral Sea may could completely dry out by the year 2020.

Water Security and Groundwater

Greater Central Asia has a fast growing population of 50 million people with another 150 million inhabitants in the surrounding areas of Eurasia (mostly in parts of Russia and China). The challenge remains to create a sustainable basis for water security in Uzbekistan and the region given the magnitude of the Aral disaster. With all of the focus on the diminution of the surface waters of the region, there is also concern with the failure to make sustainable use of the region's groundwater systems. The issue is closely related to the Aral disaster because it forced increased dependence on groundwater and because it greatly increased threats to water quality. The problem is most evident during summer months when demand for water sharply increases and the population is left to consume poor quality surface water. At such times, the health of the entire population of the region is placed in jeopardy. Infectious intestinal diseases, often caused by contaminated drinking water, are a primary cause of the high rate of childhood deaths in Uzbekistan and the Aral Sea Basin (UNESCO, 2009).

Fig. 1 illustrates the fact that just over half (54%) of Uzbekistan's water resources come from surface waters, namely the two rivers Syr Darya and Amu Darya. There is an almost equally matched but untapped resource, however, in underground water.

In the Republic of Uzbekistan and other Central Asian countries, stocks of underground waters are contained in artesian aquifers protected over geological time by impervious rock from external pollution. These waters are characteristically high in salt and "hardness" (mineralization). Technology is required to render the water potable (see Gadaev & Boboeva, Chapter 23). These underground sources, once treated, can be used to sustain human, agricultural, and industrial needs and, also over time, to restore some of the stock previously diverted from the Aral Sea Basin. The

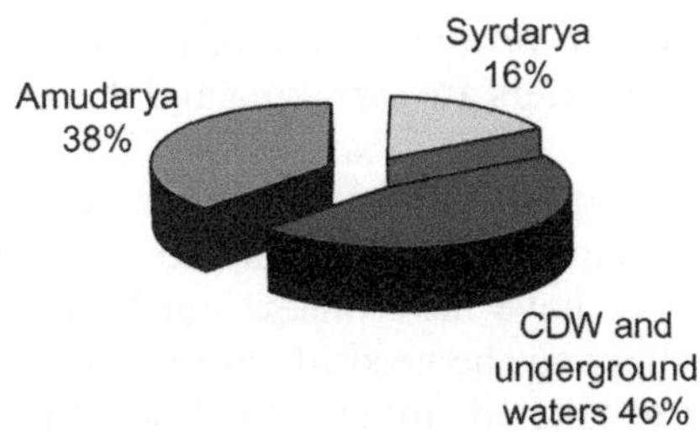

Fig. 1. Sources of Uzbekistan's Main Water Sources, Amu Darya, Syr Darya, CDW (Canaled Drainage Water), and Underground Water. *Source*: Authors.

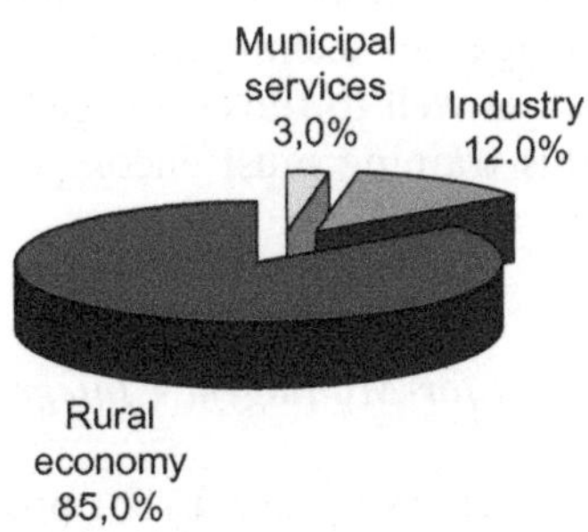

Fig. 2. Water Resource Use in Uzbekistan, 2009. *Source*: Authors.

development of these technologies should ideally be factored into a comprehensive water management system plan.

As a newly formed independent Republic, Uzbekistan must rapidly create high quality water intake, treatment, and delivery systems to balance competing water supply demands. On one hand, there is the need to halt depletion of water in the Aral Sea and, ideally, to augment the waters in that basin. On the other, Uzbekistan must be capable of supporting an exploding population and a strong economy.

Fig. 2 illustrates the current consumption and distribution of water resources in Uzbekistan. As seen, agriculture (i.e., rural economy) has received the lion's share while a paucity of water has gone to industry and particularly municipal services. Any effort at Aral Sea restoration is likely to demand a re-diversion of a large amount of this water. Even in the absence of this effort, dramatic efficiencies are required in irrigation practices and also in industry. As municipal services grow to address a rising population and the need for improved sanitation, they must adopt extreme efficiencies to limit water waste and to promote reuse of the resource. And to complicate matters, the surface waters are shrinking and the groundwater resources are getting polluted.

Historically, water has been systematically overused and poorly managed. State-of-the-art water management systems and conservation technologies, while known to experts, have nevertheless not been introduced to major effect in Uzbekistan. Beyond the need to address infrastructure deficiencies and upgrade systems to take advantage of modern technology, there is the even more primary step of capacity building. Major efforts are needed to train water resource managers in new sustainable approaches and to curtail prior water-wasteful practices. Of particular importance is training in irrigation technology. A leading cause of Uzbekistan's water crisis is its use of old and antiquated irrigation techniques. Accordingly, education and training must be extended as well to those engaged in agricultural practice. For the needed effect, this training must encompass Uzbekistan's Central Asian neighbors, as well.

The Need for Multiagency Integration

An analysis of the Aral Sea crises reveals a root problem going back to the Soviet period when the disaster was first unleashed. There was little if any cross-communication between the different agencies and specialists who created the disaster. Those foreseeing the problems were not in the right chain of communication and command to head off disaster. Today, Uzbekistan and its neighbors similarly lack an integrated agency that handles agriculture and water affairs or, alternatively, separate agencies that could perform checks and balances on each other to ensure equality of resource access within and between neighboring countries.

Such learning must also occur at the level of regional policy, where the same tension exists between addressing the well-recognized and documented disaster of the Aral Sea and the struggles of individual countries to advance economically and update their water delivery systems. Scientists from environmental organizations have brought attention to the crisis by writing about it in journals and in conference papers. Many proposed solutions have been put on the table. Some of these include the idea of a major infusion of water to the Aral Sea from an accessible source, such as the Caspian Sea, or the Volga, Ob, or Irtish rivers. Such a diversion of a major water body into the Aral Sea would restore the Aral Sea to its former size in 20–30 years at a cost of $30–50 billion (Transboundary Water and Related Energy Cooperation).

Other concepts attempt to fix parts of the existing water use/irrigation system by upgrading drainage, adapting wastewater for recycling and, also, growing salt-tolerant crops. A number of major projects promise to help the overall situation, potentially including new canal systems. USAID is funding infrastructure rehabilitation in Tashkent and the Water Users Association in Uzbekistan (USAID, 2001). The United Nations has attempted to bring Central Asian leaders together to address the health issues emanating from the Aral Sea crisis (Aral Sea, 2003). These efforts are praiseworthy. But to actually effect regional renewal and restoration, a higher order of collaboration is required among an expanded group of these organizations. The fate of the Aral Sea region is ultimately in the hands of the interstate pact that has been made between the states. If these governments are able to look toward the problem as a whole rather than their individual interests, there might be a future for water resources in the Central Asian region.

Cross-boundary agreements about water are easily confounded by the threats implied by the issues discussed. The drying of the Aral Sea has resulted in a complex of problems. These include widespread environmental degradation, the loss of fresh water and arable lands, the wind-blown distribution of sea-bed salts, total decline of a fishery that was the backbone of nutrition and economy, demands to restrict agriculture, radical declines in public health and, given all of these negative changes, the associated economic collapse (IFAS, 2009). These complex problems would be challenging in and of themselves. But unresolved, they represent a threat to the political stability of the region. The resulting security challenge for Central Asia is of major concern. The region has forged a relationship to the major world powers and retains a close link to Russia, which has a historical responsibility for the situation.

CONCLUSION

The Aral disaster underscores a broader issue. The primary challenge facing water supply and delivery specialists in Uzbekistan is to balance competing water supply and consumption demands. Among these demands is the urgency of halting further depletion of water in the Aral Sea, and, ideally, to augment the waters in that basin. But there is also the need to develop high quality water delivery systems for human use. And, of course, there is the question of what water can be left for the agricultural sector to irrigate its crops.

Some of the steps necessary to be integrated into water management in the Central Asian region include use of more efficient irrigation technologies like drip irrigation, structurally improved irrigation canals, greater use of groundwater, reuse of treated wastewater, installation of desalination plants where necessary, charging farmers to use the water from the rivers, and dramatically reducing the application of chemicals on cotton.

All of these tasks require trained experts. Capacity building is therefore a major priority. For real progress to be made in the short to medium term, technology students and professionals from Uzbekistan and other Central Asian nations need to be trained in state-of-the-art techniques of water management.

A key focus of water management must be the development of new water sources. The Republic of Uzbekistan and other Central Asian countries have vast stocks of underground waters which have been isolated from surface pollution. However, these waters are characteristically high in salt and hardness. If technologies can be applied to overcome these negative characteristics, then the underground sources can be used to sustain human, agricultural, and industrial needs and, also over time, to restore some of the stock previously diverted from the Aral Sea Basin.

A cautionary note is in order. Just as the Aral Sea was a common resource of inordinate importance to the ecology, climate, and people of the region, so is the vast underground water resource. Unless strict national and regional policies are in place, it would be easy to see this pristine underground resource polluted or overexploited and drawn down to unrenewable levels, as the Aral Sea was aboveground. One might envision, then, the equivalent of the Aral Sea disaster below ground.

REFERENCES

Abdullaev, I. (2009). *Cotton in Uzbekistan: Water and welfare*. Center for Development Research, ZEF. Retrieved from www.zef.de/module/register/media/83a8_Abdullaev%20100807.pdf

Aral Sea. (2003). *Aral Sea. Integrated water resources management for wetlands restoration in the Aral Sea Basin*. Retrieved from http://www.cawater-info.net/aral/groundwater_e.htm

Gadaev, A. N. (2006). Water and health problems in Uzbekistan; drinking water disinfection by-products. Gordon Research Conference and US EPA, Mount Holyoke College, MA.

Gadaev, A. N. (2007). Improvement of water supply services in the city of Samarkand, Uzbekistan. UNESCO Symposium on New Directions in Urban Water Management, September 12–14, Paris, France.

Gadaev, A. N., Capece, J. C., & Wood, D. (2004). Central Asian transboundary water management: Social and environmental issues. *American Water Resources Association (AWRA) annual conference*, November 1–4, Orlando, FL.

Gadaev, A. N., & Niyazov, I. (2010). Uzbekistan: Water resources management and health issues. *Ecology and Noospherology*, *8*, 16–20, Spain.

IFAS. (2009). *Environment state of the Aral Sea Basin*. Retrieved from http://enrin.grida.no/aral/aralsea/english/arsea/arsea.htm. Accessed on October 21, 2009.

Transboundary Water and Related Energy Cooperation. (2009). Retrieved from http://www.cawater-info.net/bk/water_law/pdf/lennaerts_eng.pdf. Accessed on November 4, 2009.

USAID. (2001). *Biodiversity assessment for Uzbekistan*. Retrieved from http://pdf.usaid.gov/pdf_docs/PNACN475.pdf. Accessed on November 4, 2009.

UNESCO. (2000). *Water related vision for the Aral Sea Basin*. Division of Water Sciences. Retrieved from http://unesdoc.unesco.org/images/0012/001262/126259mo.pdf. Accessed on November 19, 2009.

CHAPTER 2

GOING WITH THE FLOW: ECONOMIC IMPACTS FROM THE OVERUSE OF IRRIGATION

Ilkhomjon Niyazov, Farhod Ahrorov, with Astrid Cerny and Michael R. Edelstein

ABSTRACT

In this chapter, we examine the consequences of the substantial diversion of water to irrigation in the Aral Sea region, the principle cause of the Aral Sea disaster. The purpose of this study is to understand the relationship of the two rivers feeding Uzbekistan to the water inflow into the Aral Sea and to economic productivity in agriculture. We find that because water consumption exceeds water formation, the prognosis for future growth in agricultural output is at risk. Uzbekistan and its neighbors need to address the upstream and downstream rights to water access for the Syr and Amu rivers. The solutions to our immediate problems can be found inside Uzbekistan. Improving water efficiency and water conservation is a necessary, economical option compared to other proposals.

Disaster by Design: The Aral Sea and its Lessons for Sustainability
Research in Social Problems and Public Policy, Volume 20, 17–27

ISSN: 0196-1152/doi:10.1108/S0196-1152(2012)0000020010

INTRODUCTION

The Republic of Uzbekistan and the majority of the neighboring countries are situated in the Aral Sea internal drainage basin, the transboundary waters of which are in shared use for economic and environmental needs. The fresh waters of the rivers, lakes, and reservoirs are used for irrigated farming, industrial, and public consumption. Landlocked Central Asia relies on the massive interior lakes like the Caspian Sea, Lake Balkash, and until recently, the Aral Sea. These lakes are fed by rivers, depending on them to counteract the effects of evaporation and human fresh water needs. In the case of the Aral Sea, the sources were the two great rivers that define the Central Asian region, the Syr Darya and the Amu Darya. Despite natural fluctuations of climate and landform, these lakes would have a sustainable source of water unless one or more of these three conditions occur to disrupt the sources of supply.

First, population and its demand for water outgrow the natural ability of the water resource to regenerate. The excess demand on water sources beyond their sustainable yield has proven to be an important constraint to growth. Despite the rapid growth of population in Uzbekistan and Central Asia, this was not a major factor in the loss of the Aral Sea, nor is it the dominant continuing demand on surface water supply. However, given the potential for sustained population growth, it could become a factor in the future.

Second, competition over water between upstream and downstream users and countries may prompt damming, sluicing, diversion or excessive consumption, preventing an adequate supply to reach downstream areas. This factor has been important in our region in the past (see Chapter 9). The demise of the Aral Sea was certainly aided by the permanent diversion during Soviet times of a full 15% of the Amu Darya to the Kara Kum Canal, which grants Turkmenistan a sizable water surplus at the expense of Uzbekistan. Contests over water control on the two dominant rivers are shaping up to be a significant future concern in our region.

Third, water supply is threatened by agriculture employing crops that demand more water than can be provided from local rainfall and renewable water sources, and thus requiring irrigation. Cotton has contributed greatly to the Aral Sea disasters, as will be discussed in detail below.

WATER DEMAND EXCEEDS SUPPLY

The main sources for Uzbekistan's water resources are surface and ground waters. The total volume, or natural mean annual flow of the rivers, is

123 m^3/year. This flow comprises the combined contributions of the two main rivers, including 81.5 km^3 in the Amu Darya basin and 41.6 km^3 in the Syr Darya basin. The surface water flow formed in Uzbekistan makes up 11.5 km^3 as shown in Table 1. The table also shows that the consumption of water in the Aral Sea Basin far exceeds the water replenishment rates by a factor of approximately five.

The Aral Basin comprises the flow from the two major rivers, the Amu and Syr rivers, which are responsible for most of the freshwater available across Uzbekistan. What is instructive to note about the Aral basin, moreover, is that 59% of the total available water resources in the basin originate from surface waters outside Uzbekistan, making this imbalance all the more precarious from a downstream perspective. Transboundary import is particularly pronounced in the Amu Darya Basin, where 78% of the river's influx is from neighboring countries. Accordingly, dependencies on this transboundary water are met, to some degree, at the whim of other sovereign states. Other major sources for the entire basin are Uzbek tributaries (26%), aquifers (4%), and water draining from surface collectors (11%) (MAWR, 2010).

DOMINATION BY IRRIGATION

The next matter of importance is to examine how this water is utilized. Table 2 shows the rough breakdown of end uses for Uzbek water resources. It can be seen that 5.5% is consumed or used in service capacities, all less than half that in total for industry, fishing, and energy. In contrast, more than 90% is used for irrigation purposes and is consumed in the rural economy, a much higher percentage than found on an average for agricultural countries.

Table 1. Water Resource Availability and Consumption in Uzbekistan.

	Syr Darya Basin	Amu Darya Basin	Aral Sea Basin Total
Water resources formed/km^3	6.39	5.14	11.53
Water resources consumed/km^3	17.28	38.91	56.19

Data Source: ECOGeoscience (2009).
Source: Authors' table using cited data.

Table 2. Water Use by Branches of the Economy in Uzbekistan.

Agriculture	Service	Industry	Fishery	Energy
92%	5.5%	1.5%	0.8%	0.2%

Source: Authors' estimation based on Samylov et al. (2006).

FAILING TO LEARN FROM A LONG HISTORY OF IRRIGATION

The history of irrigated agriculture in Uzbekistan is among the oldest in the world, having its origins in the 6th–7th century B.C. This time period coincides with the flourishing of the most ancient civilizations in Central Asia. For cities to develop required the extensive development of farming. Given the arid region, irrigation from rivers was a major factor in historical and socioeconomic development. River irrigation developed some 3,000 years ago, and the technology to dig deep wells and transfer canals dates from the 3rd century CE, which facilitated expansion of oasis settlements (Golden, 2011; Liu, 2010). Agriculture on the scale to support an urban population can thus be roughly correlated in age to the founding of our major cities Samarkand and Bukhara. Over the centuries, local populations improved their skills in irrigation, water diversion from rivers, and water applications to field crops.

While our age old experience with irrigation should make us experts, in fact there has been a general failure to learn from past failures, many of which had dire consequences. Our approach to water diversion and distribution is the source of myriad problems discussed in Section Three. First, there is the issue of waste. The massive expenditure of water resources on agriculture does not mean that those resources are being used efficiently. In fact, some 70% of the water is estimated to be wasted through inefficient irrigation (see Rudenko et al., Chapter 13). The good news is that this loss can become a new source of water for the future. We only need to prevent this currently squandered resource.

Then there is the matter of contamination. Agricultural irrigation systems have caused high pollution levels in the region's surface waters (see Ahrorov et al., Chapter 3). Surface waters are the main source for agricultural irrigation. Above it was noted that 11% of Uzbekistan's water resources drain from surface collectors. Such drainage is a major source of mineralization. Runoff spreads the resulting salts and deposits, along with pesticide residues, over the soil and carries them into the rivers. Fig. 1 displays the effect of accumulating mineralization, showing that rates are much higher in the Syr Darya than the Amu Darya, but nevertheless rising rapidly in both cases. As can be seen, the rise and rate of mineralization correlates to the period of agricultural intensification, when the number of fields irrigated for cotton cultivation doubled. The last data point here is at the end of the millennium and it is unclear if the plateau evident, particularly for the Amu Darya, has continued.

Finally, irrigation introduces an endless cycle of degradation. Soils become waterlogged, accumulate salt, and become infertile. Productivity drops. Even more water is used to wash salt out of soils. The wash water then further raises the water table, returning the salts to the surface to be removed again.

Given the water quality issues, groundwater is used for municipal and industrial needs. The primary source of quality drinking water in Uzbekistan and Central Asia is groundwater, which accounts for between 85–90% of the general water consumption budget (see Gadaev and Boboeva, Chapter 23).

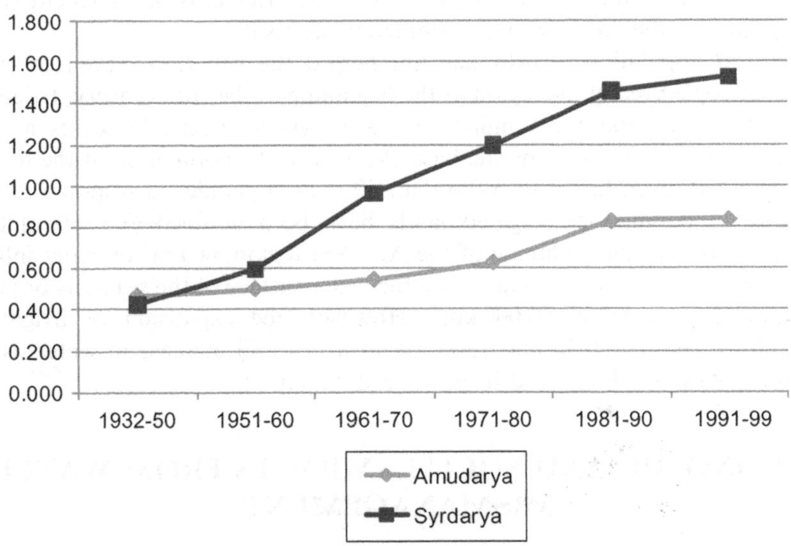

Fig. 1. Changes in Mineralization of Rivers of Uzbekistan by Percent of Water Content. *Source*: Authors' estimation.

CUTTING OFF THE ARAL'S LIFEBLOOD

The two great rivers that fed the Aral Sea run a gauntlet of irrigation canals, channels, and ditches, diverting and diminishing its flow all along their course. Historical water flow to the Aral Sea was 56 km^3 per year from both the Amu Darya and Syr Darya. It decreased to 47 km^3 between 1966 and 1970. Then, water flow plummeted to 2 km^3 between 1981 and 1983 around the time when the Aral Sea disaster really gained international attention. Today it measures less than 1.8 km^3 per year. By comparing the measurements, we can understand the implications. The data in Table 1 shows that water resource formation for the main rivers in Central Asia is 63.02 km^3/year; however, the inflow from the same rivers at their final destination at the Aral Sea is less than 2 km^3 per year (Mikhalev & Reimov, 2008; Wehrheim, Schoeller-Schletter, & Martius, 2008).

The rapid growth in water consumption is connected not only to waste through poor quality irrigation channels, but also to putting new land under cultivation – only possible with heavy irrigation – where mainly cotton and rice are grown. This expanded cropping, combined with population growth and additional employment in agriculture, has led to the current situation where the flow of water to the Aral sea from the two major river systems – the Amu Darya and Syr Darya has slowed to a trickle. In some recent years water flow to the Aral Sea has completely stopped.

Natural population growth has not helped the situation. Seven to eight million people lived in the region at the beginning of the 20th century. Irrigated lands made up about 3.5 million hectares which effectively serves as the foundation of society's economic base. At present the population of the region has enlarged seven times, to more than 50 million people. In response to the increase in population, irrigated lands have been doubled (7.5–7.9 million hectares). In the sunny climate of the Aral Sea region, 60 km^3 of water inflow per year would be needed to maintain the water surface of the sea at its original area of approximately 60,000 km^2. However, the expansion of irrigation agriculture between 1930 and 1960 meant a fourfold increase in water usage from the Aral Sea Basin and from its feeder rivers.

ECONOMIC AND SOCIAL IMPACTS FROM WATER MISMANAGEMENT

There are many perverse outcomes of the Aral Sea disaster that reflect on the systemic interconnections of the entire region. As the sea shrank ever more, the fishing industry was decimated, increasing economic dependence on agriculture in the overall Western region. Meanwhile, the desiccation of the sea has undermined the cotton harvest that was its cause. The evaporation of the Aral Sea caused salt-laden dust, which is then transported by wind, and has been measured as far away as Europe and Asia. Slowly but surely, the cotton crop yields have measurably declined in the region due to these dusts, as well as several factors directly linked to the methods of irrigation in use: erosion, water pollution, and soil salinity. At the same time, the cotton industry's increased use of pesticides and fertilizer (and their runoff) have resulted in the pollution of surface and groundwater. Significant health impacts result from the consumption of impure water and from the blowing dust. The areas closest to the Aral Sea thus have health issues which are in turn connected to social issues. As chapters 16 and 17 discuss in this volume, there is a high incidence of infant mortality, as well as many diseases, including anemia, bronchitis, and other respiratory infections, tuberculosis, kidney and liver disease, cancer, and even arthritis. The local population has thus been affected in every way, from loss of productivity in agriculture and fishing to widespread health issues.

WASTING THE ARAL CAUSED NEW PROBLEMS

The planners long ago thought sacrificing the Aral in order to grow more cotton and increase exports was worthwhile. They spared very little thought to what other consequences this might unleash, and how difficult it would be to reverse the decisions. For example, according to one report, 522,500 hectares of wetlands were damaged during the period of most intensive cotton cropping, from 1960 to 1990 (Dukhovny, 2007). These wetlands were discounted as unproductive, but actually contributed habitat for wildlife, flood control, fishing, hunting, and many other services. Considering the loss of water quality and salinization, these wetlands may have played an important role through phytoremediation in removing pollutants and restoring water quality. Most of the former wetlands are arid wastelands today.

The ecological situation in the Aral region has steadily gotten worse, and the big picture is not much better. The wrongful and inefficient use of the water taken from the Aral Sea has led to many negative consequences in Central Asia that new independent economies are not strong enough to redress.

THE DAMAGE IS DONE, NOW WHAT?

First-order solutions that need to be addressed include improving irrigation efficiency, water supply development, wastewater treatment, and watershed management. These are critical needs throughout Uzbekistan and Central Asia. It is important to develop both domestic and regional solutions.

The ecological situation is dire, and so far, the social predicament is worsening, at least in Karakalpakstan and Khorezm, the regions closest to the Aral Sea. Effective water management needs to do what it has not done before: balance the competing demands of sending more water to the Aral Sea and simultaneously meeting the water supply and quality needs of a developing economy and growing population. All of these concerns have great urgency.

CENTRAL ASIAN WATERS AND THE IMMEDIATE FUTURE

Central Asian Republics utilize the same watersheds and share many water management issues in common. It is clear that the region has a multination identity now, where regional water management and environmental protection efforts are insufficient by themselves to meet the scale of the problem that affects all Central Asian countries, whether upstream or downstream. Further multinational agreements and joint-state or joint-agency programs will undoubtedly be required. Each riparian country located in the transboundary stream basin must better manage water resources within its territory by taking into account environmental impacts as well.

Better governance and management of agriculture infrastructure is also needed. This can happen by employing our talented engineers to work with agricultural experts to introduce new plants and irrigation techniques. This can be started anytime. The stabilization of the sea floor and the problem of the airborne pesticides are somewhat more challenging.

For those who dream of restoring the Aral to its previous grandeur, it is common to return in an old idea from Soviet times on how to divert more water to the region. It is proposed to build two canals to redirect water from the Volga, Ob, and Irtysh rivers in order to restore the Aral Sea to its former size over a 20–30 year period. This US$30 billion project ironically rests on using ditches to fix problems caused by building ditches. Meanwhile, those who advocate such huge technological fixes for the Aral are not enticed by the steps that can quickly make a difference at a much lower cost. Massive water savings are possible when efficiencies in irrigation are achieved, alleviating the ongoing water management problems that plague the region. Furthermore, it can be done at much lower cost, and avoid new ecological catastrophe.

CONCLUSIONS

The problems of our region will not heal themselves. And they will not wait. We have learned from the Aral debacle that delay is our foe. We understand that so much needs to be accomplished and not everything can be done at once. If we want to rapidly manage and contain the crises from spreading even further, what do we prioritize?

First, our country must focus on water efficiency, to stop wasting water as rampantly as before. We have the skills of growing cadres of university trained water resource specialists and technicians who know how to prevent the waste of water through improved water resource management efficiency.

Second, an early focus should be groundwater with an emphasis on purification and demineralization. Current water quality fails to meet standards, yet people drink it, give it to their children and their livestock, and they water their fields with it. People deserve potable water.

Third, modern irrigation techniques need to be introduced. A big reason for the overuse of water is the mismanagement of existing irrigation systems. 7.9 million hectares are being siphoned from the lake's drainage basin, 92% of that water is being used for agriculture. This part is not easy. The high financial costs have held back any implementation of a canal modernization project. It also requires skilled people to direct and disseminate the project and a supply of trainable workers. This is challenging when so much of the population has mainly non-technical farming skills. Although the cost of lost water is conserved, the capital cost of the needed renovations is far beyond available state and international funding. The fourth option is implementation of water treatment and reclamation by all countries in the Amu and Syrdarya watersheds. Because most rural populations do not have access to safe water, these new water treatment plants would aid the local population directly and encourage improved health. Moreover, additional water resources would be created because the new wastewater treatment plants conserve fresh water and allow the reuse of treated water. Although this option is technically difficult and needs to be built from the ground up, it would provide the most relief to the population in the shortest amount of time.

Finally, a fifth option, reuse of water and deployment of drip irrigation technology, would improve water supply and irrigation.

Although all options would have their unique pitfalls, by far the most important action to take in the short-term is to work out water conservation in arid to semiarid areas, which applies to most of Uzbekistan. Improper irrigation practices are still becoming more prevalent as the situation gets more desperate. Everyone should have access to clean water, but very few people do. One of the best improvements made involved the creation of the Water Users Association in 2003. However, these associations have proven not to be powerful enough to effect change. Disagreements persist about water policy. Significant progress is required to achieve administrative cooperation on water management. These steps are intended only as a starting point. Much more needs to be done to address and attempt to mitigate the loss of the Aral Sea, if not to bring back the sea itself, at least to rectify the consequences.

A PERSONAL COMMENT ON ACHIEVING THESE GOALS

The first author, as a young person of Uzbekistan's next generation, calls on our learned and wise elders to focus your attention on our nation's water crisis. Our youth appreciate that the Aral Sea disaster was tragic and unnecessary. It reflected a failure of a generation of adults to take responsibility for a catastrophe of their own making. Such failures cannot continue. We must work together to develop realistic programs to improve the state of water, agriculture, and the rural economy in our nation. We are all responsible for the health of our environment and must deliver it to the successive next generations in good condition. Our challenge is to develop new concepts as well as techniques and technologies to solve the complex problems that my generation is about to inherit. But the focus of these efforts must be clear. Some day my own children will receive an Uzbekistan that is as healthy and ecologically sound as I can help to achieve. I know my peers, the youth of this nation, will work together to make this happen.

REFERENCES

Dukhovny, V. A. (2007). Water and globalization: Case study of Central Asia. *Irrigation and Drainage*, *56*(5), 489–507. doi: 10.1002/ird.327

ECOGeoscience. (2009). *Uzbekistan*. Retrieved from http://www.ecogeodb.com/HomePage.asp

Golden, P. (2011). *Central Asia in world's history*. New York, NY: Oxford University Press.

Liu, X. (2010). *The Silk Road in world history*. New York, NY: Oxford University Press.

MAWR. (2010). *National report of MAWR*. Tashkent. Retrieved from http://www.apipnm.org/swlwpnr/reports/y_nr/z_uz/uztb312.htm

Mikhalev, V., & Reimov, A. (2008): Land degradation in Central Asia. *Development and Transition*, 9. Retrieved from http://www.developmentandtransition.net/index.cfm?module = ActiveWeb&page = WebPage&DocumentID = 673#. Accessed May 2009.

Samylov, S. V., Abdujalilov, U. A., Ahmadjanov, B., Bekmirzaev, I., Grigoryants, A., Dotsenko, N., et al. (2006). *The national report environment and natural resources use*. Tashkent: Chinor ENK.

Wehrheim, P., Schoeller-Schletter, A., & Martius, C. (2008). Continuity and change: Land and water use reforms in rural Uzbekistan. Studies on the agricultural and food sector in central and eastern Europe (Vol. 43, pp. 89–104). Leibniz-Institut für Agrarentwicklung in Mittel-und Osteuropa (IAMO), Halle.

CHAPTER 3

POLLUTION AND SALINIZATION: COMPOUNDING THE ARAL SEA DISASTER

Farhod Ahrorov, Olim Murtazaev and Bahtiyor Abdullaev

ABSTRACT

Central Asia today is the focus of intense international attention because of its geopolitical and economic importance, its natural and human resources, its transit potential for transcontinental trade and transport, and its unsettled political situation. The last years of the Soviet period introduced increasing natural resource degradation. The deficit of water resources and contamination of the waters are two of the most severe problems within the region. Overuse of agrochemicals continues to result in the intensive pollution of agricultural lands and water resources. Many of the environmental challenges in Central Asia are closely linked to agricultural and industrial production. The most visible result of pollution is the increasing salinity of water, especially in downstream areas. The lack of domestic resources to address environmental issues requires regional solutions built upon environmental cooperation and goodwill reinforced with international assistance.

Disaster by Design: The Aral Sea and its Lessons for Sustainability
Research in Social Problems and Public Policy, Volume 20, 29–36

ISSN: 0196-1152/doi:10.1108/S0196-1152(2012)0000020011

INTRODUCTION

Central Asia is situated in the heart of the Eurasian continent with a total area of $3.9 \times 106\,\text{km}^2$ and is home to almost 60 million people. It encompasses five countries, namely Kazakhstan, Kyrgyzstan, Tajikistan, Turkmenistan, and Uzbekistan. The main natural features of the region include the highest ridges of the Pamir, the Tian Shan, and the Altai, vast deserts and steppes, important rivers such as the Amu Darya, the Syr Darya, the Irtysh, and the Ily, and large inland water bodies, such as the Caspian Sea, the Aral Sea, and a range of other lakes. The regional landscape is noted for its dry continental climate and aridity. The countries of Central Asia are the focus of intense international attention today because of their increasing geopolitical and economic importance, their natural and human resources, their transit potential for transcontinental trade and transport, and their unsettled political situation. The region is blessed with abundant natural resources, including oil, gas, coal, metal ores, and fresh water, as well as areas rich in biodiversity.

These riches did not go unnoticed by the massive neighbor to the north. Beginning in Tsarist Russia and continuing through Soviet times, the development of industry, commerce, and their associated state institutions were spearheaded by continuous modernization and integration campaigns. The Russians knew how to create modern infrastructure for the extraction and redistribution of Central Asian natural resources and commodities. By establishing their presence and building a railroad transportation network, they were able to redistribute Central Asian materials throughout the region and to other parts of what became the Soviet bloc.

This systematic development largely disregarded natural constraints, the economic costs, and the social and environmental consequences of the rapid development of the region. For example, when Soviet leaders realized, in the 1960s, that large-scale production of cotton was killing the Aral Sea and threatening fishing communities, they did not pull back from their pursuit of cotton production yields. They simply decided to import fish from the Far East for processing in the local factories that had been built on the Aral Sea as a short-term measure and to prepare for the diversion of Siberian rivers into Central Asia as a long-term plan.

In many other instances as well, the lack of knowledge and awareness in our society for resource conservation and innovation resulted in staggering risks to sustainable development. Some prominent examples that are now known worldwide include the radioactive contamination of the nuclear test site near Semipalatinsk in Kazakhstan, the opening of the so-called "Virgin

Lands" in north-western Kazakhstan for grain production, the uranium and hard-rock mining in Kyrgyzstan, the polluting industries of north-east Kazakhstan, the aluminum plant in Tursunzade, Tajikistan, and last but not least, salinization of land and mineralization of water resources in Uzbekistan.

ENVIRONMENTAL ISSUES PROLIFERATE IN UZBEKISTAN

Western Uzbekistan witnessed increasing natural resource degradation during the later years of the Soviet period due to massive irrigation and drainage system development as well as the conversion of vast tracts of desert into irrigated agricultural land (Gleick, 2000). From the perspective of ecological balance, a deficit of available water resources and the contamination of the diminished waters are severe problems within the region.

Of the major anthropogenic sources of water pollution in western Uzbekistan, pollution caused by agricultural activity is of greater significance than either of the two other predominant sources, industrial contamination, and household and municipal waste disposal in urban and rural areas. Our concern is, therefore, primarily with the former. Looking at agricultural activity more closely, overuse of agrochemicals results in the concentrated pollution of agricultural lands and water resources. Reports indicate that certain amounts of applied fertilizers: nitrates – up to 25%, phosphate – 5%, pesticides – up to 4%, are leached from the irrigated fields into drinking water reservoirs. The concentration of contaminants in drinking water reservoirs exceeds standard values for domestic drinking water supplies by 5–10 times.

Downstream regions of Uzbekistan along the Syr Darya and the Amu Darya basins thus exhibit clear signs of land and water degradation, resulting in a decline in crop yields. This degradation has threatened the food security not only within the areas where degradation is happening, but also in Central Asia as a whole (Klotzli, 1994). Since 1961, the water level of the Aral Sea has been receding progressively at the rate of 20–90 cm/yr. Accelerated salinization and desertification of the land coincides with water pollution in the Amu Darya and Syr Darya deltas. The former bed of the Aral Sea is now an area of dust, pesticide residues, and salt. The decline of the Aral Sea also causes the climate regulation of its basin to change, causing additional environmental stresses. Meanwhile, the overall regional water deficit is expanding over time. These are the local, ground level direct impacts of water management decisions made long ago. Added to that are new reasons for concern which include the rate of population growth in Central Asia, Afghanistan's increasing demands on the water supply, and the intensified desertification process across the region, that in turn fuels climate change.

A primary, current domestic concern is the measurable pesticide residues which are accumulating in the soil. These are unfortunately leading to considerable changes in agricultural yield. Since 1991, regional agricultural yields have reportedly declined by 20–30%, causing annual losses in agricultural production of as much as \$2 billion (Asian Development Bank, 2004).

The decline in yield has other contributing factors, several of which are felt cross-regionally. Between 1990 and 2000, the share of land in Central Asia with high groundwater levels increased from 25% to 35% of the total irrigated area. The area of salinized territories in the Amu Darya basin increased by 57% and in the Syr Darya basin by 79% from 1990 to 1999. These consequences are not limited to Uzbekistan. More than 30% of irrigated land is salinized in Tajikistan and up to 40% in Kyrgyzstan. Erosion is even more widespread, affecting some 51% of agricultural lands in Kyrgyzstan and some 97% in Tajikistan. In Central Asia, as elsewhere, people depend not only on cultivated land, but also on natural and seminatural ecosystems for their food and various other ecosystem services. Today, the useful productivity of such ecosystems is under significant threat, most notably the degradation of the marine and coastal ecosystems of the Aral Sea.

While Uzbekistan domestically relies on its agricultural land as a dominant producer in the region, the delicate balance of land and water is being felt in other ways. Other regionally important ecosystems – such as grasslands, river deltas, and the mountains are also declining at an alarming rate. Sparsely productive saline deserts, *solonchaks*, have developed over an area of 4 million hectares affected by drainage waters. Flooded meadow soils in the deltas have dried up and transformed into *takyr* desert soils on over 1 million hectares. In recent decades the area of natural lakes in the Amu Darya and Syr Darya deltas declined from 640 and 833 km^2 to 80 and 400 km^2, respectively, leading to the disappearance of once profitable fishing and muskrat hunting and the complete loss of marshy areas for commercial use. In the Syr Darya delta, the biological productivity of commercially valuable reeds decreased by a factor of 30–35.

When we consider that agricultural pollution is the biggest offender, and that poor water management led to all of the problems listed above, we still have to consider the technological culprits. The expansion of canalization, with wide channels funneling river flow into a network of irrigation canals, has led to inefficiency losses due to evaporation in the canals, and caused the flow reduction in the rivers themselves. The ensuing drainage discharges worsen the water quality. Since the early 1960s, the irreversible consumption of the river waters doubled. At present levels, this consumption is four times higher than during the 1930s and 1940s. The irrigation system is the crux of the problem.

As we are taking measurements in the field, accumulating evidence confirms that environmental impacts on both land and water in Uzbekistan and across the region are many, worsening, and in some places, tragic. Since the 1960s, the water quality in Central Asia has drastically deteriorated. The main reason for this has been the discharge of heavily polluted water through drainage systems currently making up to 15% of the river flow volume of the Aral Sea Basin (Kipshakbayev & Sokolov, 2002). The most visible result of pollution is the increasing salinity of water, especially in downstream areas. In the Republic of Karakalpakstan, for example, river water is unsuitable for drinking 10 months a year due to excessive mineral residues. Since the 1960s, mineralization of water in the lower reaches of the Amu Darya and Syr Darya has at least doubled, and water has also become unacceptable for drinking or harmful to health in the medium reaches of Syr Darya.

Effluents from municipal and industrial sewers and runoff from waste disposal sites and mining industries are other significant sources of pollution. The decline in industrial production of the 1990s did not significantly change the high levels of water pollution. Water contamination by metals (including mercury, copper, zinc, and chromium), organic chemicals, and other substances also poses significant risks. In Uzbekistan, only about 2% of the population lives in an area with good water quality, while some 50% live in areas with bad or very bad quality water.

Identifying the problems is important, but even more important is to set about the task of solving them. Unfortunately, there is very little domestic capacity to cope with these systemic problems of water management. According to the Environmental Sustainability Index (ESI) recently calculated by the World Economic Forum and several influential environmental think tanks (Esty, Levy, Srebotnjak, & de Sherbinin, 2005), Central Asian nations rank low compared with much of the world in their ability to protect the environment over the next several decades. As the results

Table 1. ESI Rankings for Central Asian Countries.

Central Asia							Reference Countries	
Country	ESI	Environmental systems	Environmental stresses	Human vulnerability	Institutional capacity	Global stewardship	Country	ESI
Kazakhstan	78	33	15	71	130	129	Russia	33
Kyrgyzstan	80	50	41	78	112	110	Belarus	47
Tajikistan	134	59	22	144	131	115	Azerbaijan	99
Uzbekistan	142	65	86	122	147	141	Ukraine	108
Turkmenistan	144	81	130	88	146	127	Iran	132
Regional average	116	57.6	58.8	100.6	133.2	124.4	China	133

Source: The Environmental Performance Measurement Project (http://www.yale.edu/esi/).
Note: The ESI ranks 146 countries; a higher rank indicates better performance in a particular area (the highest rank is 1 and the lowest is 146). Ranks in the shaded cells fall below the worldwide median.

displayed in Table 1 reveal, the main reasons for their poor prospects stem from weak institutional capacity and a limited ability to reflect regional and global impacts in managing national environmental resources.

CONCLUSION

Central Asia faces a critical challenge in the peaceful, productive and sustainable use of its natural resources. Many of the environmental challenges in Central Asia are closely linked to agricultural and industrial production. The massive diversion of water for irrigation has resulted in the widespread destruction of ecosystems, especially in the Aral Sea and the river deltas. Mismanagement of irrigation infrastructure has caused land salinization, swamping, desertification, and disturbed ecosystems. Environmental impacts such as climate change are in turn affecting the formation of river flow and availability of water throughout the region. Environmental pollution aggravates water scarcity by making water resources unsuitable for agricultural or domestic use. Excessive use of fertilizers leads to mineralization of soils and contamination of water stocks. Similarly, the extraction, transportation, and use of all forms of energy have had significant environmental impacts. These impacts are caused by oil drilling, coalmine development, and the building of pipelines, dams, and thermal power stations.

Despite providing basic protection, Soviet-era institutions are no match for modern environmental challenges. Moreover, since we now know what technological and managerial decisions caused the loss of the Aral Sea and the depletion of the Amu and Syr rivers, it is time to implement new technologies and management systems.

The Central Asian republics inherited numerous rules and standards that were inflexible, largely unsuitable for today's predominant market economies, and did not allow for local or democratic input in decision-making. The environmental problems in Uzbekistan and the region need serious and immediate attention both domestically and internationally. Environmental cooperation across the region is one important factor and external assistance is another. Failure to take action soon will only aggravate the potentially serious consequences for food security and human development in the region.

REFERENCES

Asian Development Bank. (2004). *Program concept for pipeline entry of the global environment facility for Central Asian countries initiative for land management*. Retrieved from www.adb.org. Accessed in March 2005.

Esty, D. C., Levy, M., Srebotnjak, T., & de Sherbinin, A. (2005). *Environmental sustainability index: Benchmarking national environmental stewardship*. New Haven, CT: Yale Center for Environmental Law and Policy.

Gleick, P. H. (2000). The changing water paradigm: A look at twenty first century water resources development. *Water International*, *1*(25), 127–138.

Kipshakbayev, N. K., & Sokolov, V. I. (2002). Water resources of the Aral Sea Basin: Formation, distribution, usage. In *Water resources of Central Asia*. Tashkent: Interstate Commission for Water Coordination.

Klotzli, S. (1994). *The water and soil crisis in Central Asia. A source for future conflict?* ENCOP Occasional Paper No. 11. Center for Security Policy and Conflict Research Zurich/Swiss Peace Foundation, Berne.

CHAPTER 4

DEATH AND REBIRTH ISLAND: SECRETS IN THE U.S.S.R.'S CULTURE OF CONTAMINATION

Michael R. Edelstein

ABSTRACT

The history of Vozrezdheniye Island in the Aral Sea is recounted, both to establish the related hazards and to acknowledge the underlying social process it shares with the Aral Sea Disaster, itself. A Closed City, part of the secretive world of Soviet arms manufacture and testing, carried out a hidden agenda of experimentation in the heart of the vast blue sea. Even as the Aral Sea was polluted and starved of water from without, parallel authoritarian regimes were poisoning its heart at the center. In this way, the military industrial complex became an actor with parallel duplicity to the public planned economic sector in charting the death of the Aral Sea. Ironically, the very attraction of Rebirth Island as an isolated weapons testing ground was undermined by the actions of the Soviet government to desiccate the Aral Sea, thus uniting Vozrezdheniye to the main shore and destroying its seclusion. In both instances, the case study is a further indicator that the people of Karakalpakstan were considered marginal and expendable, underscoring environmental injustices that continue long after the demise of the Aral Sea.

Disaster by Design: The Aral Sea and its Lessons for Sustainability
Research in Social Problems and Public Policy, Volume 20, 37–51

ISSN: 0196-1152/doi:10.1108/S0196-1152(2012)0000020012

INTRODUCTION

It is instructive that the prologue of Ken Alibek's (1999) book *Biohazard*, the crucial expose about the Soviet bioweapons program written by its former head, begins with a recollection that truly must haunt him. A hundred monkeys, tethered to poles in a field "on a bleak island in the Aral Sea," choking frantically as they succumb to a biological weapon dropped into their midst.

Vozrezdheniye Island is translated into English as Rebirth or Renaissance Island, an irony that cannot be ignored. There may be no part of the dying Aral Sea that symbolizes death more. In their classic study of the Soviet Union, *Ecocide in the USSR*, Feshbach and Friendly term Vozrezdheniye as one of the most dangerous places on earth (1992). Its story is important as a part of the Aral history and as an exemplar for the actions of authoritarian systems that caused the larger loss of the Aral Sea.

The Culture of Contamination

Elsewhere in this volume, it is made clear that the downstream Aral Region has received an inordinate concentration of pesticides and other agricultural poisons. In a direct way, the Aral Sea became a sink, or the final place of deposit, for all chemicals added to the farmlands that branched off the two rivers feeding it, the Amu Darya and Syr Darya. And now, converted to desert, the Aral Kum's eroding floor redistributes particles of poison around the region and beyond (Breckle & Wucherer, 2012). This tragedy itself was the result of the Soviet Union's authoritarian society and its structural dependence on the pseudo-validity of interim term planning loosely based on science and carried out through technocratic principles where the ends justified the means. Having determined a goal, steps needed to attain that goal would be achieved at all costs and without question. The system was reinforced by fear, since dissent was not tolerated (Figes, 2007). The effect was to imbue the public operating structures of society with a near military discipline in which orders were not questioned. In my prior work, I have described the Soviet system as a "culture of contamination" (Edelstein & Tyschiachniouk, 2007).

Closed Cities

The tragedy unfolding in the fields of Vozrezdheniye Island was a different manifestation of that same culture. President Eisenhower famously termed it the military industrial complex in his illuminating retirement speech (Eisenhower, 1960). And, in the Soviet Union, the term had at least as much validity as in the United States. In both cases, spawned by crisis, secret authoritarian societies were developed to carry out the development, production, and delivery of weapons. In the United States, the Manhattan Project created during World War II to win the race for development of atomic weapons created just such a secret world. Parallel universes existed for chemical and biological weapons as well. By nature, such weapons demand authoritarian control systems (Mumford, 1964).

The Soviet Union followed these same paths, creating an elaborate secret world of Closed Cities, analogous to the American atomic cities created under the Manhattan Project. There, scientists, staff, and security lived and worked, carefully walled off from the outside world. In Russia today, such cities continue to exist and operate. Perhaps the most illuminating description published about life in the Closed Cities was offered by Kutepova in my co-edited volume *Cultures of Contamination* (Kutepova & Tsepilova, 2007). She writes of the psychology of the internal authoritarian environment that it combines control by fear with the seduction of power.

> Closed Cities have existed under top-secret conditions since their inception. They house the nation's most hazardous facilities. Their actions have been historically destructive of the environment and surrounding population .... the special mentality of Closed Cities mixes fear with the power that [it]... holds over the rest of the world. The community is closed and intimidated, yet its members hold a conscious superiority arising from their ability to develop and produce weapons of mass destruction. The main ideological postulate of the...population thus reads, "No one can threaten us, but instead, everyone should be afraid of us because we handle something terrible for the whole world."

The Closed City and its military agenda require seclusion, secrecy, and cover, and it relies on the fact that outsiders will be inattentive, intimidated, powerless, or economically dependent on the enterprise. As a military complex, the rules governing appropriate risk taking would not be acceptable to the general public, yet these rules apply to Closed City civilians. The scientific purpose of many Closed City enterprises, moreover, operates in an environment of experimentation and testing that exhibits a great deal of uncertainty and unpredictability. Such research, scrutinized by the public, would be encumbered with elaborate demands for precaution. The purpose of the Closed City system is to avoid such encumbrances by whatever means possible in order to address the urgent demand for success in meeting whatever goals have been set.

THE SECLUDED ARAL SEA ISLANDS

Given the deep history of Silk Road travel through Central Asia, it is an interesting statement of the vastness of the Aral Sea that Vozrezdheniye Island was not discovered until 1848.[1] Then the Russian ship *Konstantin* charted a group of three islands that were called the Tsarskiye or Tsar Islands. One island, Konstantin, like their ship, was named after Grand Duke Konstantin Romanov, chair of the Russian Geographic Society. The Soviets later renamed one of two other Komsomolskiy Island. The largest was renamed Vozrezdheniye (Rebirth) Island.

Forced collectivization of farms was introduced to the Soviet Union in 1930 as part of Stalin's "great break," or effort to break from the old society. By forcing the peasantry onto collectivist farms, a cheap and secure supply of food could be produced to fuel industrial production mandated in a new Five Year Plan that would require a major mobilization of workers. The ruthlessness with which the exaggerated goals of the plan were to be pursued was evident when, in January and February 1930, 60 million peasants in 100,000 villages were forced off their own farms and into collectives. The most independent and successful farmers were deemed a source of opposition to the effort and were denounced as "kulaks" and sent into exile where they could mount no resistance. In January 1930, 60,000 "malicious kulaks" were sent to labor camps and their families exiled to the most remote areas of the country, Karelia, Siberia the Urals, and Kazakhstan (Figes, 2007). Recognizing the value of its remoteness, Vozrezdheniye became a forced place of exile for some of these kulaks (Bozheyeva, Kunakbayev, & Yeleukenov, 1999).

This employment of the island's extreme geographical isolation won Vozrezdheniye a special but unfortunate place in later Soviet history. Conditions were perfect for secret activities requiring solitude. The surrounding sea served as a giant moat keeping out intruders. Vegetation was sparse. The summers were blazing hot (up to 140°F/60°C) and the winters bitter cold, extremes that discouraged microorganism survival. The soil was sandy. Just as human escapees could not, neither wind, animal, nor insect – were likely to reach the mainland from there (Global Security Organization, 2012; Middleton, 2005).

THE FIELD TESTS

On its way to becoming labeled one of the most dangerous places on earth, Vozrezdheniye was selected for field testing of early biological weapons.

Soviet interest in biological weapons had emerged during the Russian Civil War, when disease played a major role on the battlefield. By 1936, an expedition of 100 scientists had traveled to Vozrezdheniye Island to conduct experiments with a disease called tularemia. The testing ended abruptly, however, when the top scientists were arrested. A 1942 grave marker on the island for a young scientist killed from contact with a biological agent indicates activities there during the Second World War. During the war, the Soviets deployed biological weapons against the Germans in such battles as Stalingrad with considerable collateral damage (Alibek, 1999).

During the Cold War era, great public attention was paid to building "the bomb" and delivering it. The chemical weapons' programs worldwide came out of the closet and went commercial with the new war on pests and the expansion of agriculture. The third pillar of weapon production, biological weapons, remained secretive even to its very existence; programs remained under wraps. The Soviets developed a very extensive and aggressive program. Alibek notes that while the Americans developed bioweapons for diseases for which there was a cure, the Soviets sought to produce weaponized diseases for which no cure was in existence. Biological weapons were manufactured in many places in the U.S.S.R., but Alibek writes that the Soviet Union became the first biological weapon superpower after completion of the world's first industrial-scale biological weapons factory at Stepnogorsk in what is now Kazakhstan. The factory was capable of producing two tons of anthrax a day formulated to be magnitudes stronger than prior forms. Based on testing of the new weapon at Vozrezdheniye in 1987, he reported that 5 kg of Anthrax 836 was capable of fatally infecting half the people living in a square mile of territory (Alibek, 1999).

In the postwar era, Soviet bioweapons manufacturing and testing facilities were concentrated in Kazakhstan and Uzbekistan. Bioweapons testing intensified at Vozrezdheniye Island on a large scale when Aralsk-7, a closed military city and bioweapons test site, was built on Vozrezdheniye and Komsomolskiy Islands in 1954. The scientific research laboratory was built on Vozrezdheniye while the military garrison was divided between Vozrezdheniye and Aralsk (Alibek, 1999; Global Security Organization, 2012).

Vozrezdheniye was served by a seaport and an airport. A military settlement was built in the northern (Kazakh) third of the island where officers, some with families, and soldiers lived. The housing compound supported as many as a 1,000 inhabitants. Besides the military garrison, this number also included Russian scientists, technicians, and their families. The settlement had barracks, residential houses, an elementary school, a nursery school, a cafeteria, warehouses, and a power station. Soldiers received hazard pay to serve there (Alibek, 1999; Global Security Organization, 2012).

Aralsk-7 had a contingent of 1,000 Soviet troops. Six hundred of these were based at Aralsk and 400 on Vozrezdheniye (Miller, Engelberg, & Broad, 2002; Middleton, 2005). The contingent on Vozrezdheniye was trained in germ warfare (Miller et al., 2002) and was most likely led by officers from the elite military academy at Kostroma. Kostroma trained officers in tactical use of nuclear, chemical, and biological weapons and their troops were also involved in deactivation and disinfection activities. Kostroma officers have been engaged in addressing problems created by special weapons systems and, most famously, played a key role as liquidators during the Chernobyl disaster (Smirnova & Edelstein, 2007).

While residences were placed in the northern part of Vozrezdheniye, the research labs and testing fields were placed on the lower two-thirds (Uzbek) part of the island. In the most southern area, a vast field of telephone poles had been sunk on a grid with a kilometer distance between each. Each pole held an air pollution monitor. It was to these poles that the ill-fated test subjects would be strapped during the experiments (Alibek, 1999).

It is not clear whether earlier scientific teams were kept on Vozrezdheniye year round. But during the 1980s and early 1990s, a team of 150 scientists and security staff was dispatched (presumably from another Closed City) for four or five months of each summer to test the military's stock of biological weapons. Already by this time, the Aral Sea was in an advanced state of decline, the water polluted by agricultural runoff. Other than lizards and rats, there was no evident fauna on the island. The few local residents lived in earthen huts without running water and suffered from malnutrition and hepatitis. The time spent in the barracks was reported to be interminably boring, relieved only by drinking and sexual escapades. Winds from the eroding dry seabed evident around the island blew dust everywhere. The scientists jokingly referred to the place as the kingdom of the cockroaches (Alibek, 1999).

A large number of biological disease agents were tested at Vozrezdheniye from the 1970s, if not before. These included tularemia, Q-fever, brucellosis, ganders, and conventional plague (Alibek, 1999; Miller, 1999), and new highly resistant strains of weaponized plague and anthrax (Miller, 1999), and also typhus, botulinum toxin, Venezuelan equine encephalitis, and smallpox. There was a focus in the early times on diseases of animals that could be passed to humans and later on with microbial strains with military applications having high virulence and environmental resistance. Work was also done with genetically altered diseases in the later period designed specifically to overcome existent cures (Bozheyeva et al., 1999; Miller 1999; Miller et al., 2002). The facility, which operated until 1991, was used to test the lethality, dispersion, and delivery options for biological weapons (Koblentz, 2011). Simulated bacteria were also used to study the dissemination of aerosol particles in the atmosphere (Global Security Organization, 2012).

The agents were tested on animals tied to the poles in "the killing field." There horses, sheep, donkeys, baboon and particularly monkeys, and a variety of small animals (rabbits, white mice, gerbil, and hamsters) were sacrificed (Miller, 1999). While the scientists ate a boring Soviet diet, Alibek recalls their jealousy at the Animals' last suppers.

> We used to say that the most fortunate inhabitants of the Soviet Union were the condemned monkeys of Rebirth Island. They were fed oranges, apples, and bananas, and other fresh fruits rarely seen by Soviet Citizens ... our test subjects had to stay healthy until their last breath .... (Alibek, 1999, p. 17)

And, in fact, the research was all about discerning when that last breath would occur and how it would be taken. While the victims were thus confined, the experiment of the day would be carried out. It might vary the agent, its variant, the potency, the form of the agent (liquid, powder), the means of delivery (bio-bomblets, etc.). Depending on the conditions, they would survive the initial exposure and, carried back to the laboratory, the progress of their death would be carefully studied. The test subjects were often vaccinated with the cure for the agent to be used in order to test new super-strains designed to defeat known treatments. Success, for the scientists, was defined by achieved deaths (Alibek, 1999). When later caught having violated the terms of the Biological Weapons Treaty it had signed only shortly before intensifying its work at Vozrezdheniye in the 1970s (Bloggett, n.d.), the Soviet Union would portray the activities at Vozrezdheniye as aimed only at developing defensive measures for its troops in battle. In fact, the laboratory developed methods of biological defense and decontamination that were later used for troops, for example, in Afghanistan (Global Security Organization, 2012). However, Alibek (1999) makes clear that the primary research was on offensive biological weaponry.

ROLE OF SECRECY

The essense of any closed military system is secrecy. There were reasons to prevent local residents of the Aral region from understanding the hazards involved. And, of course, Vozrezdheniye was part of a weapons race during both hot and cold war periods. The Soviet bioweapons program had been moved from the defense establishment to a new agency at the start of the cold war. This agency, Biopreparat, operated a network of 40 facilities, all given cover stories to appear as peaceful facilities but all providing cover for civilian work on bioweaponary (Miller, 1999).

Operations in Vozrezdheniye were highly secretive. The islands were continually circled by speeding patrol boats. Teams sent there were extremely discrete. Norms of secrecy encapsulate Closed Cities as a group and are enforced by clear rules. Civilians as well as military assigned to Closed Cities understand and obey these rules at risk of punishment for violations (Kutepova & Tsepilova, 2007). It is likely that a similar Closed City rule set would have applied here. Even the design of the facility sought to make it secretive. The port and airfield on Vozrezdheniye were kept at minimal levels and the residences and labs looked unremarkable. When activities could not easily be passed of as normal, however, elaborate rules were needed. For example, when 500 monkeys were needed as test subjects in 1982 for a weaponized vaccine resistant version of the disease tuleremia, they were acquired in Africa and brought to the island using a fictionalized chain of custody and disguised itinerary to cover up the actual destination of the shipment. The tests were then conducted and the monkeys killed. After the results were flown back to Moscow, everything on the island was swept clean and disinfected and all indications of any habitation during that period erased. Other than statistics in the research report, no vestage of the monkeys remained. After Vozrezdheniye ceased to be used for testing in 1992, all records of its existence were destroyed (Alibek, 1999).

Now, secrecy is a veil that is both hard to pierce but also difficult to preserve. And there was intense interest on the part of the Americans during the Cold War in tracking the Soviet bioweapons program. German agents had reported bioweapons testing at Vozrezdheniye, as reflected in U.S. intelligence reports. By 1958, CIA confirmed the presence of a possible bioweapons facility based upon a U-2 surveillance flight over the island (U2 4035 on August 5, 1957). The aerial images clearly showed every detail of the island, but some of what the images portrayed actually threw the Americans off track. First, they assumed that a Russian bioweapons facility would mirror the ones built by America, and there were a number of observations that did not match their expectations. Not only were transport facilities too primitive, but prevailing winds led directly to Konstantin Island, which was inhabited. Residential areas with playgrounds were located on Vozrezdheniye. By American standards, such factors would preclude bioweapons testing there (Koblentz, 2011).

Clearer images later arrived via American Corona satellite (1962–1970). The Americans were then able to match the design of the facility to that of a Japanese bioweapons facility in Manchuria overrun by the Russians in WWII (Koblentz, 2011).[2] It was not until the deputy head of the Soviet bioweapons program, Kanatjan Alibekov, defected to the United States that the full extent of the site came to be appreciated, however. Alibekov, who changed his name to Ken Alibek, also disclosed what was below ground at the site. Of particular interest, was Vozrezdheniye's unfortunate status as "the world's largest burial ground for weapons grade anthrax" (Miller et al., 2002, p. 178). The anthrax was transferred quietly to Vozrezdheniye in 1988 under the orders of the Soviet Union's last President, Mikhael Gorbechev. As part of his effort to create openness (*glasnost*) and transition (*perestroika*), Gorbechev faced many skeltons in the Soviet closet. One of these was a super form of anthrax, created under Alibek's supervision at the bioweapons facility he headed in the Kazakh city of Stepnogorsk, designed to defeat available antidotes. Given the clear violation of the Biological and Toxins Weapons Convention of 1972, the Soviets looked for a quick disposal solution that would hide its existense forever (Miller et al., 2002). They lacked the option of autoclaving the anthrax because the quantities were just too vast (Blodgett, n.d.), so the weaponized disease agent was placed on trains to the Aral region and transferred by boat to Vozrezdheniye. In some cases, the anthrax traveled more than a thousand miles. There, some 200 tons of anthrax, more than enough to kill the world's population, was covered in bleach to kill the spores and buried in the ground (Miller, 1999; Miller et al., 2002).

MATTERS OF RISK AND HAZARD

Earlier it was noted that U.S. intelligence initially discounted the aerial surveillance of Vozrezdheniye in part because civilians were nearby and downwind of the apparent testing fields. It was assumed that a biological weapons testing facility would not blatantly place civilians at risk. In fact, the Soviets tolerance for risk of public exposure was much higher than anything the Americans would contemplate.[3] The American concern was not without basis, however. A wind shift during a bioweapons test in 1960 indeed forced the evacuation of downwind Konstantin Island and led to the permanent relocation of the population (Koblentz, 2011).

Then there was the community on Vozrezdheniye itself, less than a mile away from the test range to the south. A kindergarten and playground underscored that children were present. Moreover, unlike the scientists, themselves, none of the children or families brought to the island were apparently inoculated for any of the agents tested and some weapons had no vaccines. The former head scientist in charge of Vozrezdheniye explained that the tests were not done unless the prevailing wind was facing south away from the residential area (Bozheyeva et al., 1999; Global Security Organization, 2012; Miller et al., 2002). It is at first hard to understand what seems to be a cavalier attitude toward risk of loved ones. However, I found a parallel in interviews I conducted in the 1980s with scientists and wives of scientists working at the Hanford Nuclear Complex in Washington State (Edelstein, 2007). It became clear that Closed Communities have a strong scientific esprit de corps that is expressed as confidence by families in the ability of the scientists to be protective and by the scientists themselves in their own abilities to avoid risk.

The Soviet bioweapons projects were not without incident. Regional wildlife was affected. In 1976, a mass fish death occurred in the Aral Sea. In June 1986, entire flocks of sheep lost their wool in the region in response to outbreaks of plague. In May 1988, over the course of only about an hour, approximately 500,000 saiga antelope suddenly dropped dead in the Turgay steppes northeast of the Aral Sea, an event that also prompted the emergency evacuation of people in the region. No clear cause was ever disclosed for the disaster (Bozheyeva, 1999; Global Security Organization, 2012; Miller et al., 2002). Rodents on Vozrezdheniye were found to exhibit unexpectedly high levels of plague. After the fall of the U.S.S.R., plague outbreaks were reported in several areas throughout Central Asia in 1991 (Alibek, 1999).

Humans were not free of impact. At Vozrezdheniye, the toll began early on, when a young female scientist died of pathogen exposure in 1942 and was buried on the island. We have also noted the 1960 release that forced the evacuation and abandonment of the island Konstantin. In 1971, an outbreak of smallpox at Aralsk may have been attributable to a scientist involved in smallpox experimentation (Blodgett, n.d.). In 1972, a sudden wind shift occurred during a release of plague; two fishermen out on the Aral Sea were found dead drifting in their boat, struck down by the disease. A visitor to the island was sent to the anti-plague station in Aralsk when she contracted smallpox (Global Security Organization, 2012).

By the 1980s, Aralsk was a former fishing and canning center that had been abandoned by the receding sea and supplying food and supporting the closed city became the only work available to remaining local residents (Alibek, 1999). Activities on the secret island caused serious concerns among local residents because of repeated epidemics and the mass deaths of animals and fish in the area. Individual cases of infectious disease also occurred in people who spent time on the island (Global Security Organization, 2012). Residents of the area were reported to be suffering from rare cancers (Miller et al., 2002).

The potential for impacts beyond the region were also acknowledged. For example, there were reports of Vozrezdheniye dusts blown as far as Turkey and Europe. Whether capable of carrying pathogens, such distant points of contact certainly contradicted notions that Vozrezdheniye was environmentally isolated. All of these concerns became matters for public discourse the moment that public discourse became possible. During the Perestroika period, Kazakhstani public activists made Vozrezdheniye a focus of public activism. It was a major concern, specifically, for three civil society organizations, the Nevada-Semipalatinsk international anti-nuclear movement, the nongovernmental organization Tamshy in Aralsk, and the Aral-Asia-Kazakhstan public movement (Bozheyeva, 1999).

CEASING TO BE AN ISLAND

By the early 1990s, the desiccation of the Aral Sea resulted in significant changes. At Muynak, the fish processing factory which was once on the banks of the Aral Sea was now 80 km from the beach; it began importing Pacific Ocean fish to maintain operations. The same problem was seen as the size of Vozrezdheniye expanded from 200 to 2,000 km^2. with the shrinkage of the sea. Operations there began to become impaired. It became more expensive to import things because the port had to be continually moved to the new more distant shoreline, complicating ground transportation and loading and unloading operations. And, of course, the emergence of a land bridge eliminated any security benefits. Even aside from other pressures to close Vozrezdheniye, these factors made the site a liability (Bozheyeva, 1999; Global Security Organization, 2012).

Military personnel began leaving Vozrezdheniye Island in 1991, after scientists departed and labs were shut. Public representatives from Kazakhstan were not allowed to visit the island until the Soviets were ready to abandon it, during 1991–1992. On their part, Uzbekistan reported not knowing that Vozrezdheniye Island was a biological testing site and disposal area until informed of it by the United States sometime in the late 1990s, according to Ambassador Shavkat Khamrakulov (Tucker, Khamrakulov, & Karimova, 2002). As a result of such lapses, there were sizable periods during which the island was unprotected (Blodgett, n.d.). Miller summarized the situation aptly.

> Kazakh and Uzbek officials knew that the Russians had viewed their countries as expendable, but until the Russians left, in the early 1990s, they had no idea of the magnitude of the ecological disasters confronting them. (Miller et al., 2002, p. 171)

On January 18, 1992, independent Kazakhstan's Supreme Soviet issued the edict "On Urgent Measures for Radically Improving the Living Conditions of Aral Area Residents," officially closing the military facility on Vozrezdheniye. Russian President Boris Yeltsin's Edict No. 390 of April 11, 1992, declared the end to the offensive bioweapons programs and declared Vozrezdheniye closed. By 1995, facilities were to be dismantled and the island decontaminated and transferred to Kazakhstan. An American inspection in August 1995 found these conditions met with one exception – surviving anthrax spores were found, a particular concern given the merger of the island into the mainland (Global Security Organization, 2012). A U.S. program was begun in 2002 to find remaining Anthrax (Tucker et al., 2002).

Various sources have expressed doubt over whether biological weapons were adequately contained at Vozrezdheniye. For example, Alibek acknowledged on CNN: "It is clear, when you destroy tons and tons of their weapons, it wouldn't be possible to kill everything. And now, what we know is this island is contaminated."

During the 1970s and 1980s, Soviet scientists found unexpectedly high levels of weapons-grade bubonic plague bacteria in rodents in the north of Vozrezdheniye. Some predicted that plague would be perpetuated by fleas. Other persistent reservoirs involving animals and insects might have been affected downstream, providing a means for spreading disease into Uzbekistan. Moreover, confidence that ultraviolet radiation protected the site faded once it was realized that bacterial and fungal spores grow below ground (Bozheyeva, 1999). Anthrax spores have been known to survive in residence for 100 years (Gale, 2012). The fact that viable pathogens may remain underground is a concern as oil and gas prospecting occurs nearby (Bozheyeva, 1999; Blodgett, n.d.). And entry to the island by scrap collectors and other scavengers has proven to be a continuing breach of security; animals can also easily cross and return onto the former island (Bozheyeva, 1999). To address these eventualities, both Kazakh and Uzbek scientists presently monitor the island for plague and other diseases, beyond anthrax, that might have been introduced to the island during Soviet times and could spread to the mainland through rodents (Kassenova, 2008).

There is also the possibility that any of the thousands of scientists, engineers, and technicians that worked on Vozrezdheniye Island may have taken samples from the island or might be using knowledge from the program to help emerging bioterrorist programs. CIA reportedly suspected one scientist of passing the Aralsk strain of smallpox to Iraq in 1990 (Blodgett, n.d.).

CONCLUSION

One might begin by commenting on the cold war period's dedication to surpassing the enemy in killing capacity. Mutual Assured Destruction (MAD) is the antitheses of any concept of sustainability. In the post-Cold War period, the potential proliferation of "weapons of mass destruction" might instead be termed Self Assured Destruction (SAD). The evolution from MAD to SAD is in no small part an outgrowth of the obsession with weapons development epitomized by the activities on Rebirth Island.

The Vozrezdheniye bioweapons program underscores the willingness of the Soviet system to carry out military programs that risked civilian populations and the environment. As an interesting parallel, the forced effort to intensively produce monocropped cotton was similarly coerced. There are other interesting parallels between Vozrezdheniye and the race for killing capacity that has characterized modern agriculture, itself an outgrowth of chemical warfare development. Whatever unknown threats persist in the Aral region from biological weapons production, the cumulative threat from pesticide use is known to have had a deleterious effect on the population of the Aral region (Lall, chapter 16). And, there is the important commonality between the development of a bioweapons program in a remote yet not totally unpopulated area and the intensive use of pesticides on monocrop farm fields to which farm workers and families are routinely exposed.[4] Both share a willingness to sacrificing people to reach plan goals. In both instances, this case study is a further indication that the people of Karakalpakstan were considered marginal and expendable and serves to underscore the environmental injustices that continue in their post-Aral period.[5]

Ironically, the very attraction of Rebirth Island as an isolated weapons testing ground was undermined by the actions of the Soviet government to desiccate the Aral Sea, thus uniting Vozrezdheniye to the main shore and destroying its seclusion. It was almost as if this final contortion of a dying Aral Sea sought to erase once and for all any remaining traces of this disaster within a disaster.

NOTES

1. By early accounts, the Aral Sea is referred to as the Sea of Islands, although it was even earlier called the "blue sea" (Breckle & Wucherer, 2012).
2. The actual satellite and U2 photos of Vozrezdheniye and the comparison of the Japanese and Soviet facilities can be seen at http://www.globalsecurity.org/wmd/world/russia/Vozrezdheniyerozhdenly.htm.
3. Although, as I document elsewhere in writing about the U.S. Hanford Nuclear Reservation in Washington State, there were collateral exposures of civilian populations in the United States as well from secret weapons programs. These were denied until the 1990s (Edelstein, 2007; for Americans and risk, see Edelstein, 2004).
4. See Wilson, chapter 10.
5. Edelstein, chapter 9, develops the theme of environmental injustice further.

REFERENCES

Alibek, K. (1999). *Biohazard.* NY: Dell.

Blodgett, B. (n.d.). A case study on Vozrezdheniye Island/Aralsk-7: Is there an unknown biological threat? Blodgett's Historical Consulting. Retrieved from https://sites.google.com/site/blodgetthistoricalconsulting/a-case-study-on-Vozrezdheniyerozdeniye-island-aralsk-7. Accessed on July 1 2012.

Bozheyeva, G., Kunakbayev, Y., & Yeleukenov, D. (1999, June). Former Soviet biological weapons facilities in Kazakhstan: Past, present, and future. Middlebury College, Center for Nonproliferation Studies, Monterey Institute of International Studies. Retrieved from http://cns.miis.edu/opapers/op1/. Accessed on July 1 2012.

Breckle, S., & Wucherer, W. (Eds.). (2012). *Aralkum – A man-made desert: The desiccated floor of the Aral Sea (Central Asia).* NY: Springer-Verlag.

Edelstein, M. R. (2004). *Contaminated communities: Coping with residential toxic exposure* (2nd ed.). Boulder, CO: Westview Press.

Edelstein, M. R. (2007). Hanford: The closed city and its downwind victims. In M. R. Edelstein, M. Tysiachniouk, & L. V. Smirnova (Eds.), *Cultures of contamination: Legacies of pollution in Russia and the U.S.* (Vol. 14, pp. 253–307). Research in Social Problems and Public Policy. NY: Elsevier.

Edelstein, M. R., & Tyschiachniouk, M. (2007). Psycho-social consequences due to radioactive contamination in the Techa River Region of Russia. In M. R., Edelstein, M. Tysiachniouk, & L. V. Smirnova (Eds.), *Cultures of contamination: Legacies of pollution in Russia and the U.S.* (Vol. 14, pp. 185–204). Research in Social Problems and Public Policy. NY: Elsevier.

Eisenhower, D. D. (1960). *Military-industrial complex speech, Dwight D. Eisenhower, 1961.* Public Papers of the Presidents (pp. 1035–1040). Retrieved from http://coursesa.matrix.msu.edu/~hst306/documents/indust.html. Accessed on July 27, 2012.

Figes, O. (2007). *The whisperers: Private lives in Stalin's Russia.* NY: Picador.

Gale Encyclopedia of Espionage & Intelligence. (2012). Retrieved from http://www.answers.com/topic/Vozrezdheniyerozhdeniye-island-soviet-and-russian-biochemical-facility. Accessed on July 1.

Global Security Organization. (2012). *Vozrezdheniye Island/Renaissance/Rebirth Island.* Retrieved from http://www.globalsecurity.org/wmd/world/russia/Vozrezdheniyerozhdenly.htm. Accessed on July 12, 2012.

Kassenova, T. (2008, July 18). Biological threat reduction in Central Asia. *Bulletin of Atomic Scientists.* Retrieved from http://www.thebulletin.org/web-edition/features/biological-threat-reduction-central-asia. Accessed on July 1, 2012.

Koblentz, G. D. (2011). *Living weapons: Biological warfare and international security*. Princeton, NJ: Princeton University Press.

Kutepova, N., & Tsepilova, O. (2007). Closed city, open disaster. In M. R. Edelstein, M. Tysiachniouk, & L. V. Smirnova (Eds.), *Cultures of contamination: Legacies of pollution in Russia and the U.S.* (Vol. 14, pp. 147–164). Research in Social Problems and Public Policy. NY: Elsevier.

Middleton, N. (2005). Welcome to Anthrax Island. *The Guardian*, Wednesday, April 20. Retrieved from http://www.guardian.co.uk/science/2005/apr/21/russia.international-news. Accessed on July 1, 2012.

Miller, J. (1999). At bleak Asian site, killer germs survive. *New York Times*, June 2. Retrieved from http://www.nytimes.com/library/world/global/060299anthrax-island.html. Accessed on July 1, 2012.

Miller, J., Engelberg, S., & Broad, W. (2002). *Germs: Biological weapons and America's secret war*. NY: Simon and Schuster.

Mumford, L. (1964). Authoritarian and democratic technics. *Technology and Culture*, *5*(1), 1–8.

Smirnova, L. V., & Edelstein, M. R. (2007). Chernobyl: A liquidator's story. In M. R. Edelstein, M. Tysiachniouk, & L. V. Smirnova (Eds.), *Cultures of contamination: Legacies of pollution in Russia and the U.S.* (Vol. 14, pp. 361–372). Research in Social Problems and Public Policy. NY: Elsevier.

Tucker, J. B., Khamrakulov, A. S., & Karimova, A. (2002). Biological decontamination of Vozrozhdeniye Island: The U.S.-Uzbek agreement. *Event transcript.* Washington: CNS. Retrieved from http://cns.miis.edu/archive/cns/programs/dc/briefs/011802.htm. Accessed on July 1.

CHAPTER 5

POTENTIAL CLIMATE AND HYDROLOGICAL CHANGES IN THE ARAL SEA REGION

James R. Miller, Imtiaz Rangwala and Debjani Ghatak

ABSTRACT

This chapter describes the climatic setting of the Aral Sea region, investigates how the climate might change during the 21st century, and discusses potential impacts on water resources. Temperature and precipitation fields are analyzed to describe the mean climate for the Aral Sea region. Composite analysis has been employed on the precipitation field from the Global Precipitation Climatology Project (GPCP v2.2) to assess the spatial pattern of changes in precipitation during the last several decades. Furthermore, temperature and precipitation projections available from the 2007 Intergovernmental Panel on Climate Change report are synthesized to examine the nature of climate change during this century.

Cold season precipitation has increased during recent decades, particularly over the mountainous terrain east of the Aral Sea. Climate models also project increases (5–20%) in winter precipitation during the 21st

---

Disaster by Design: The Aral Sea and its Lessons for Sustainability
Research in Social Problems and Public Policy, Volume 20, 53–64

ISSN: 0196-1152/doi:10.1108/S0196-1152(2012)0000020013

century; however, several models suggest decreases (0 to −15%) in precipitation during summer. Despite the increases in cold season precipitation, the large increases in temperature (4°C) during the 21st century are likely to cause increased evaporation which could exacerbate the regional water budget deficit. This may constrain the water supply in the region, particularly during summer and autumn when water demand is highest. To fully understand the impacts of future climate change on regional water resources, hydrologic models that include anthropogenic management of water will be required.

INTRODUCTION

The Aral Sea was the fourth largest lake in the world before 1960. It is located in an arid-temperate region where there is considerable water stress. Annually about 100 mm of rain would fall on the surface area of the lake at its previous size, and about 1,000 mm would evaporate from the surface (Micklin, 1991). This large imbalance between precipitation and evaporation meant that the water budget of the Aral Sea could be maintained only if there was another source of freshwater input. That input came from two major river systems, the Amu Darya that entered from the south and the Syr Darya that entered from the northeast. After 1960, major withdrawals from the rivers, primarily for irrigation, reduced water inflow to the Aral Sea, and the surface area was reduced to its present size. The environmental and social disaster that followed is documented in other chapters in this book.

The Amu Darya and Syr Darya rivers receive most of their water from the rugged mountainous regions located east of the Aral Sea. They are part of the world's loftiest mountain-plateau chain with the headwaters of the Syr Darya in the Tian Shan Mountains and the Amu Darya in the Pamirs and Hindu Kush. After leaving the mountains, both rivers travel through arid and semi-arid lowland regions before reaching the Aral Sea. Thus, the Aral Sea region has a diversified geomorphology consisting of mountain, plateau, and desert. Fig. 1 (from Nezlin, Kostianoy, & Lebedev, 2004) shows the region of interest and the path of the two river systems from their headwaters in the high mountains through Uzbekistan and into the Aral Sea.

The two river systems are the major source of fresh water for the region. Agriculture is supported by large withdrawals of water for irrigation. Some of the withdrawn water may return to the river through drainage channels

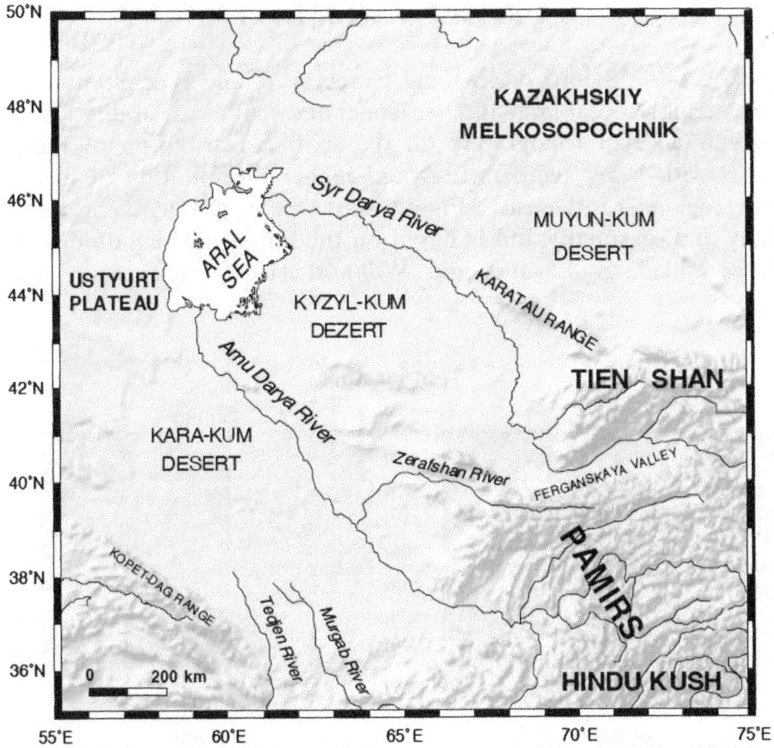

Fig. 1. Aral Sea Region. The Figure is Reproduced from Nezlin et al. (2004).

and infiltration into the underlying aquifers. This water is essentially recycled and available again for use downstream. However, most of the withdrawn water does not return to the rivers. Since demand fluctuates during the year, with the majority of withdrawals occurring between late spring and early fall, the impact on the downstream flow is significant.

In this chapter, we provide the climatic setting of the Aral Sea region by first examining the present climate and how it has changed in the last 30 years. The focus is on temperature and precipitation, and for precipitation, whether it falls as rain or snow. We then use global climate model projections to see how climate might change during the 21st century and discuss potential implications for water resources.

PRESENT CLIMATE

Long-term observations of seasonal temperature and precipitation for the Aral Sea region reveal large inter-seasonal and spatial variability. There is a strong influence of topography on the spatial distribution of these two variables with lower temperatures and higher precipitation in the mountainous regions to the east. Fig. 2 shows how the temperature varies spatially and seasonally and is based on the long-term climatology (1900–2008) compiled by Legates and Willmott (1990). The mean seasonal

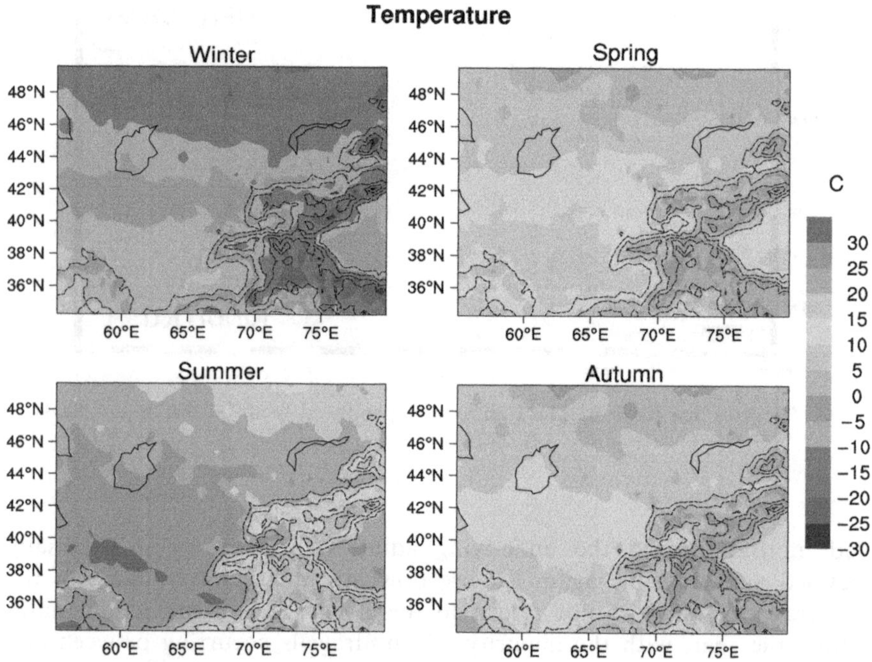

Fig. 2. Long-Term (1900–2008) Seasonal Temperature (°C) Climatology for the Aral Sea Region Based on the University of Delaware Dataset (Legates & Willmott, 1990). *Note:* All temperatures are positive in summer, spring, and fall except for negative values in the high mountains in the southeast. In winter almost all temperatures are negative except in southwest corner (35N-40N; 55E-70E) and a small area centered at 35.5N, 73E. *Note*: A color version of the figure is available online.

temperatures in the region vary between −30°C to greater than 30°C (i.e., −22 to 86°F). The low temperatures occur in the mountains during winter, and the high temperatures occur in the plains south of the Aral Sea. Daily temperatures can be significantly higher or lower than the seasonal means.

Precipitation plays a dominant role in the regional water budget. Fig. 3 shows that the highest precipitation occurs in the mountainous regions in the east and southeast. These mountainous headwater regions receive their maximum precipitation during winter and spring while the summer is dry. The primary source of moisture is from the cyclonic systems that originate in the westerlies. Moisture input from the Indian monsoon is very limited but does occasionally reach the south-eastern Pamirs in summer (Aizen et al., 2009).

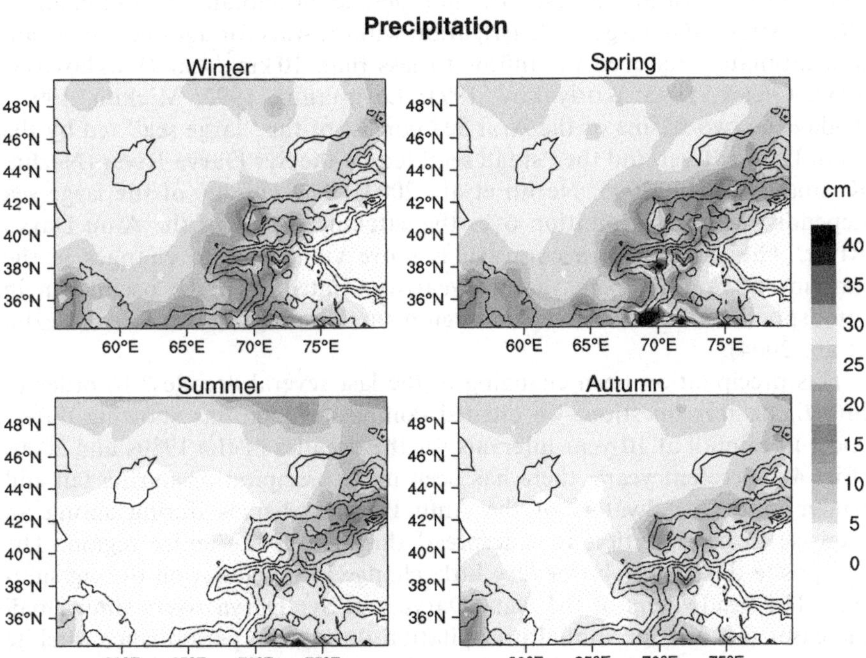

Fig. 3. Long-Term (1900–2008) Seasonal Precipitation (cm) Climatology for the Aral Sea Region Based on the University of Delaware Dataset (Legates & Willmott, 1990). *Note:* A color version of the figure is available online.

River flow is dominated by spring snowmelt, so it is important to know how precipitation varies spatially and temporally in the mountains. However, there have been inconsistencies among previous studies of precipitation in the mountainous headwater regions of the two river systems. Legates and Willmott (1990) found mean annual precipitation in the high mountains to be approximately 300 mm. According to Van der Leeden (1975), annual precipitation is 596 mm in Tadzhikistan (Amu Darya headwaters) and 470 mm in Kirghiz (Syr Darya headwaters). The higher values are closer to the estimates of Micklin (1991) who indicated a range from 800 to 1,600 mm in the mountainous regions.

For the present climate, the annual runoff into the headwaters of the two rivers is about 110 km^3 with flow in the Amu Darya being about twice that in the Syr Darya (Micklin, 1991). Prior to 1960 half of this water was lost before reaching the Aral Sea, primarily through evapo-transpiration, infiltration, natural ecosystems, and deltaic floodplains (Shiklomanov, 1979). After 1960, large-scale irrigation withdrawals for agriculture began, and ultimately reduced the inflow to less than 10 km^3/year (Dukhovnyy, 1984; Gleick, 1993; Kotlyakov, 1991; Levintanus, 1992; Micklin, 1991). Today the two basins of the Aral Sea consist of the "large sea" fed by the Amu Darya River and the "small sea" fed by the Syr Darya River (Nezlin, Kostianoy, & Li, 2005; Nezlin et al., 2004). The volume of the large sea depends on the precipitation over the catchment area of the Amu Darya River. The small sea, in contrast, is more vulnerable to changes in the amount of rainfall from year to year associated with the oscillation in atmospheric circulation over the region surrounding the Aral Sea (Nezlin et al., 2004).

Has precipitation been changing in the last several decades? In order to investigate this question, we created composite diagrams showing differences in rainfall at 10-year intervals for the decades of the 1980s and 2000s (Fig. 4). In recent years, there has been more precipitation during fall and winter in the headwaters of the Amu Darya, whereas during spring an increase in precipitation is widespread throughout the entire region. The composite difference shows very little change in precipitation during summer. The headwaters of the Amu Darya and Syr Darya rivers show small increases ($\sim$12%) in annual precipitation during the period since satellite data became available, although large variability from year-to-year is apparent in the record (Fig. 5). This trend is consistent with the 6–8% increase in precipitation over a longer period (1900–2005) for central Asia reported in the Intergovernmental Panel on Climate Change 2007 report (IPCC AR4 report: IPCC, 2007).

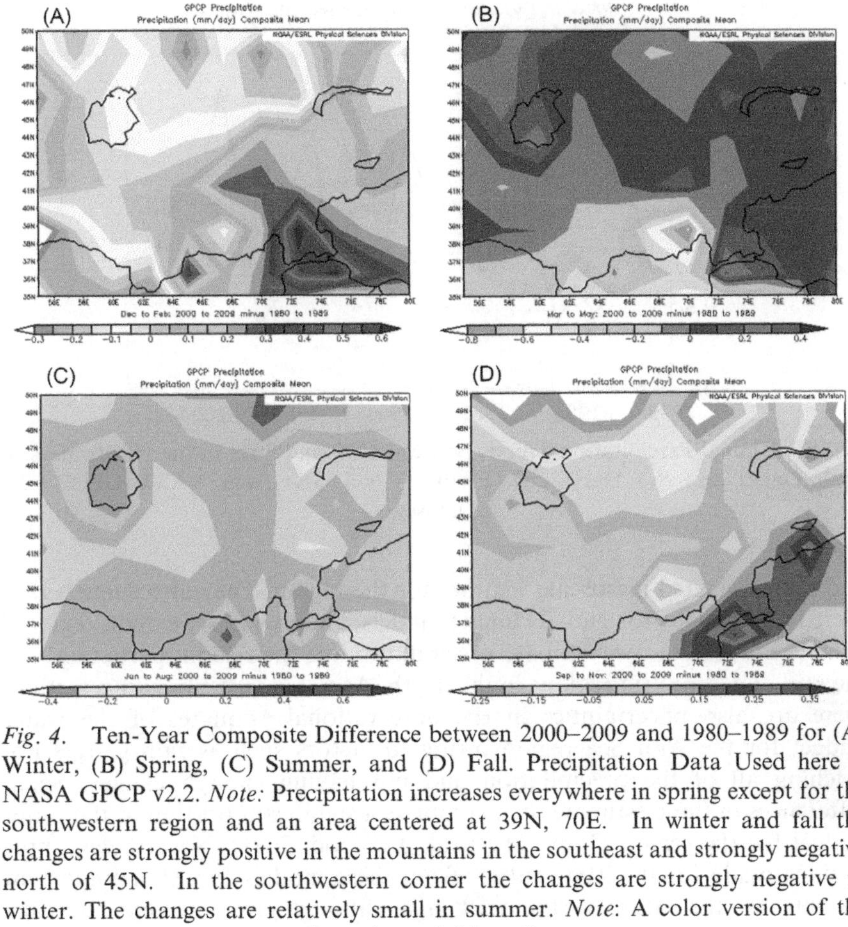

Fig. 4. Ten-Year Composite Difference between 2000–2009 and 1980–1989 for (A) Winter, (B) Spring, (C) Summer, and (D) Fall. Precipitation Data Used here is NASA GPCP v2.2. *Note:* Precipitation increases everywhere in spring except for the southwestern region and an area centered at 39N, 70E. In winter and fall the changes are strongly positive in the mountains in the southeast and strongly negative north of 45N. In the southwestern corner the changes are strongly negative in winter. The changes are relatively small in summer. *Note*: A color version of the figure is available online.

FUTURE CLIMATE

Future projections of climate change are based on global climate model simulations. These climate models are generally better at simulating potential future changes in temperature than in precipitation. For Central Asia, Christensen, Hewitson, Busuioc, Chen, and others (2007) suggest that climate models are typically poor in simulating future climate, primarily because the models do not adequately represent the complex topography

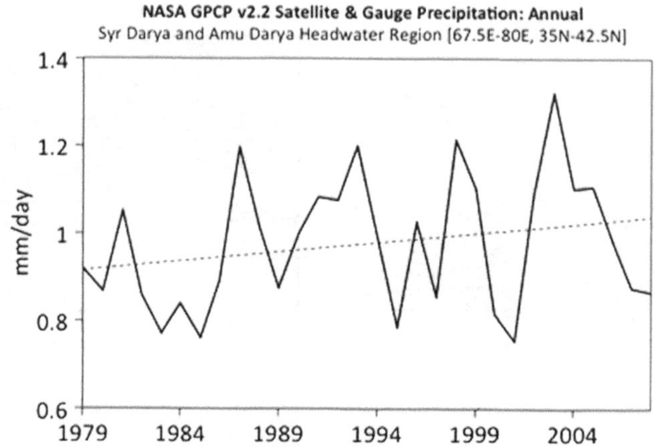

Fig. 5. Annual Precipitation Trends between 1979 and 2008 in the Headwaters of Amu Darya and Syr Darya (67.5E-80E, 35N-42.5N) from NASA GPCP v2.2 Dataset.

and the associated mesoscale weather for the region. They also suggest that there is a tendency for global climate models, and to a lesser extent, regional climate models, to overestimate precipitation for the region, particularly in the arid and semi-arid regions in the north. As noted in the previous section, there are also uncertainties in the observational estimates of the water budget for the Aral Sea region owing to factors such as rain gauges not catching all of the precipitation and not having enough gauges. Small differences in these estimates can comprise a significant portion of the water budget for the region (Troy, Sheffield, & Wood, 2011). Nevertheless, projections made for the 21st century from these models can offer important insights into the nature of climate change, although given the above caveats, one should be careful not to put too much emphasis on the magnitude of change, particularly for precipitation.

Here we present climate projections for the region based on the 21 global climate models used in the IPCC AR4 report. These models considered the SRES A1B scenario for 21st century greenhouse gas emissions; this is one of the moderate greenhouse gas emission scenarios considered by the IPCC. Fig. 6 shows changes in temperature and precipitation by the end of the 21st century for annual, winter (DJF), and summer (JJA) as well as the inter-model agreement for changes in precipitation.

Fig. 6 shows that the mean-annual temperature increases by 3.5–4°C by the end of the 21st century with slightly greater increases projected during

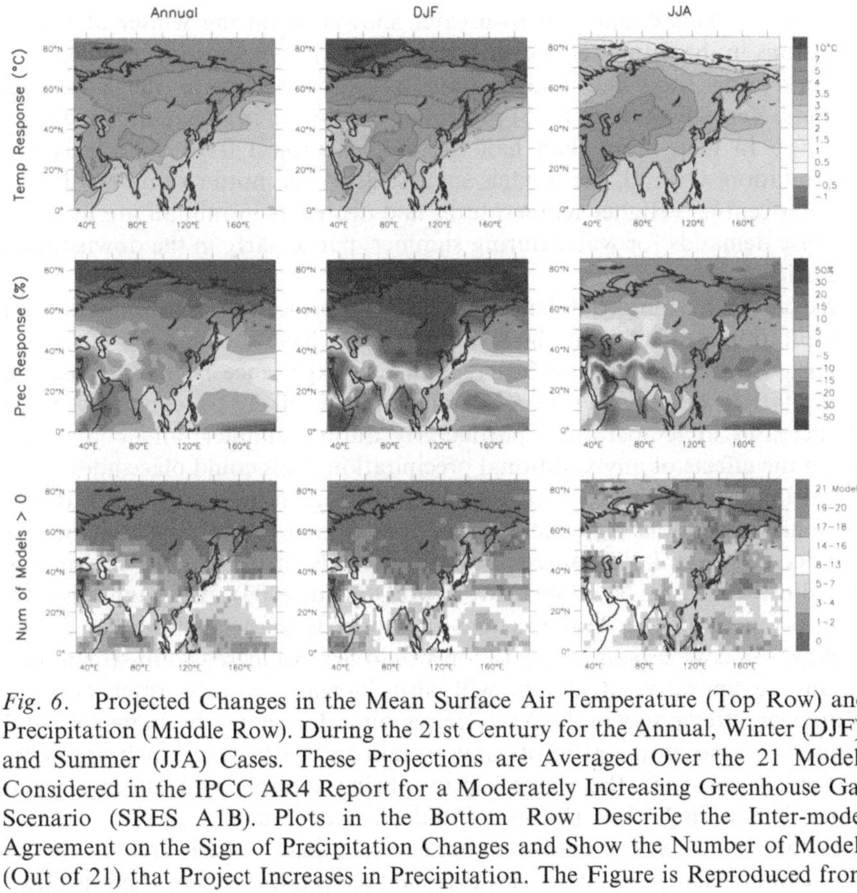

Fig. 6. Projected Changes in the Mean Surface Air Temperature (Top Row) and Precipitation (Middle Row). During the 21st Century for the Annual, Winter (DJF), and Summer (JJA) Cases. These Projections are Averaged Over the 21 Models Considered in the IPCC AR4 Report for a Moderately Increasing Greenhouse Gas Scenario (SRES A1B). Plots in the Bottom Row Describe the Inter-model Agreement on the Sign of Precipitation Changes and Show the Number of Models (Out of 21) that Project Increases in Precipitation. The Figure is Reproduced from Christensen et al. (2007). *Note:* All the temperature responses are positive. Precipitation responses are positive everywhere except in the following locations: 25N-40N, 30E-70E and 25N-35N, 160E-180E for annual; a band starting at 30N on the western boundary and extending southeast to 15N, 100E and then extending northeast to 30N on the eastern boundary for DJF; and between 35N and 50N and extending to 80E from the western boundary for JJA. *Note*: A color version of the figure is available online.

summer and winter months. For precipitation, there are small increases (~5%) in mean-annual precipitation in the headwaters of the Amu Darya and Syr Darya rivers. For winter, more than 90% of the models show 5–20% increases in precipitation, particularly in the rivers' headwaters. This could lead to greater snowpack during winter at higher elevations in the future. However, increases in air temperature could lead to earlier spring snowmelt and/or more rain than snow in the transition seasons. Very few models (<5) project any increases in precipitation during summer. In fact, more than half the models project 0–15% decreases in precipitation. Overall, the models suggest drier and hotter summers during the 21st century. Higher temperatures and decreases in rainfall are likely to increase demands for water during summer, particularly in the downstream and arid regions.

We can have some confidence in the model projections because observed precipitation trends during the last few decades have shown signs of similar change. Annually, the Aral Sea region may experience slight increases in precipitation, however, evaporative demands are likely to increase because of increases in temperature, particularly during summer, and could outweigh the effects of any additional precipitation. This could place additional constraints on the future water supply in the region. For northern Eurasia, there are indications that increases in cold season precipitation in the last few decades may have caused increasing trends in the annual streamflow (Rawlins, Ye, Yang, Shiklomanov, & McDonald, 2009). Although this may have occurred in the Aral Sea region, too, it is uncertain whether future increases in cold season precipitation would lead to increases in streamflow because evaporative demands will also increase in a warmer climate. Moreover, streamflow in the Aral Sea region is dominated by snowmelt, and future decreases in the ratio of snow to rain and earlier snowmelt are likely to further constrain the water supply during summer and autumn when water demand is highest (Barnett, Adam, & Lettenmaier, 2005).

In the past 50 years, land-use changes have had a much greater impact than climate change on water resources in the Aral Sea region. That is likely to continue. Therefore, it is important to develop appropriate hydrologic models, including anthropogenic effects, to fully understand the impacts of future climate change on regional water resources. Ferrari, Miller, and Russell (1999) were able to successfully use a global climate model[1] to project river flow into the Aral Sea that was consistent with the actual observed reduced flow. To model land-use changes for the Amu Darya River system, they sequentially incorporated the following into the global climate model: the effects of groundwater flow, expansion of wetland area, deltaic flooding combined with additional evaporative loss in the floodplains at the entrance to the Aral Sea, and finally human behavior (i.e., water withdrawals for irrigation). The same approach can apply to other large river systems in arid and semi-arid regions.

SUMMARY

Climatologically arid and semi-arid regions surround the Aral Sea. Mean seasonal temperatures range between plus and minus 30°C. Most of the precipitation falls well to the east in the rugged mountains that form the western boundary of the Tibetan Plateau. The freshwater budget of the Aral Sea must be maintained by the precipitation in these mountains, and historically, that has occurred primarily in the form of spring snowmelt that fed the Amu Darya and Syr Darya rivers. Major increases in irrigation to support agriculture after 1960 disrupted the balance in the freshwater budget of the Aral Sea and led to its demise and related environmental disasters.

The future of the region depends primarily on land-use and water-use policies, but will also depend on how climate might change during the 21st century. Current global climate models are limited in making future projections in this region, in part because of the complex topography. Global climate models project that temperatures will increase by 3–5°C and that precipitation may increase during winter and spring and decrease in summer. However, any increases in precipitation in this arid region could be overwhelmed by increases in evaporation, thus elevating the water stress in an already stressed region. Improvements in global and regional climate and hydrologic models to incorporate changes in human land-use will help to refine future projections of climate and hydrological changes.

NOTE

1. The model was developed by Russell, Miller, and Rind (1995) in conjunction with the river routing scheme of Miller, Russell, and Caliri (1994).

ACKNOWLEDGMENTS

This work has been supported, in part, by the National Science Foundation grant AGS-1064326.

REFERENCES

Aizen, E. M., Aizen, V. B., Mayewski, P. A., Joswiak, D. R., Kaspari, S., Sneed, S., … Finaev, A. (2009). Stable-isotope and chemical time series from Fedchenko Glacier Firn Core (Pamir). *Journal of Glaciology*, *55*(190), 275–291.

Barnett, T. P., Adam, J. C., & Lettenmaier, D. P. (2005). Potential impacts of a warming climate on water availability in snow-dominated regions. *Nature*, *438*(7066), 303–309.

Christensen, J. H., Hewitson, B. C., Busuioc, A., Chen, A., & others (2007). Regional climate projections. Climate change 2007: The physical science basis. In S. Solomon, D. Qin, M. Manning, Z. Chen, & others (Eds.), Contribution of Working Group I to the 4th assessment report of the Intergovernmental Panel on Climate Change, Cambridge University Press, Cambridge (pp. 847–940).

Dukhovnyy, V. A. (1984). The problem of the Aral Sea and nature protection measures. *Problemy Osvoyeniya Pustyn*, *6*, 3–14.

Ferrari, M. R., Miller, J. R., & Russell, G. L. (1999). Modeling the effect of wetlands, flooding, and irrigation on river flow: Application to the Aral Sea. *Water Resources Research*, *35*(6), 1869–1876.

Gleick, P. H. (1993). Aral Sea Hydrologic Data, 1926–1990. In P. H. Gleick (Ed.), *Water in crisis: A guide to the world's fresh water resources* (pp. 313–314). NY: Oxford University Press.

IPCC. (2007). The physical science basis. In S. Solomon, D. Qin, M. Manning, Z. Chen, M. Marquis, K. B. Averyt, M. Tignor & H. L. Miller (Eds.), Contribution of Working Group I to the Fourth Assessment Report of the Intergovernmental Panel on Climate Change, Cambridge University Press, Cambridge, UK.

Kotlyakov, V. M. (1991). The Aral Sea basin: A critical environmental zone. *Environment*, *33*(1), 36–39.

Legates, D., & Willmott, C. (1990). Mean seasonal and spatial variability in gauge – Corrected global precipitation. *International Journal of Climatology*, *10*, 111–127.

Levintanus, A. (1992). Saving the Aral Sea. *Journal of Environmental Management*, *36*, 193–199.

Micklin, P. P. (1991). *The water management crisis in Soviet Central Asia* (120 pp.). Pittsburgh, PA: The University of Pittsburgh, Center for Russian and East European Studies.

Miller, J. R., Russell, G. L., & Caliri, G. (1994). Continental-scale river flow in climate models. *Journal of Climate*, *7*, 914–928.

Nezlin, N., Kostianoy, A., & Li, B. L. (2005). Inter-annual variability and interaction of remote-sensed vegetation index and atmospheric precipitation in the Aral Sea region. *Journal of Arid Environments*, *62*(4), 677–700.

Nezlin, N. P., Kostianoy, A. G., & Lebedev, S. A. (2004). Interannual variations of the discharge of Amu Darya and Syr Darya estimated from global atmospheric precipitation. *Journal of marine systems*, *47*(1), 67–75.

Rawlins, M. A., Ye, H., Yang, D., Shiklomanov, A., & McDonald, K. C. (2009). Divergence in seasonal hydrology across northern Eurasia: Emerging trends and water cycle linkages. *Journal of Geophysical Research*, *114*, D18119.

Russell, G. L., Miller, J. R., & Rind, D. (1995). A coupled atmosphere-ocean model for transient climate change studies. *Atmosphere-Ocean*, *33*, 683–730.

Shiklomanov, I. A. (1979). *Anthropogenyy izmeneniya vodnosti rek, Gidrometeoizdat*. USSR National Committee for International Hydrological Decade, Summary of the USSR Activities on the I.H.D. Programme for Five Years (1965–1969). Leningrad, 1969.

Troy, T. J., Sheffield, J., & Wood, E. F. (2011). Estimation of the terrestrial water budget over northern Eurasia through the use of multiple data sources. *Journal of Climate*, *24*(13), 272–3293.

Van der Leeden, F. (1975). *Water resources of the world: Selected statistics* (568 pp.). Port Washington, NY: Water Information Center.

CHAPTER 6

THE SIGNIFICANCE OF BEING DOWNSTREAM: UZBEK CONCERNS OVER THE ROGUN DAM

Murad Askarov

ABSTRACT

In this chapter, the Permanent Representative of the Republic of Uzkekistan to the United Nations, His Excellency Murad Askarov, sets forth his nation's concerns over transboundary rivers in Central Asia and the protection of Uzbekistan's rights as the most downstream nation in the region. A key focus of the chapter is the continuance of Soviet-era dam projects in the headwaters of the region's two principal rivers, the Amu Darya and the Syr Darya. In particular, the chapter focuses upon the potential adverse impacts associated with the completion of the Rogun hydropower project by Tajikstan in the headwaters of the Amu Darya. International opposition to the project is summarized. And the unequivocal opposition of Uzbekistan is made clear.

On July 20, 2011, the UN Security Council engaged in a discussion on the impact of climate change for security, defined in the broadest sense of this

Disaster by Design: The Aral Sea and its Lessons for Sustainability
Research in Social Problems and Public Policy, Volume 20, 65–74

ISSN: 0196-1152/doi:10.1108/S0196-1152(2012)0000020014

concept. As nation after nation spoke, the vast majority of the speakers explicitly highlighted the growing lack of water as one of the major challenges faced in the many regions of our planet.

Today, nobody doubts the immutable but unfortunate truth that natural disasters in our world are becoming more frequent with every passing year. There is increasing negative impact as the climate is more stressed. During 2010 we have witnessed earthquakes of unprecedented destructive power in Haiti, Japan, Chile, and New Zealand, floods in Pakistan and Australia, and severe drought in the United States, China, and the Horn of Africa.

There are warning signs that the same phenomenon also exists in Central Asia. The glaciers of Pamir and Tian Shan mountains – the main sources of water for the Amu Darya and Syr Darya rivers – are melting (UNEP, 2008). In 2010 our region has also experienced severe drought, adversely affecting the whole complex of socioeconomic development parameters included in the Millennium Development Goals. That year there was a significant decrease in food production due to further environmental degradation. The trend toward the trend toward environmental crisis continues. Of protracted concern is the ever worsening situation in what once was the fourth largest intracontinental basin. The Aral Sea has disappeared, becoming Aralkum – a huge desert.

In his address at the Millennium Development Goals Summit the President of the Republic of Uzbekistan Islam Karimov (2010) stated that

> The tragedy of Aral Sea, which practically during the lifetime of one generation has turned from once one of the unique and most beautiful seas into a drying and vanishing water reservoir, stands as a stunning example and evidence of our irresponsible attitude toward environmental problems. The problems of the drying up Aral Sea are the problems of millions of people living in this region, who hope for assistance and appeal to such a high-profile organization as the United Nations.

In this context, a reasonable and rational use of water and energy resources is one of the most serious challenges for the Central Asian region. Wrongly directed management of these key resources not only hinders development, but also further aggravates environmental disaster. Today, the way we address the Aral region's challenging array of ecological, socioeconomic and demographic problems has global implications. The Secretary-General, Ban Ki-moon, witnessed this for himself during his visit to the Aral Sea in 2010. He said "It is clearly one of the worst environmental disasters of the world ... It really left with me a profound impression, one of sadness that such a mighty sea has disappeared. It was a vivid testament to what [...] happens [...] when we waste our common natural resources, when we neglect our environment, when we mismanage our environment" (UN News Center, 2010).

In preparation for the sixth World Water Forum held in Marseille this year, an international conference was convened in Uzbekistan in May 2011.[1] International academicians and eminent experts acknowledged that immense efforts aimed at conservation and sustainable use of water resources are being undertaken in Uzbekistan. Associations of water users are being created as part of country's large-scale reforms in agriculture. Cotton production is significantly reduced, and today cotton fields occupy only about 25% of the irrigated areas. Despite the intensive development of the national economy, water intake from rivers was decreased by more than 30% compared to early 1990s. More than $1 billion has been allocated to land improvement in recent years. Due to these efforts, the condition of almost 740,000 hectares of irrigated land was improved. Uzbekistan has widely adopted modern approaches to solving the problem of water scarcity, including integrated water resource management.

Uzbekistan, home to almost half of the entire population of Central Asia and the nation most severely affected by the Aral Sea catastrophe, has a moral right to insist that water resources in the region be coordinated in a coherent, rational, and reasonable manner. There is a legal basis for this expectation, as well. Basic UN legal instruments for regulating relations in joint use of water should be fully implemented in Central Asia, including the UN conventions on the Protection and Use of Transboundary Watercourses and International Lakes (March 17, 1992) and the Law of the Non-Navigation Uses of International Watercourses (May 21, 1997).

These documents clearly establish the principle that all decisions on the use of Transboundary Rivers, including the decision to construct hydropower facilities, shall not in any way prejudice the environment and infringe upon the interests of neighboring states. Guided by these universally recognized international legal principles and norms, Uzbekistan is categorically against the construction of gigantic dams in the upper reaches of transboundary rivers in Central Asia that excludes the interest and consent of the downstream countries. Nevertheless, such projects are being contemplated.

President of Uzbekistan Islam Karimov stated our concerns clearly in his address at the Millennium Development Goal Summit in September 2010:

> In these circumstances, any attempt to implement projects that were developed 30–40 years ago, during the Soviet period, for construction in the headwaters of these rivers of large-scale hydro facilities with gigantic dams – especially when you consider that the seismicity of the zone for the forthcoming construction ranges from 8 to 9 points – all

this can cause irreparable damage to the environment and is a reason for the most dangerous man-made disasters.

One concern over dam construction is its impact on food security. Since 2008, the UN has consistently raised alarm over growing food insufficiency. Climate change and water scarcity lead to less food production – the basis of life itself – without which it is meaningless to talk about security and stability.

It is clearly stated in the General comment No. 15 of the Committee on Economic, Social and Cultural Rights that, "The human right to water entitles everyone to sufficient, safe, acceptable, physically accessible and affordable water for personal and domestic uses."

Furthermore, Paragraph 31 of the General comment No. 15 obliges States parties "To comply with their international obligations in relation to the right to water, States parties have to respect the enjoyment of the right in other countries. International cooperation requires States parties to refrain from actions that interfere, directly or indirectly, with the enjoyment of the right to water in other countries. Any activities undertaken within the State party's jurisdiction should not deprive another country of the ability to realize the right to water for persons in its jurisdiction."

Uzbekistan believes that fresh water should be used primarily for food production and human consumption. This is especially true for Central Asia, which shares common water resources. In such a state of mutual dependency, each state should act with full responsibility by exercising rational use of the water resources required for both life and security by both individual countries and the region as a whole.

Another basis of concern is the issue of catastrophic risk. The International Commission on Large Dams, which Uzbekistan joined in 2011, has found that large dams are risky and dangerous, especially those located on Transboundary Rivers. Such concerns are particularly justified when one considers the proposed Rogun Dam on the Amu Darya River, which is the primary focus of my concern in this chapter.

Construction of the Rogun Hydropower Plant (HPP) upstream on the Amu Darya River entails significant risk of technological failure with social, economic, and environmental consequences. The project is opposed by experts, international organizations, and countries located downstream.

There are strong reasons for such opposition. The design of the Rogun HPP dam dates back to the Soviet era, 35–40 years ago, when gargantuan projects were preferred. The engineering for the project does not meet current requirements, standards, construction norms, and rules. Eminent professionals and experts agree that the project fails to meet current requirements for dam safety.

These standards and requirements have been informed by negative experiences with large hydropower plants built in the last few decades. These projects include "Three Gorges" in China, "Sayano-Shushenskaya HPP," and "Boguchanskaya HPP" in the Russian Federation, and also large hydropower plants built in South America. Given these more stringent requirements, there is now a basis for permit denial, revocation, or amendment of large-scale projects, even when construction is underway.

During engineering for the Rogun HPP project, Soviet specialists failed to find adequate technical solutions for a number of major unresolved issues. In particular, measures are lacking to assure the integrity of the dam by mitigating the inevitable effects of filtration, high mobility in the rock masses in the area of construction, and adverse structural impacts caused by huge masses of water on the 100 meters thick salt layer beneath the base of the dam. The precarious nature of the situation was further illustrated by the 1993 failure and collapse of a coffer dam at the site due to flood damage and by several other subsequent accidents.

Then there is the proposed construction of a dam that is problematic in every way. First, there is the unprecedented height of 335 meters. Second, the location is in a seismically active zone. It is sited in mountain ranges newly formed by the Vakhsh tectonic fault, itself interconnected with the chain of faults underlying the Southern Tian Shan range. This is a region that has been described as having the highest seismicity in all of Central Asia. Periodic earthquakes have been measured here reaching 10 points on the Richter scale and routinely scoring above 9.

According to the Tajik News Agency *Asia Plus*, "On 10 July 1949, in the area around the city of Khait, an 11-Richter scale magnitude earthquake occurred. More than 50 thousand people were buried under the destroyed buildings and a huge mudslide" (Shodiev & Mannonov, 2012).

Richard Stone (2009), staff writer for *Science magazine*, writes that "The possible trigger for such a catastrophe would be another major earthquake shaking the region. A large temblor is virtually certain: The Pamirs are a seismic hot spot, and a quake of magnitude 7 or greater rattles the faults around Sarez every century or so."

These facts negate the statements by Tajik officials that 9-point magnitude earthquakes are impossible nowadays.

Regional earthquakes have already proven catastrophic. An earthquake in Tajikistan in the first half of last century claimed the lives of more than 100,000 people. In 1911, an earthquake of more than 9 points led to the formation of the Usoi natural landslide dam and Lake Sarez, with its capacity of 20 billion cubic meters of water.

At present this area is entering a stage of heightened seismic activity. According to the U.S. Geological Survey, up to eight earthquakes are recorded weekly in the Pamir-Hindu Kush mountain chain, where the Rogun HPP construction site is found. On average, a strong earthquake occurs in Tajikistan every 4 years and destructive ones every 10–15 years. Based on the analysis of available data, a number of experts predict high-magnitude destructive earthquakes within the next 10 years in this mountain range. Even Tajik scientists and seismologists confirm these expectations.

In addition, the construction of a huge dam will require moving and stacking 80 million cubic meters of soil, which, along with the collection of 14 billion tons of water in the bowl of the reservoir, will create additional pressure on the mountain range. This will certainly lead to an increase in the seismic vulnerability of the region. It is safe to predict that the construction of Rogun HPP will significantly increase the frequency and intensity of earthquakes in this region.

The Rogun HPP dam will be highly vulnerable to destruction from natural and human causes. Should that occur, scientists and engineers estimate that giant waves will be caused by the dynamic pressure of 14 cubic kilometers of water trapped in the reservoir, a human-made tsunami more than 100 feet in height, which will rush down the river Vakhsh at high speed, as fast as 500 kilometers per hour. A chain reaction of devastation will then occur. A second dam, the Nurek HPP, will be completely destroyed, along with all other hydroelectric stations and waterworks in the Vakhsh cascade, and flooding will engulf the towns of Nurek, Sarban, Kurgantyube, and Rumy. Furthermore, continuing on its destructive path, the flood wave will demolish and cover dozens of other towns and villages in Tajikistan, Uzbekistan, and Turkmenistan, causing incalculable and tragic consequences as well as deaths of many hundreds of thousands of people.

Of comparable concern are the consequences of the Rogun HPP project for the Central Asian region's long-term sustainability and its irreversible threats to the environment and in socioeconomic sphere.

The construction of this gigantic hydro facility will upset the delicate ecological balance in the region and have a devastating impact on water resources and the broader environment. In order to impound a water reservoir with a volume of 14 cubic kilometers, the significant part of the flows of the Vakhsh River will have to be blocked for at least 8–10 years. Cutting off its flow will completely disrupt the long-term regulation of water flows in the region and will raise the already increasing water scarcity to a disastrous level.

Rogun Dam construction will also completely disrupt the structure of natural water flows. A sharp flow decrease during the vegetation season will cause a severe summer water scarcity and drought. An increase of flow during the autumn–winter period will result in disastrous winter floods in countries that lie downstream of Amu Darya.

Meanwhile, changes in the hydrological regime of the Amu Darya will intensify drops in the water level during dry season that already constitute a loss of 15%. Drainage from lakes and wetlands in the lower reaches of Amu Darya will dramatically accelerate the formation of new salt marshes and saline "takyrs," serving as major cells of salt transfer to adjacent farmlands, reducing soil fertility and crop yields. As a result, the environmental disaster of the Aral Sea, already felt at a planetary scale, will worsen further.

Such environmental effects have socioeconomic consequences, as well. These changes will totally destroy the economic foundation of agricultural production, and, along with it, the prevailing lifestyle of more than 10 million people living in the oases of the Lower Amu Darya in Uzbekistan and Turkmenistan. These victims will be doomed to drought, famine, and, in the end, forced relocation.

Calculations predict a direct economic loss on the order of 20 billion dollars for downstream countries, including Uzbekistan, as a result of the construction of Rogun HPP (Prosta Vostoka, 2012). There will be no means to recover these losses. Beyond this cost, there will be multiple economic and social problems associated with the forced relocation of vast numbers of people. These forced migrants will be left in a situation of irreparable water shortage. As a result, social unrest will increase sharply in the region.

As can be seen, the total economic cost of this project's implementation is incalculable. Beyond such sums, how – in what currency and numbers – can we estimate the suffering and misery of millions of people? It is obvious that the construction of this dam is not only contrary to technical standards and economic logic, but to common sense.

The threats presented by the construction of Rogun HPP have served to raise legitimate concerns by the international community. Eminent international organizations objecting to this project include the UN, the World Water Council, and the International Union for Conservation of Nature. They are joined by members of the European Parliament, the United States Congress, the parliaments of Belgium and other countries, and scientific and research centers in countries that include Japan, United States, Netherlands, and Korea. Broad discussions of the problems associated with the project have been held over the past few years at various levels at various UN agencies, as well as at international scientific forums and conferences. These events have helped to clarify the issues and formulate the principle positions of the international community about the need for deep and thorough, impartial and independent, international assessment of the project.

It is important to note that the opinion of independent experts is fully consistent with the established norms of the international law on use of transboundary water resources. There are a number of UN conventions on this matter (the UN Convention on the Protection and Use of Transboundary Watercourses and International Lakes, adopted on September 18, 1992; the Convention on the Law of International Watercourses, adopted on May 21, 1997) that clearly define requirements for mandatory consideration of the interests of all affected Parties. Consequences for those who will be impacted by transboundary water facilities must be fully considered before projects are initiated that will cause transboundary effects.

Additionally, the Convention on Environmental Impact Assessment in a Transboundary Context (entered into force in 1997) requires a full accounting of environmental consequences from the construction of such facilities and a need for thorough assessment of impact involving all stakeholders.

Taking into account the opinions of international organizations and agencies and on the basis of international law, the World Bank in 2010 decided to carry out an independent international assessment of the Rogun HPP construction project and allocated $20 million for its implementation. The project was undertaken by companies from France, Switzerland, and other European countries and is expected to be completed in early 2013.

For its part, in 2010, the Government of Tajikistan voluntarily committed itself before the World Bank to halt the construction work on the site until the completion of an independent technical–economic and socio-environmental impact assessment of the project. It simultaneously pledged to the International Monetary Fund to stop collecting funds from its citizens intended for financing the construction of Rogun HPP.

In violation of these commitments, however, Tajikstan has unilaterally and flagrantly violated its agreements to halt the construction activities at the Rogun HPP site. The Tajik side continues to pursue its "obsession" with building Rogun HPP. Abundant evidence suggests that Tajikistan continues a wide range of construction activities on the hydro plant site that were initiated in the 1970s. Currently, such extensive work includes construction of drainage tunnels, an engine room, quarries, and other components of the facility. This work is being done in private and concealed from the international community.

Naturally, questions arise. What is the ultimate goal of the activities carried out by Tajikistan? What is the aim of building Rogun HPP? Why, despite numerous objections and in violation of international law, is Tajikistan trying to expedite the completion of the construction of the plant's first phase?

The answer is obvious. Tajikstan's ultimate goal is to complete the construction of the first phase of the hydroelectric station. It would then present the World Bank and the entire international community with a fait accompli, hoping to force its acceptance and legitimacy. The current reviews of impacts would be rendered moot. This outcome is exactly what the community of experts and international organizations is opposed to.

This perilous situation could be easily avoided. As many international experts attest to, the problem of providing reliable electricity for Tajikistan can be solved by building a number of small hydroelectric power plants instead of the huge Rogun Dam. This alternative approach comes at a much lower cost and can be completed in a shorter period of time. It avoids risking large-scale catastrophic human-caused threats and the secondary environmental and socioeconomic impacts for both Tajikistan and its neighbors. It is vital that the World Bank consider this alternative.

In the meantime, the Tajik's current efforts to continue construction of the Rogun HPP complex while hiding their efforts from the international community and downstream countries located on the Amu Darya can lead to severe unintended and irreversible consequences that cannot be corrected. Obviously, this eventuality cannot be ignored.

NOTE

1. "Towards the 6th World Water Forum – joint actions towards water security." Final Outcome Document of Tashkent Conference, Tashkent, Uzbekistan, March 13, 2011. http://www.cawater-info.net/6wwf/conference_tashkent2011/final_resolution.htm

REFERENCES

Karimov, I. (2010). Address by the President of the Republic of Uzbekistan Islam Karimov at the Millennium Development Goals Summit, September 20. Retrieved from www.press-service.uz/en/#en/news/show/vistupleniya/address_by_h.e._mr._islam_karimov_presid/)

Pravda Vostoka. (2012). Once again on the issue of construction of Rogun Hydroelectric Power Plant. *Pravda Vostoka*, July 12. Retrieved from http://mfa.uz/eng/press_and_media_

service/dates/water_resource/once_again_on_the_issue_of_construction_of_rogun_hydroelectric_power_plant.mgr. Accessed on July 12, 20212.

Shodiev, K., & Mannonov, A. (2012). Khait. The lost city. *Asia Plus New Agency,* July 13. Retrieved from http://www.news.tj/ru/newspaper/article/khait-poteryannyi-gorodwww.news.tj/ru/newspaper/article/khait-poteryannyi-gorod. Accessed on July 15, 2012.

Stone, R. (2009). Peril in the Pamirs. *Science, 5026*(December), 614–617. doi: 10.1126/science.326.5960.1614.

UNEP. (2008). *Global glacier changes: Facts and figures*. Retrieved from http://www.grid.unep.ch/glaciers/. Accessed on July 15, 2012.

UN News Centre. (2010). *Shrinking Aral Sea underscores need for urgent action on environment.* UN News Centre, April 4, 2010.

SECTION TWO
THE ARAL DISASTER IN HISTORICAL PERSPECTIVE

Having established multiple disasters now radiating from the core loss of the Aral Sea, we turn to the volume's theme of *Disaster by Design* in Section Two. Neither the trunk nor the branches was a natural process; all took root in the hands of humans intent on altering nature to serve humans.

In Chapter 7, "A Last Movement for a Lost Sea," Yusup Kamalov, longtime defender of the Aral Sea, explains how the Aral's demise was designed and planned and implemented without public recourse. In their willingness to reshape nature, Soviet planners proved themselves technocrats without bounds. As the Aral story so blatantly illustrates, the Soviet willingness to destroy nature displayed a total disregard for the ecological integrity of natural systems and the people living in them. But, once compromised, Kamalov asks, how are these human and ecological rights regained? This is a fundamental question of environmental justice. Kamalov provides a first person account of the evolution of the disaster as a key player trying to address its ramifications. He makes clear that the mindset that created the disaster to begin with has not yet changed. Environmental justice will go unserved until that change occurs.

We next pose a question of perennial interest. Is the Aral disaster specific to the conditions of decision making found in the Soviet Union or is it more generically a problem of the modern era. To raise this issue, we turn to geographer Howard Horowitz, whose Chapter 8, entitled "Aral Sea Analogs in the American West," offers just this perspective. Drawing upon three cases studies from California involving the diversion of waters that threatened sizable inland lakes, Horowitz walks through the historical, geographic and social perspectives on each.

We get a sense, from this comparison and the long-term perspectives of Kamalov and Horowitz, about the conditions that produced the crisis and then forged eventual outcomes. The similarities in the organizational roots of the disastrous actions are indeed common to both situations. But the conditions that forged the outcomes diverge. Horowitz shows that, when the countervailing force of civil society – with its basis in rights and a functional legal system – is enabled, disaster can actually be averted. And we see that the period of democratic activism in the USSR during the period of perestroika approached the conditions of public discourse and activism necessary to save the Aral Sea. There was lacking only a mechanism to force the authoritarian state to shift direction. The public movements of the time had not risen to a level of checks and balances.

Finally, in Chapter 9, "Disaster by Design: The Multiple Caused Catastrophes of the Aral Sea," Michael Edelstein sets forth the key themes of this volume. First, there is the disappearance of the Aral Sea, not as some miraculous act of nature, but as a calculated consequence of human action. Edelstein draws from keen observers' now priceless historical accounts to chart this course. We see the destruction of the Aral Sea as more than just an unintended, unanticipated outcome of policies to send water elsewhere, but as a systematic exercise in what Feshbach and Friendly (1992) term "ecocide." The effort encompassed a planned economy and the bureaucracy that enforced it, decisions to expand agriculture into marginal "virgin lands," scientists who had lost any semblance of objectivity, a prevailing sense of control over nature, a disregard for those victimized by events and an authoritarian society capable of extreme acts against its own population.

Michael R. Edelstein
Editor

REFERENCE

Feshbach, M., & Friendly, A., Jr. (1992). *Ecocide in the USSR: Health and nature under siege.* New York, NY: Basic Books.

CHAPTER 7

A LAST MOVEMENT FOR A LOST SEA☆

Yusup S. Kamalov

ABSTRACT

This chapter addresses the Aral Sea disaster, not as an acute event demanding emergency response, but with a long-term view aimed at determining the causes of the disaster and providing analysis and troubleshooting to prevent such catastrophes in the future.

Emergency response requires quick action and no cost is spared. Such actions look very bold and are very often used to enhance the public image of politicians, businessmen, and leaders of various organizations who have rendered assistance.

The second direction, described here, is a more complicated and thankless task. The investigation faces obstruction from those who caused the disaster or who can use the crisis situation to their advantage. It is hard to get financing for long-term solutions. This chapter describes the key causes of the death of the Aral Sea. I then describe our engagement in the

☆This chapter draws from a number of unpublished works by the author, including "The Concept of The Union for Defense of the Aral Sea and Amu-Darya: On Water Usage in the Aral Sea Basin." Proceedings of the International Conference "Market and Water Resources." Nukus, 2000, Uzbekistan.

Disaster by Design: The Aral Sea and its Lessons for Sustainability
Research in Social Problems and Public Policy, Volume 20, 77–88

ISSN: 0196-1152/doi:10.1108/S0196-1152(2012)0000020016

hard work of looking for ways to correct errors and provide innovative solutions that would satisfy the majority of stakeholders.

The UN Declaration for Human Rights on the availability of clean water gives a new context for the case made in this chapter. Here I argue that all people along the rivers of Central Asia (and elsewhere) share the same rights to have enough high-quality water and to fully enjoy a healthy environment, regardless whether they live upstream or downstream.

INTRODUCTION

The Aral Sea, a pearl of the earth, was made to dry up by the diversion of its waters. It has since died as an ecosystem. I have myself witnessed both its vibrant life and its agonizing death. And I continue to work for its restoration. A native of Nukus, capital of the Karakalpakstan Autonomous Republic, I well remember the healthy Aral Lake region that brought such prosperity and pleasure to our lives. I went to Moscow from 1969 to 1975 to study and was away from the region when first understandings of the crisis emerged. But I had a special source of information. My father, Sobir Kamalov, was the chair of the Karakalpak branch of the Uzbekistan Academy of Sciences. And he became involved in the issue from the beginning. He explained the situation to me, about the shrinking Aral Sea and the diversion of water for irrigation.

After I graduated from the Moscow Power Engineering Institute, I returned to Nukus to work in that field. I spent 2 years working at the power plant. At that time, our home became a center for conversation about the Aral Sea. Many scientists involved in the issue came from Moscow. They always visited my father, and he always invited them home. Everyone talked a lot about the Aral Sea issue particularly and, in general, about the environment. This was the atmosphere I grew up in. I grew up as an environmentalist.

I was back working in Nukus at the power plant, as I indicated, when the Karakalpak Academy of Sciences opened what they called "the computer center" linked to one at the Moscow Academy of Science. It was actually a research center including a lot of interesting laboratories. One was for mathematically simulating ecology. I became so involved that I started to study mathematics. I was sent to Moscow to work there at the Laboratory of Mathematical Ecology at the Academy of Sciences. I did not stay to write my doctorate, returning instead to Nukus to work for the computer center and Laboratory of Water Resources as an engineer collecting data and information on water resources in Karakalpakstan. I tried to build mathematical models that would describe our situation.

The scientists I met at that time were totally against the disappearance of the sea. They were for saving it and solving the problems using environmentally friendly means. But, even though these were leading scientists of the time, unfortunately, the orders from the top were of a different persuasion. The officials were prepared to divert Siberian rivers to Central Asia. The scientists opposed this step.

A DELIBERATE ACT OF VIOLENCE AGAINST NATURE

The disaster was actually set in motion long before. Ancient Central Asian traditions of thrifty water use have been overturned by a series of nature "conquerors." Going back to the reign of the last czar, before 1917, the Russian scientist Voeykov argued that the Aral Sea was just "a useless evaporator." He proposed that the Amu Darya and Syr Darya rivers be fully diverted for irrigation. One would think that the plan of the czar to terminate the Aral Sea would have inspired the communists to protect it. That is not what occurred. After the Civil War, the communists implemented the czar's plan. It was as if they thought of the sea as a heroic citizen charged with a duty. One head of the USSR Water Affairs Ministry even said the "Aral Sea must die like a soldier in battle!"

These mistakes were magnified when, in the late 1920s, the Central Asian States were formed and borders outlined without consideration of ecology. Today we confront daily the barriers to rational water planning of having our two rivers shared by five countries. But from the beginning, to divide the rivers was a conqueror's strategy.

The authoritarian, unapproachable nature of water conquerors was in keeping with the fact that, up until the 1950s, the USSR Water Affairs Ministry was a subdepartment of GULAG, which ran the brutal prison camps. There was no public debate. One example of this was the route of the Karakum canal, the principle user of water from the Amu Darya River. Had it been connected to the delta of the Amu Darya, as Stalin planned, it would have been logical as that is where most of the population and the major water deficit are. However, when Stalin died, the project was relocated to the mouth of the Amu Darya, where the route must traverse a sandy water course, with the result that almost half the water is lost. The government required total submission to its plans, stifling independent comment. It distributed water free of charge, assuring careless management. New technology increased water delivery. The combination of these factors was that people learned not to consider water precious. Appealing to patriotism, the most truculent projects were championed as necessary to "build communism." War and then "cold" war excused any excess, with the idea that it could be corrected later on if necessary. Thus rationalized, water lost its value; it was dispersed for free, distributed in open and unlined ditches, flooded onto fields, and allowed to become polluted. Sacrificing the Aral Sea was the biggest blunder, by far, but its illogic had much company.

PERESTROIKA AS A TIME OF POSSIBILITIES

The problem of the Aral Sea was one of the USSR's hidden secrets exposed to the world during "Perestroika." Articles appeared in the media and experts from around the world began to visit. With the opening up of society, there was a period of big hope for the future, that the right decisions would be made regarding water resources, democracy, and so on.

In 1989, working with people of like mind, we formed an organization called UDASA, the Union for Defense of the Aral Sea and Amu Darya. At this time, we were able to create an absolutely grass roots organization from the ground without any orders from the top. Our first chairman was the famous writer Orazbay Abdirakhmanov. He initiated the creation of the UDASA. It was even he who named the organization. I felt the name was too long and that we could omit the "Amu Darya." But he said people would get used to the longer name. The main goal of UDASA has been to convince people to protect the Aral Sea and the rivers as the healthy environment for all living beings. As we say in our Mission Statement, "All the concerns for human well-being without a healthy natural environment are empty words!"

It was a time of enthusiasm. There was a lot of hope. People at this time were still economically secure and had free time to be involved. We were meeting each other almost every week, talking about possible ways to solve the problem. And sometimes more than 300 people came to the same meeting. It was a real grass roots organization. We first registered as an organization in April 1990 at the Nukus mayor's office. People supported us very much, and we collected a lot of money. We published our first newspaper, the *Janlan, Aral!* (*Revive, Aral!*).

My thinking at the time was simply to implement a market system and users would pay for water and then everything would be solved. If they were thinking about their wallets, then they would implement water-saving technologies. From the beginning, even from 1988, I was thinking about uniting all water management systems all along the river from glacier to sea. We understood Uzbekistan could not manage the water by ourselves because it was coming from outside. These are still my thoughts.

During the period we battled to save the sea, in the late 1980s and throughout the 1990s, many publications opposed the cotton monopolies in Central Asia and especially in Uzbekistan. However, it is a legend that cotton caused the Aral Sea crisis. This many-voiced choir of opposition to cotton culture overpowered contrary voices, including mine. We sang a different tune. The real issue was not cotton; it was the need to significantly improve water management, stop the anachronistic delivery methods, and engineer for efficiency. We did not push to reduce the cotton quotas. If you replaced cotton with rice, it uses much more water. We were not totally against cotton, we were against losing the Aral Sea by wasting its precious water on nothing.

What We Lost

Of course, we failed to save the Aral Sea. The resulting problems for the Karakalpak Region and neighboring areas of the Aral Sea disaster zone are many. And each primary problem generates a secondary one. The substantial fishing industry in the region is gone, and with it the economy, resulting in unemployment. There is a shortage of irrigation and drinking water throughout the shrinking deltas of the Amu Darya and Syr Darya. Over-irrigation has so saturated the soils of the larger region that salt has been transported to the surface. The salty soil must now be washed by flushing through even more water. Polluted drainage water filled with excess salt flows back into the Amu Darya. Imagine this insanity!

Water shortages and logging have diminished the forests along the rivers up into the Pamir mountains. Hundreds of lakes in the deltas of both rivers disappeared while hundreds of new artificial ponds have emerged in upstream and middle stream areas of the rivers. Fauna and flora in the basin of the Aral have been decimated. More than 40 species of fish disappeared with the loss of the sea and diminishing lakes and rivers. The health of both people and livestock in the river deltas has rapidly deteriorated. And the entire process has intensified climate change in the Aral Sea region stretching all the way up into the mountains of Pamir and Tian Shan. Summers are dryer and hotter and winters are colder along the former Aral Seaside; in the mountains, summers are more humid and winters milder; more rain falls year round. Glacier melt has intensified. A factor in the loss of these particular glaciers is salt transport by winds from the bottom of the former Aral Sea which essentially decreases the temperature at which the glaciers melt.

LOST HOPE AND UNREALISTIC DREAMS

Unfortunately, as the catastrophe progressed, many people lost hope. Many people thought that somebody else would solve the problem, not themselves. Somebody will come and give the order. They had hope, due to the many conferences held around the sea, that these conferences would be able to change something. UDASA's goal is just to prove that change only comes when people themselves take initiative and influence the bureaucracy.

People were just waiting for somebody to make a decision and take action. Of course, to paraphrase Marx, such an extraordinary idea as diverting rivers attracts the attention of people. My father once said that "two Siberian Rivers and three Volgas would not be enough to satisfy our irrigators." In any case, if such new water would be diverted to Central Asia, it would not be used to refill the Aral Sea, but to make new irrigation and cotton fields. Nobody will return the Aral Sea without economic profit. This is a reality.

A lot of us remember what it was like to live during the Soviet times when it was believed that anything was technically possible. Here's a snapshot: If a decision could be made to divert another river even from another planet, they would start doing it. Now people realize that this is just a story for kids. Another myth is about the possibility to move the entire population of the delta regions of the two rivers to appropriate regions around Tashkent or Kazakhstan where there are good living and ecological conditions. This is also a myth. No one is waiting for us.

The movement has since waned. Now there are only a few people helping with our organization. We don't have money. Many people are in a very bad economic situation. Many have left for abroad, to Russia, to Kazakhstan. Some have died. Many people have lost hope that something might be changed. Now there are only five or six people. Those involved come and meet and discuss questions and meet guests. There are not too many at all anymore.

We continue to defend the region even though the Aral Sea is lost and will not return in my lifetime. If it took 50 years for the Aral to die, then restoration would not take less time. For example, we have protested the discharge of contaminated waters to the Amu Darya system. Take another example: when we found that the Ministry of Agriculture and Water Resources was allowing polluted water from the fields in three districts of Karakalpakstan to drain through the nature preserve, Baday Tugay, where all human activity is strictly prohibited, we went to court to stop this practice. We lost, even though the violation was obvious. This instance, like all the others, demonstrates that even our judges and prosecutors know only the reality within a totalitarian country where presumption of the State's prerogative dominates, always. One born as a slave does not adopt freedom!

THE CONTINUING OBSESSION WITH LARGE PROJECTS

The new state of Uzbekistan, following the initiatives of President Karimov, brought the problem of the Aral Sea to the attention of the world society. As a result, the International Fund to Save the Aral Sea (IFAS) was created along with its subsidiaries the International Coordination Water Commission and the Catchment Water Management Units for the Amu Darya and Syr Darya. IFAS received support from the World Bank, UNEP, and the Global Environmental Foundation for a series of projects aimed to address portions of the situation but never its core causes or essential remedy. The problem is that IFAS, the nations involved with it, and their sources of global support do not focus on what is best for the sea and the ecosystems.

The International effort has remained obsessed with large projects, just like the Soviets were! Thinking about steps to restore the Aral has tended toward the grandiose scale. A key plan was to use part of the flow of Siberian rivers, especially the Ob River. The first stage of the project was already financed and built. It is a myth that this effort really was about saving the Aral Sea. Nearly all the water taken from Siberia was destined for irrigation with only 2–3 km^3 directed to the Aral, just enough to flush out rivers and lakes in the deltas. Only a trickle of diverted water would reach the sea, much too little to be meaningful.

Forgetting, it seems, all of the ignored warnings that occurred in the past about the consequences of diverting water from the Aral Sea, the same mistakes would be repeated again. The new canal bringing Siberian water would become a center for settlement, population growth, and intensive water demand. Of course, the canal's long-term viability would not be assured given climate change (e.g., drought in Siberia), increasing water demand there, and the potentially divergent future interests of Russia. The new water would invariably be more expensive, driving up agricultural production costs and making produce from the region less viable in the market place. And then there is the lost opportunity to make wise and sustainable decisions like investing in drip irrigation, which would render the existing waters in the region sufficient in and of themselves. Similarly, investing funds to cover the already existing canals would produce an immediate 50% benefit in water recovery.

The idea of diverting a river's flow was not an unusual one for human beings at the beginning of the 20th century. However, we human beings of the 21st century fully understand that such a huge construction might lead to "unexpected" consequences. Why did I use a quotation mark for the word "unexpected"? Because for every such project, there is someone who anticipates these consequences, and whoever it is they are dismissed as skeptics. In this same way, the consequences of the Aral project were predicted long before the beginning of water diversion from either river. But even now some people claim that nobody warned them about these consequences. The authoritarian regime ignored all apprehensions of those with doubts. Their closed minded view led to the catastrophe. Similarly, supporters of the redirection of the Siberian rivers have no hope for success.

ENVIRONMENTAL INJUSTICE

We have to negotiate with our neighbors and win their support for a basic proposition. People around the sea have the same rights as people in the middle and upstream of the river, not only to survive but also to have natural objects around them, a healthy, beautiful environment. For now, we do not even have rights to keep the sea!

This injustice is made clear when one looks at continuing projects in the region. Take, for example, the artificially created Aydarkul–Arnasay lakes system-built in 1969 in Bukhara, Syr Darya, and Navoiy regions of Uzbekistan. You will not believe me that the length of this artificial lake is 350 km. And it looks like this lake has a right to exist, but the Aral has no right to exist! Also, there are artificial lakes in Turkmenistan, Golden Age Lake, and Sarykamysh Lake. They have a right to exist as an artificial lakes, but the Aral has no right! It is very strange. It is a matter of human rights and of natural rights. Humans do not accept nature's rights yet, so we assert the human right to have the Aral Sea as property. It existed for many centuries until the activity of people upstream in the middle and upper river destroyed it. And now it looks like we, the people of the Aral Sea region, have no rights to exist. Perhaps if we would take water from these three artificial lakes and send it to the Aral Sea, we could have the Aral at a high enough level to keep up hope for the future. Now we fill artificial lakes from the rivers and canals, but why not return the water to the Aral Sea? It is logical from one side; from the other side it is like paying back nature's borrowed property. The people on the top and even on the ground are not thinking in such terms. They think only about today. But what about human rights for the Aral region?

TRYING TO BUILD ON HARD LESSONS LEARNED THE HARD WAY

A number of lessons emerge from our experience

1. The need for responsible users

As noted, the problem of water in Central Asia is management not amount. And I refer to management aimed at protecting the natural systems that deliver the water and the interests of the region's people. There is more water than necessary. Annual combined flow of the Amu Darya and Syr Darya is equal to $100\,km^3$, enough to supply 6 billion people on the planet with 45 L of potable drinking water a day per person. The obsession we find with redirecting Siberian rivers toward us might rise to the level of a matter of conscience were it the case that any conscience had materialized among those determining the course of these events.

In Uzbekistan, after independence, we have failed to create a viable local mechanism for dividing and managing allotted water resources. These were carefully controlled even if misallocated during the Soviet period. No real water market has been created here, as it has in some neighboring countries, in part because the nongovernmental organizations created to manage water lack authority and are ineffectual. Farmers represent the main portion of Uzbekistan's population. Fearful of the government, restricted from moving due to language barriers, and lacking industrial work to move into, the growing farm population is very dependent on the status quo. Producers without liberties are not interested in saving resources. Corruption and a lack of management capability also are barriers. The result is that little is currently being done to conserve water.

2. State planning

The situation is much worse today. Nevertheless, we keep the same approach, using too much land and soil. It is still an outcome of the same practice of water planning We always have such a plan that Karakalpakstan has to grow this much cotton, and we need a good report at the end of the year that we met this plan and overachieved the goals by 10% or 20%. Everything was for the plan and not for revenues or the real amount of cotton grown. We are still in socialism in Uzbekistan. The Soviet type of socialism continues.

3. Cotton quotas

We still do not blame cotton. Cotton is the only agricultural product that is a profitable export for Uzbekistan. The Uzbek government maintains a state monopoly on producing and selling cotton. Even selling cotton in internal markets is prohibited. The profit is, therefore, entirely taken by the government, which buys cotton at bottom price from farmers and sells it abroad at the world market price. At times, the difference between these rates reaches 1,000%! As an outcome of this monopoly, cotton growing is unprofitable for farmers. And they begin to resist, refusing in every possible way to produce cotton and replacing it with rice and other crops, which are more profitable. To force their hand, government issued a "state order" obligating every farmer to grow a certain amount of cotton. This minimum quota for cotton growing, involving a mandated percentile of total crop yield, is further required in order for the farmer to receive bank loans necessary for buying such requisites as seed and fertilizer.

Cotton has been grown in arid central Asia for a very long time. Cotton does not require that much water to grow. Moreover, a drought-resistant type of cotton was bred that uses five times less water. In actuality, the replacement of cotton by rice and wheat caused water use to increase by two to five times, aggravating the situation. In short, it was possible for Uzbekistan to have cotton and the Aral Sea, too, if only things had been handled differently.

Cotton must be taken out of state monopoly. Freedom must be given to farmers to choose a more profitable crop and the market to sell products. However, current realities hamper this change. Under socialism, resources of the state subsidized costly projects that now require continued support.

For example, the Karshi canal carries Amu Darya water 132 m upslope to irrigate the Karshi Steppe, using 450 MW of electricity to do so. Agriculture on the Karshi steppes could not compete in a free market without this public subsidy. Likewise, the Amu-Bukhara channel similarly requires pumping water for irrigation. Either support must continue or the production will cease.

4. Lack of a free water market

To discuss the continued dependence upon water-intensive crops, such as cotton, as the lead exports of Uzbekistan, Turkmenistan, and to some degree of Tajikistan, one must acknowledge the conditions that deprive farmers of the choice of what crops to grow. In these centralized economies, natural resources belong to the government and farmers are required to report what they grow. In reality, farmers own neither the land nor the crop. As noted, a farmer cannot refuse to grow cotton and shift to different products or practices such as keeping livestock. Farmers are discouraged from charting their own course and, instead, reduced to following regulations. Government is responsible for supplying agricultural production with a required amount of water; farmers, themselves, are left without motivation to conserve.

The problem here is that with the government involved in everyday issues of the economy, Central Asian countries are no different from giant corporations caught on a profit-focused tread mill. Profit, represented in monetary value, serves as the only criterion of management. Suppression is common and the farmer becomes the government's slave. Meanwhile, to fulfill the conditions to save the ecosystem, to strengthen the health of the nation, and to fulfill international duties, material production must be independent of the government. Such independence can be provided only a free market economy, in which independent producers compete with each other, and the government guarantees their equal opportunity, a healthy environment, and acceptable living standards.

How do we create a mechanism of appropriate management? Two "basin water organizations" or BVOs were created under the International Fund for Saving the Aral Sea, one for each river. Their functions are limited to surveillance of distribution of water volumes according to previously agreed volumes. They lack the authority to restrain government or water consumers. Yet, such BVOs, in my opinion, are prototypes of future public-interest corporations managing the water market in Central Asia. They would be "owners" of water ecosystems, selling water resources to different "free buyer" consumers and directing earned profits to maintain water ecosystems from high mountain glaciers all the way to the Aral Sea for the well-being of the ecosystems. Creation of this market and placing it beyond the sovereignty of any one state will lead to a real integration of the Central Asian economy and provide a financial base for the stable development of the region. The real issue is whether governments can agree.

CONCLUSION

I have never lost hope yet. I am a pathological optimist. If I lose hope, then I lose life. There must be some hope in any case. Now the situation is worse than before, but there is some good change. There is a lot of information now. We are much better informed than before, during Soviet times. And other nations in the region are experimenting with land tenure. Kyrgyzstan and Kazakhstan developed new approaches in agriculture; they gave land access as private property. Consequently, productivity on the land is higher and users pay for water. Even in Tajikistan, they have a method to pay for water and they received attention on this issue from the international community. It gives us some hope that this can be solved. Climate is also changing and it will push us to save some water, at least to think about ways to solve the problem.

Now that the tragedy of the Aral Sea has nearly unfolded in total, we seek to find an adequate answer to the problem of the Aral Sea to serve as a model for broad principles for harmonious relations between the nature and the society. To achieve this goal, we must define the main causes of the Aral Sea crisis as they occur not only in the Aral Sea Basin, but in the whole world. UDASA is the union of like-minded associates, who are confident that the Aral Sea is the heritage of the whole of civilization, of the whole planet, a pearl of Nature. We have to be here to save at least our own nature. Every nation has its own fortune, and it is ours to be here and try to save our nature. If we cannot save the Aral Sea, even with assistance from the international community, then how can we expect to save our planet?

THE ARAL MUST BE SAVED AND RESTORED!

CHAPTER 8

ARAL SEA ANALOGS IN THE AMERICAN WEST

Howard Horowitz

ABSTRACT

The hydrologic losses due to net evaporation in the Aral Sea have interesting analogs in the interior-drainage basins of the American West. Each of the three places discussed here – the Salton Sea, Owens Lake, and Mono Lake – has its own unique historical and geographic circumstances, but the story of each place has certain parallels to the Aral Sea disaster. Each place experienced dramatic water losses during much of the 20th century, but the emergence of environmental science and law in recent decades has caused significant policy changes. The Salton Sea is still declining, and modest efforts by state and federal agencies to halt the decline are inadequate. A proposal to build dikes to save part of the water body and convert the rest to salt evaporation ponds cited the Aral Sea as a model for the Salton Sea's future. The dry Owens Lake bed yields windblown dust that exceeds the Clean Air standard for fine particulate matter (PM 10), so Los Angeles is now required to release additional water back into the basin to create more shallow wetlands. In Mono Lake, a negotiated settlement has reversed the water loss while protecting vital interests of all parties, and a substantial ecological restoration plan is being implemented. The history of the American

Disaster by Design: The Aral Sea and its Lessons for Sustainability
Research in Social Problems and Public Policy, Volume 20, 89–103

ISSN: 0196-1152/doi:10.1108/S0196-1152(2012)0000020017

analogs to the Aral Sea, especially the success story of Mono Lake, may indicate potential pathways to progress in reducing problems caused by large-scale water diversion.

INTRODUCTION

Although the geographical area impacted by the shrinkage of the Aral Sea greatly exceeds the scale of hydrologic losses elsewhere, several interior drainage lakes in the American West have undergone significant hydrologic losses for similar reasons during the last 100 years. Another category of analogs with the Aral Sea watershed include the deltas of overdrawn river systems. The numerous American diversions that have resulted in the desiccation of the Colorado River delta in Mexico have caused ecological and political problems that have analogs with the diversions of the Amu Darya and Syr Darya in Central Asia.

Like the Aral Sea, the Salton Sea continues to suffer from severe hydrologic loss consistent with inadequate efforts to remediate the problems. Owens Lake, a smaller body of water in the Great Basin fed by runoff from the Sierra Nevada, was mostly desiccated within 15 years after the Owens River diversion to Los Angeles began in 1913. In recent years, to reduce health hazards from windblown dust, the California air pollution control agency has required that some water be released into the stream, creating some wetland acreage in the former lakebed. Mono Lake, an ecologically unique Great Basin lake similarly fed by mountain runoff from the Sierra Nevada, also lost most of its inflow by the diversion of its principal feeder streams to Los Angeles, beginning in 1941. The lake level dropped 45 vertical feet over the next 48 years, exposing unique columns of tufa but raising the salinity and threatening both resident and migratory bird populations. However, after a series of court battles and negotiations, the decline of Mono Lake has been reversed, due to a remarkable effort by an alliance of scientists, environmental lawyers, and citizen activists.

The geographic and hydrologic parallels of the Mono Lake and Aral Sea situations are close, which suggests that the lessons of the Mono Lake recovery could also be applied to the Aral Sea. Unfortunately, the political contexts are different, and the forces that have brought about the Mono Lake recovery are not yet sufficiently developed in the Aral Sea region. As a rare success story that stands out among failures, Mono Lake will get the largest share of this analysis of Aral Sea analogs in the American West.

GEOGRAPHIC CONTEXT

The Aral Sea, the Salton Sea, Owens Lake, and Mono Lake are all water bodies fed by interior drainage. Since they have no outlet, the inflow of snowmelt from the surrounding mountains is offset by evaporation from the surface, leaving salts to build up in the water column. Each of these water bodies is a remnant of much larger ones that existed 10,000–15,000 years ago, when vast volumes of frozen water in the late Pleistocene ice sheets began to melt. Those much bigger and deeper lakes rose until they reached an outlet. The outlets would carry dissolved salts away towards the ocean, and the big lakes would become freshwater. Many of these freshwater lakes were enormous and deep, including Lake Bonneville in Utah, Lake Lahontan on the east side of the Sierras, Lake Manley in Death Valley, and Lake Missoula in Montana. The Great Basin in its pluvial maximum contained at least 27.8 million acres of lakes, which is eleven times more water surface than the 2.5 million acres of today (Grayson, 1993). These lakes gradually declined in depth and shrank in size as the climate warmed and feeder stream flows were reduced, until the surface levels dropped below the outlets, and the lakes became interior drainage once again (Fig. 1).

Some of the late Pleistocene lakes filled grabens, which are deep structural basins underlain by faults linked to tension slowly pulling apart the earth's crust. These lakes gradually declined in depth and shrank in size over the last 10,000 years as the climate became warmer and drier, until the surface levels dropped below the outlets and the lakes became the much smaller and saltier remnant water bodies of today.

Other huge Pleistocene lakes were caused by ice lobes that blocked and dammed big rivers, backing up lakes far bigger and deeper than any modern man-made reservoir. Lake Missoula was perhaps the biggest of them, filling the Clark Fork valley of the Columbia River to a depth of over 2,000 ft above where the city of Missoula, Montana sits today. Ancient levels of this lake surface are revealed by wave-cut terraces on the mountain slopes far above the city. These ice dams were destined to collapse catastrophically as the glaciers melted and shrank, causing floods so gigantic that they dramatically reshaped land for hundreds of miles downstream. J. Harlan Bretz, an early 20th-century geologist, first startled and then convinced the earth science establishment that the "channeled scablands" of Eastern Washington were created by these giant floods, which occurred repeatedly during the Pleistocene as ice sheets advanced and retreated.

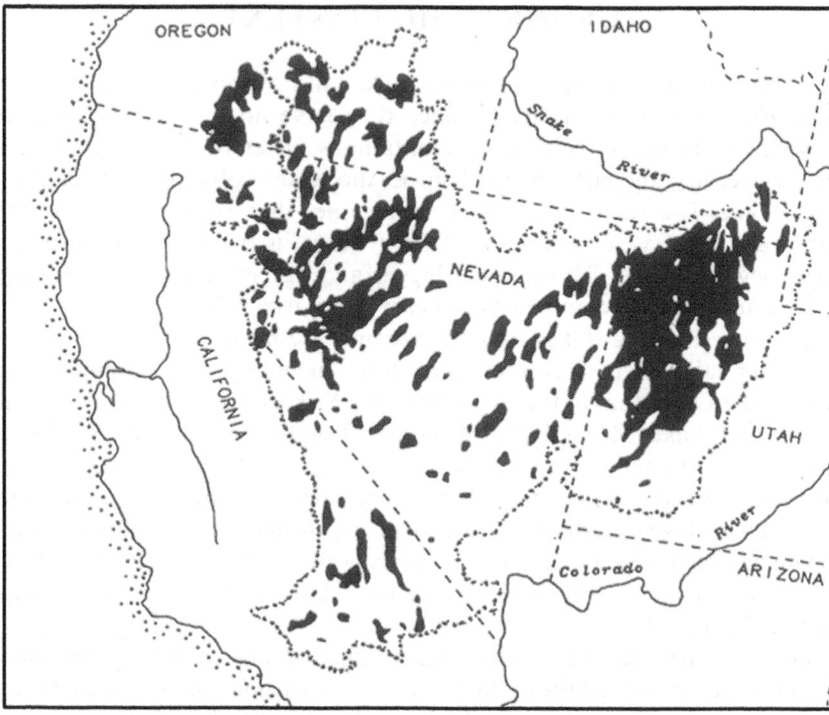

LATE PLEISTOCENE LAKES IN THE GREAT BASIN

Fig. 1. Map of Region – Pleistocene Lakes. *Source:* Howard Horowitz – adopted from Grayson (1993).

The Columbia River watershed today includes vast topographic and hydrologic variation from region to region. In the drier interior sections, there are many analogs between problems in the Aral Sea watershed and the Columbia watershed with regard to irrigated agriculture, industrial pollutants, groundwater overdraft, and fisheries' decline. These problems are not, however, unique to the Aral Sea or the American West, but are shared by many populated arid and semi-arid regions.

There are too many similarities between the interior basins of Central Asia and interior basins of North America regarding specific attributes and places to attempt to identify them all. This chapter focuses exclusively on the interior drainages that are the closest analogs to the Aral Sea Basin: the Salton Sea, Owens Lake, and Mono Lake.

CASE ONE: THE SALTON SEA

Despite the protestations of the Salton Sea Authority that it "is NOT a man-made lake," the current "historical" version of this water body is indeed the result of human error. In 1903, a channel cut into the lower Colorado River near the Mexican border by enterprising farmers trying to divert some water into the Imperial Valley ended up capturing the river's flow. The new course of the Colorado River flowed down into the Salton basin, which sits about 300 ft below sea level. The big silty river continued to fill it for 18 months, creating the Salton Sea. Levee-builders finally contained the river back into its former channel, but by then the Salton Sea covered 400 mi^2, making it the largest lake in California.

The Salton Sea Authority correctly points out that Colorado River has migrated into this deep down-warped basin many times before. This basin is on the western side of the Colorado River delta, and has been occupied by a succession of large freshwater lakes since Pleistocene times. Whenever the lake level rose high enough to find an outlet south toward the sea, it became a freshwater lake. Whenever the river migrated elsewhere in the delta, inflow would virtually cease and the lake would gradually dry up to become a huge alkali flat, as it was prior to the farmers' 1903 canal work.

In the early decades of the 20th century the Salton Sea was primarily freshwater, and it was stocked with warm-water fish that multiplied well, supporting a growing commercial fishery and sport fishery. Recreational and retirement communities grew around the shores, many of which have been abandoned in recent decades. The salinity gradually rose until it exceeded that of the ocean, and the heat and lack of shade proved to be oppressive. Periodic massive fish kills and prolonged episodes of strong odors drove most people away. In 2003, reporter Alec MacGillis described Bombay Beach as follows:

> Rows on rows of shotgun houses and mobile homes stand with missing doors, windows, and walls. Not a human occupant in sight. At sea's edge, several homes, old trucks, an armchair and a hardware store sink halfway into the sand, caked in salt that looks like the ash of a nuclear winter.

Despite the periodic fish killing, the nutrient-rich waters have continued to produce prodigious populations of warm-water salt-tolerant fish such as tilapia. These fish attract huge numbers of migratory birds traveling along the Pacific Flyway. The Salton Sea remains today one of the most important locations in North America for bird migrations (Fig. 2).

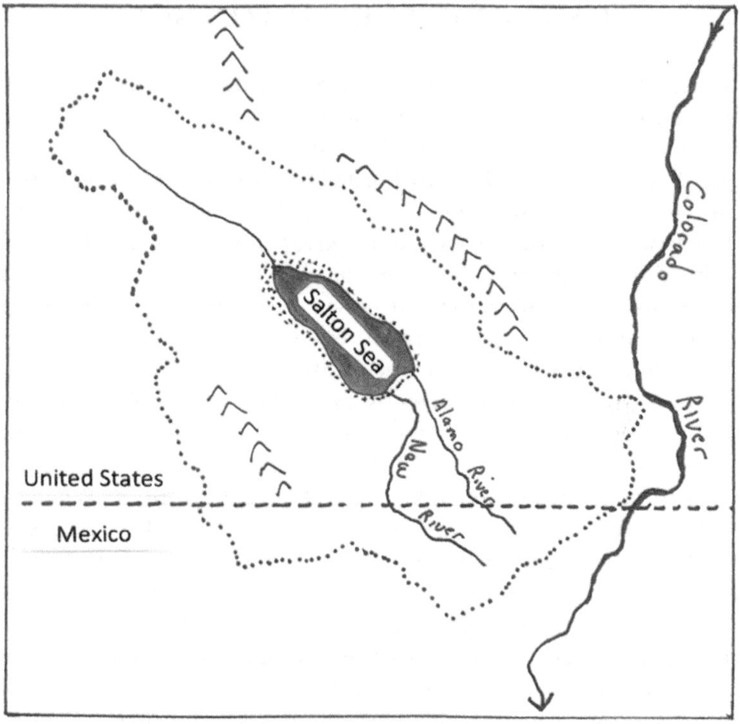

Fig. 2. Salton Sea Basin. *Source:* Howard Horowitz.

In order to maintain as much freshwater inflow as possible, irrigation waters from the Imperial Valley, the Coachella Valley, and the Mexicali Valley were routed into the Salton Sea. The New River and the Alamo River flow north 60 mi from Mexico into the Salton Sea, providing a source of marginal quality water. Not surprisingly, these warm fertilizer-laden inflows periodically produce severe episodes of eutrophication. Despite various efforts to increase freshwater inflow, excess evaporation has caused the sea to shrink from over 400 mi^2 to less than 376 mi^2 in one century, and the future water balance prognosis is not good.

In 1993 the Salton Sea Authority was chartered as a joint power agency by the State of California "for the purpose of ensuring the beneficial uses of the Salton Sea." It comprises many cooperating agencies, including water districts, county governments, and Indian tribes, with ex-officio members from various federal and state agencies. Note that the legal agreement that chartered the Salton Sea Authority in 1993 includes the following key clause:

> "Beneficial use" includes the primary purpose of the Sea as a depository for agricultural drainage, storm water and wastewater flows; for protection of endangered species, fisheries, and waterfowl; and for recreational purposes (Salton Sea Authority, 2012).

The 2003 California water transfer deal settled prolonged legal wrangling between Imperial Valley irrigation farmers and San Diego urban water consumers, but created new problems for the Salton Sea. The settlement shifts control of hundreds of thousands of acre-feet of irrigation water from agriculture to the city of San Diego over a period of 15 years. The policy reasons for transferring water from agricultural use to urban residential use may be compelling, but nonetheless this water will be lost as inflow to the Salton Sea. This water transfer deal also included an estimated $300 million for restoration projects, and the State of California "has assumed responsibility for mitigating negative impacts from the transfer of water on the Sea" (Salton Sea Authority, 2012).

The authority actually has a modest budget, a small full-time staff, and appears to have little real "authority" to effectively address the Salton Sea's problems. However, they have established a technical advisory committee, and have been trying to plan restoration pilot projects of various kinds. These pilot projects range from desalination to algae beds to solar ponds and others, but the Authority freely admits that they are all tokens that can never match the scale of the Salton Sea and its decline.

One proposed long-term "restoration" project has been "endorsed" for further funding and study by the Salton Sea Authority (although this endorsement carries no commitment to completion). It would involve the construction of dikes that would separate the Salton Sea into two basins: inflows would be routed to the north basin for aquatic restoration, while the south basin would become evaporation ponds and shallow wetlands. This plan, if implemented, would be clearly analogous to the Aral Sea situation. Despite the restoration funds and the agency rhetoric, these plans exist only on paper, and the prognosis for the Salton Sea is not good.

CASE TWO: OWENS LAKE

The extraordinary emergence of Los Angeles as an iconic American destination in the early 20th century triggered an enormous population explosion in a region with limited local water supplies. A handful of aggressive and visionary entrepreneurs recognized the need for bigger and better water supplies, and planned a bold strategy to acquire the Owens River and build a gravity-flow aqueduct to divert the water southwest to the city. The Owens River watershed is fed by snowmelt from the eastern slope of the Sierra Nevadas to the west, and from the western slope of the White Mountains to the east (Fig. 3).

One of these entrepreneurs, William Mulholland, ran the Los Angeles Department of Water and Power (DWP) for over 20 years; another, Harrison Otis, published the *Los Angeles Times*; a third man, Fred Eaton, used charm, guile, and deception to buy land with strategic water rights in the valley, which he later sold to the city for big personal profits.

Since the American West was settled by pioneers and homesteaders in the mid-1800s, the basic water law has been "the doctrine of prior appropriations." It assigns ownership of water rights on the basis of chronological

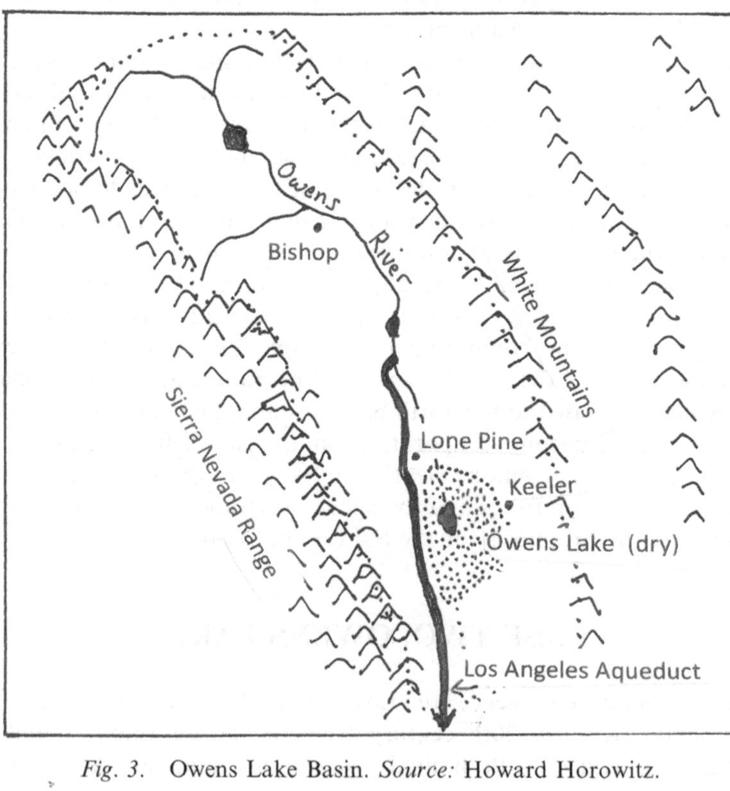

Fig. 3. Owens Lake Basin. *Source:* Howard Horowitz.

order. The first property owner along the river who puts water to "beneficial use" gets to own the rights to the water he uses. Landowners who make later property purchases may have to let the water pass through their property to be used by property owners with earlier claims

The late Marc Reisner's (1986) classic book *Cadillac Desert* provides a vivid description of the "Water Wars" that engulfed the Owens Valley as local landowners gradually came to recognize the danger posed by the Los Angeles water agents. Some local farmers with priority water rights were tricked into selling their lands to city agents disguising themselves as sheep ranchers. Other city agents used deception, money, or bare-knuckled force. The first bond issue to finance the construction of the Los Angeles Aqueduct was passed by voters in 1905, with help from a newspaper-supported campaign of misinformation about drought. By 1913, the aqueduct was operational, and the Owens River was diverted to the aqueduct intake about 50 mi north of Owens Lake. Over the next 15 years bitter Owens Valley residents used dynamite to bomb the aqueduct repeatedly, and in 1924 armed angry locals "occupied" the aqueduct intake. Of course these "water wars" did not accomplish anything except for arrests and prison terms for the local vigilantes (Karhl, 1982).

Within about 15 years, by the late 1920s, Owens Lake was reduced to a mostly dry lake bed, and has remained a largely desiccated alkali flat ever since (Sauder, 1994). One hundred years ago, its waters were a major feeding stop for migratory birds along the Pacific Flyway. Over the decades, some birds continued to visit the wetlands and springs that still occupy portions of the former lake, and more recent court rulings have required water releases to expand those wetlands. This has enlarged the wetlands substantially and has increased the annual bird count numbers as well.

The driving force behind the required increase in flows to the lake was the recognition of the health hazard posed by windblown dust, especially the threat posed by fine particulate matter (smaller than 10 μm in diameter, regulated by the Clean Air Act as PM 10) (Reheis, 2006). In 2008 the Los Angeles DWP reached a settlement with the state Air Pollution Control District (APCD) that required the release of enough water back into the Owens watershed to create 27 mi^2 of shallow wetlands. The 27 mi^2 of wetlands happens to equal to one-fourth of the former 108-mi^2 area of Owens Lake prior to the aqueduct diversions, although that is coincidental to the numbers fixed by the settlement.

Although some progress has been made with wetland restoration and bird habitat enhancement, various conflicts as well as litigation between the Los Angeles DWP and environmental regulatory agencies continue into the present moment. The most recent issue of the Mono Lake Newsletter describes current battles involving both Owens Lake air pollution rulings and Mono Lake water diversion rulings. The interesting point is that clean air requirements are managing to bring about an expansion of wetlands in the once-desiccated Owens Lake bed, although the progress is still modest.

CASE THREE: MONO LAKE

Mono Lake is a unique saline water body that occupies a volcanic basin located in eastern California at the base of the dramatic Sierra Nevada fault scarp, directly east of Yosemite National Park. Mono Lake, along with Pyramid Lake and Walker Lake in Nevada, is a remnant of enormous Pleistocene-era freshwater lakes. Each of the three remnant lakes has its own remarkable history: Pyramid Lake is a sacred water body at the heart of a Piute Indian Reservation, Walker Lake is surrounded by a military weapon depot, and Mono Lake became the focus of a colossal environmental battle to prevent the Los Angeles DWP from causing its desiccation, in the same way Owens Lake had been left to dry 50 years earlier (Fig. 4).

The legendary water czar William Mulholland was still a California power player in the mid-1920s, and he realized that the Los Angeles Aqueduct water from Owens Lake would not be sufficient to accommodate the region's continuing explosive suburban development. He managed to successfully secure water rights drawn from two different sources, the Colorado River and Mono Lake's feeder streams, and they both were tapped to feed the region's endless thirst for water and growth.

By 1941 the Los Angeles Aqueduct was extended further north to intercept four of the five streams that carried snowmelt down from the Sierras into the Mono Lake basin. Although this aqueduct extension required some pumping, unlike the gravity-driven Owens Lake portion, it still delivered water that was cleaner and cheaper than the Colorado River water. For the next five decades, the Mono Lake watershed supplied about one-fifth of Los Angeles' water. During those decades, the lake surface level dropped at an increasing rate, and Mono Lake appeared to be doomed to the same destiny as then-dry Owens Lake. By 1990 Mono Lake was 45 ft lower than it was when stream diversions started, and the shore had receded far enough to expose 22 mi^2 of dry lakebed (Mono Lake Committee, homepage).

Mono Lake supports a truly unique ecosystem that is extremely simplified yet highly productive. Even before the feeder stream diversions began, the

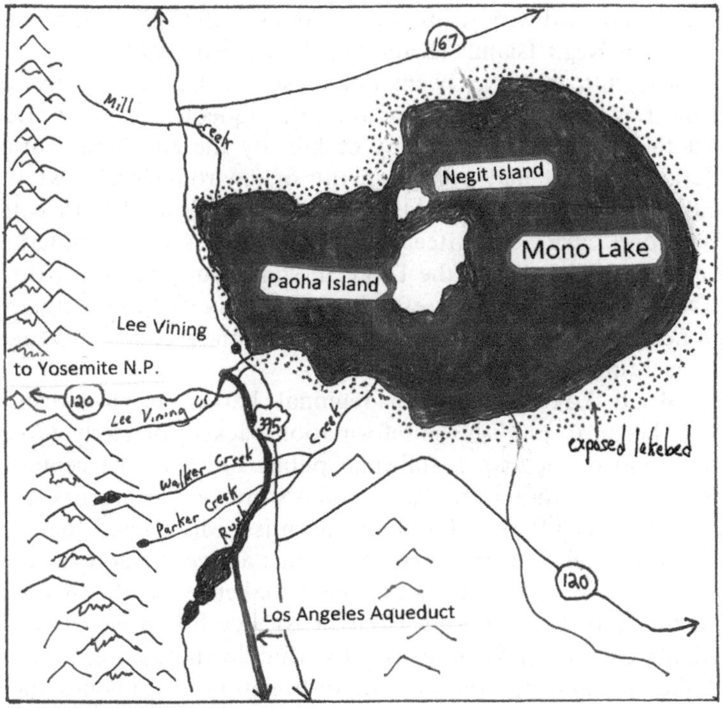

Fig. 4. Mono Lake Basin. *Source:* Howard Horowitz.

lake water had such high concentrations of salts and minerals that fish could not survive in it. Instead, Mono Lake supports a food web based on saline algae that feed dense numbers of brine shrimp and millions of alkali flies. These species, in turn, feed massive populations of migratory birds, and Mono Lake is one of the most important stopovers on the Pacific Flyway. Yet as the lake shrank, salinity increased, and biologists became increasingly concerned about the viability of the entire ecosystem. For example, Negit Island, a critical nesting area for California gulls, became accessible to coyotes, raccoons, and other predators as dropping water levels exposed a pathway from the mainland. The nesting gull population on Negit Island was quickly impacted, and they abandoned the island, dropping from 38,000 pairs down to 10 pairs in just 2 years (Steinhart, 1980). The California gulls returned to nest in 1985 when the water level rose enough, but the island was abandoned again in 1991 when the lake level dropped because of drought.

Nesting gulls returned with rising waters, and the 1997 bird count estimated 38,000 pairs on Negit Islands (Kaufman & Shuford, 1997).

In the early 20th century, Owens Lake residents had no effective options to fight the Los Angeles DWP (bombing the aqueduct was not effective!). They lost their water and their way of life. By the late 20th century, the Mono Lake residents had a larger array of environmental laws to work with, and a broader base of public sympathy. They used both with great skill. The Mono Lake Committee was formed in 1978 and immediately used science and law to challenge the DWP in state court, based on the state's Public Trust Doctrine, which asserts the state's duty to "protect the people's common heritage of streams, lakes, marshlands, and wetlands" (Mono Lake Committee, Mono Lake Story).

"Save Mono Lake" became a regional battle, as the Mono Lake Committee began to publish the information-packed quarterly *Mono Lake Newsletter*, and opened up a full-time public office in the center of Lee Vining (the nearest town to the lake, and a stopover for tourists driving to Yosemite National Park). Fisheries scientists discovered native trout populations in several diverted feeder streams above the aqueduct intakes, and this finding opened a new set of legal challenges to the diversions. In addition, the state Air Pollution Control Agency found elevated levels of fine particulate matter (PM 10) caused by windblown dust from the exposed Mono Lake bed. During over 10 years of litigation, these issues and others were contested, and in 1989 all water diversions were halted temporarily until a settlement could be reached.

In 1994, under great legal pressure to yield, the DWP agreed to a negotiated settlement that allows only limited diversions of water to Los Angeles. The volume of water diverted has been dramatically reduced (from over 80,000 acre-ft/year before the settlement to 15,000 acre-ft/year today, which can increase to no more than 30,000 acre-ft/year once the lake level rises to the target elevation of 6,392 ft above sea level.) That target lake level is 16 ft above its lowest point, but 25 ft below the level in 1940 before diversions began. Precise target levels will be difficult to maintain, since they are influenced by wet and dry cycles of precipitation as well as by limits on water diversion. In the two decades since the diversion limits have been in effect, Mono Lake has risen back up about 8 ft, which is halfway to the target level. The settlement is complex, but it effectively saved Mono Lake, and has already restored the isolation of Negit Island from land-based predators.

Legal skirmishes continue between Los Angeles and environmental activists throughout the region, as the city continues to assert priority water rights under the doctrine of prior appropriations. However, these battles are relatively minor compared to the basic settlement framework now accepted by all parties. Facing the loss of almost one-fifth of its water, the city embarked on a major water conservation program that has successfully reduced per-capita demand by nearly enough to compensate for the reduction in Mono supply.

CONCLUSION: POLICY ERROR AND INCREMENTAL DISASTER

The water policies that caused the Aral Sea disaster do have many similarities to water policies that had parallel harmful consequences in the American West. Some of these similarities involve river diversions and groundwater overdrafts. The water is used for agriculture, or for residential use, or for industrial use, but once-mighty upstream rivers become gradually reduced to desiccated beds downstream at the delta. The Colorado River and several smaller North American rivers share this fate with the Amu Darya and Syr Darya in Central Asia, and also with interior drainages in parts of Africa and Australia.

The closest analogs between the large-scale Aral Sea water crisis and the smaller-scale American West interior drainage lakes crises can be found through examining the changes in the water bodies themselves, and the policy priorities that drive those changes.

The Salton Sea, Owens Lake, and Mono Lake have unique individual histories, but they all share the common feature of shrinkage due to evaporative loss following the diversion of fresh water inflows.

The loss of the Aral Sea has been gradual and incremental. Considering the massive size of the area impacted, it may be regarded as rapid, but the sea has been shrinking since large-scale cotton irrigation began more than a half-century ago. By the 1980s, American geographers were describing the problem in academic journals (e.g., Micklin, 1988). Yet, the recognition of the problem did not result in any substantive policy changes, and the Aral Sea has continued to shrink. Effective water policy change in the Aral Sea Basin has not been achievable for a multitude of reasons: political fragmentation, economic orthodoxies, inefficient irrigation infrastructure, and conventional approaches to agriculture are among them.

Of the three American West interior drainage basins examined here, the Salton Sea most closely resembles the Aral Sea. The Salton Sea occupies a below-sea level basin in the Colorado River delta, and has a geological history of alternating periods of inundation and desiccation. They both share the long-term prospect of continued gradual shrinkage over time, as inflows from irrigated agricultural regions do not offset evaporation from the water surfaces. The consequences of decline are shared by both the Aral Sea and the Salton Sea, although the evaporative loss is far more advanced in the Aral Sea. Each of them features numerous abandoned communities, hazardous dust in the wind, and desperate plans to save some water by letting other places go dry.

The Owens River was literally captured by the Los Angeles DWP and diverted directly down to the city, causing Owens Lake to dry up almost completely within 15 years. Although some remnants of spring-fed wetland remained, providing patches of bird habitat, the lake was essentially abandoned to desiccation. As a sacrifice zone created by powerful political forces for the benefit of a distant area's economic growth, the wasting of Owens Lake was not an incremental process, but a rapid one. Owens Lake was already lost by the time Soviet administrators increased Central Asian river diversions to expand the cotton production. It became a miniature analog of what the Aral Sea has been gradually becoming.

The emergence of environmental legislation in the 1970s provided modest opportunities for restoration, and the mitigation of fine particulate dust (PM 10) has recently forced the return of some water back to Owens Lake. Current restoration programs call for shallow wetlands to occupy about one-fourth of the former lake area. The use of clean air laws to require larger releases of water into the ecosystem was possible in California, with relatively aggressive air quality control policies, but is unlikely to be implemented in many other places.

Mono Lake is the only one of the Aral Sea analogs to have suffered gradual decline because of deliberate diversions, and to experience significant hydrological and ecological recovery through the implementation of substantial policy changes. This has also fed economic recovery, as the beauty and wonder of the Mono Lake basin draws increasing numbers of tourists. Mono Lake State Park was slated for closure with 70 others due to budget cuts in spring 2012, but after a public outcry the park was kept open.

The negotiated agreement of 1994 that has "saved" Mono Lake was made possible by the citizen activists and scientists who were able to challenge the previously dominant power structure by using new environmental laws, at both the state and federal levels. The final settlement was indeed negotiated, not court-imposed, although the Mono Lake Committee's ability to have a strong negotiating hand was based on favorable court rulings along the way. Los Angeles does retain some diversion rights, between 15,000–30,000 acre-ft/year, but most of the water flowing from upstream in the mountains once again follows the four natural stream courses back into Mono Lake. The key element in the settlement was the long-term commitment to a stable and sustainable lake level. Scientists, engineers, lawyers, and environmental activists managed to agree that this would be sufficient to maintain a vibrant Mono Lake ecosystem, and still leave something for Los Angeles.

The difficulty and magnitude of the Aral Sea problems are far greater than those found in any of the American West analogs. However, there are comparisons to be observed, and lessons to be drawn, from each of these places and stories. The only hope for long-term recovery of these damaged ecosystems lies with the emergence of recognition that failure to implement changes will be more disastrous than a serious commitment to policy change.

REFERENCES

Grayson, D. K. (1993). *The desert's past: A natural history of the great basin.* Washington, DC: Smithsonian Institution Press.

Karhl, W. (1982). *Water and power: The conflict over Los Angeles' water supply in the owens valley.* Berkeley, CA: University of California Press.

Kaufman, R., & Shuford, W. D. (1997). Population size and reproductive success of California Gulls at Mono Lake, California, with Emphasis on the Negit Islets." Contribution no. 726, Point Reyes Bird Observatory, Stinson Beach, CA.

MacGillis, A. (2003). A Salty Tale of the Salton Sea. *Baltimore Sun,* August 20, Page 2A.

Micklin, P. (1988). Dessication of the Aral Sea: A water management disaster in the Soviet Union. *Science, 241*(4870), 1170–1176. doi: 10.1126/science.241.4870.1170.

Mono Lake Committee. (1978–2012). *Mono Lake newsletter.* (34 volumes, 4 quarterly issues per year.) Lee Vining, CA. Retrieved from http://monolake.org

Reheis, M. (2006). Owens (Dry) Lake, California: a human-induced dust problem. USGS (U.S. Geological Survey.) *Impacts of climate change and land use in the southwestern United States.* Retrieved from http://geochange.er.usgs.gov/sw/impacts/geology/owens

Reisner, M. (1986). *Cadillac desert: The American west and its disappearing water.* New York, NY: Viking Penguin Press.

Salton Sea Authority. (2012). *18 Years of progress.* Retrieved from http://www.saltonsea.ca.gov/salton-sea-authority.html

Sauder, R. (1994). *The lost frontier: Water diversion in the growth and destruction of owens valley agriculture.* Tucson: University of Arizona Press.

Steinhart, P. (1980). The city and the Inland Sea. *Audubon, 82*(5), 98–125.

CHAPTER 9

DISASTER BY DESIGN: THE MULTIPLE CAUSED CATASTROPHES OF THE ARAL SEA

Michael R. Edelstein

ABSTRACT

In this chapter, the title theme of "Disaster by Design" is explored and justified. Even from early times, the Aral Region was subject to alterations of natural conditions due to human intervention, often deliberate and designed. After the final conquest by Russia, the region became a fixed colony as part of the Soviet Union, ripe for exploitation characteristic of the Soviet approach to nature broadly and to stigmatized areas specifically. The Aral region was selected for irrigated cotton and other cultivation even though the consequences for desiccation of the sea, desertification, and salinization were understood. The decision was so calculated that even a cost–benefit analysis was offered to show that the Aral fishery was worth but a fraction of the cotton potential. The destruction of the region was made possible by a Soviet system of central planning and peripheral control. The brief glimmer of hope for the region evidenced during glasnost *was the only moment where the Aral's fate was*

Disaster by Design: The Aral Sea and its Lessons for Sustainability
Research in Social Problems and Public Policy, Volume 20, 105–151

ISSN: 0196-1152/doi:10.1108/S0196-1152(2012)0000020018

not sealed. The outcome is a model of ecological disaster by design, an environmental injustice, and an indication of the abusive nature of authoritarian power.

All of this country uses water from the channels of the Oxus, which causes great depletion of the river, that is why it does not run to the Caspian Sea/Sarykamysh as during past times. Very soon all this country will be ruined and become desert because of the lack of water in the river Oxus.

—Anthony Jenkinson (1558)[1]

Even for the matter-of-fact minds of our day, there is, apart from the remoteness of its situation, a veil of mystery enveloping Lake Aral, which is sufficiently provocative of the interest of the most phlegmatic observer. That the most ancient classic historians should have ... omitted all mention of an inland sea which is one hundred times larger than the Lake of Geneva, is an enigma stimulating enough to demand a solution. Nor is the attraction of so curious a circumstance lessoned in considering the fact that, though the existence of the lake remained actually unknown to Europeans until comparatively very recent times, the Arabians who conquered Central Asia had no difficulty in observing and recording its exact dimensions more than ten centuries ago.

—James Wood, The Shores of Lake Aral (1876, p. 110)

INTRODUCTION

In this chapter, I examine the Aral Sea disaster as a caused disaster. A good starting place is to reflect on Central Asians' historical control over waterways going back to ancient times. Central Asians rendered the concept of "natural waterways" a euphemism long ago.

ANCIENT DESIGNS, LIFTING THE VEIL OF MYSTERY ON THE GREAT DISAPPEARING SEA

Like the fog that once rolled off its blue waters, a veil of mystery indeed envelops the Aral Sea. Wood, writing in 1876 as quoted above, wonders how this huge body of water escaped notice by European explorers.[2] Was this some act of magic? Or did its invisibility relate to the observations made by Jenkinson in 1558 regarding the chameleon ways of the Amu Darya (referenced here by its Greek name Oxus), sometimes connecting to the Caspian Sea, the even larger inland sea to the west of the Aral Region fed by the Volga River. The answer to this question will help inform our understanding of the Aral Sea disaster today and the role of human activity in designing and implementing this catastrophe, the focus of this chapter.

The past appearance and disappearance of the Aral Sea is indeed surrounded by considerable mystery. Sergei Tolstov, a Russian archeologist who carried out a systematic investigation of ancient civilizations in the Aral Sea region from the 1930s through the 1950s, went to great lengths to consider whether the Amu Darya was ever connected to the Caspian Sea instead of the Aral Sea. His work, fortunately, was recently translated and reprinted by UNESCO. Flying low over the desert in an early airplane, he and his team explored the course of a long idle river bed called the Uzboi located on the Ust Yurt Plateau to the southwest of the Aral Sea. Tolstov found the evidence he was looking for. The Uzboi was once connected to the Amu Darya via Lake Sarykamysh. But how then did the river manage to switch to the Aral Sea instead?

Tolstov concluded that the Amu Darya's connection to the Caspian Sea in ancient times was broken some 2,500 years ago and, except for a relatively brief reunion during the middle ages, the Amu Darya fed the Aral Sea instead. To form his theory, Tolstov had to interpret his field findings in light of the written historical record. That record, as a bonus, offers an explanation for the creation of the Aral Sea. Let us allow Tolstov to explain in his words how the original separation between the Caspian Sea and the Amu Darya occurred resulting in the formation of the Aral (Tolstov, 2005, p. 295).

> I remind the reader of the legend by Herodotus about the Akes river (Oks-Amu-Darya), which irrigated the valley belonging to the Khorezmians. Some king blocked the river in the mountain passage with dams that caused the formation of a lake (obviously the Aral Sea) and the drying up of some areas including the territory of Uzboi .... Another version of the story has an ancient Khorezmian king rerouting the Amu Darya. Did the Khivaian Khorezmians route the water to Khiva, cutting off the Uzboi which connected the Amu Darya to the Caspian Sea?

So around the time that agriculture started in Khorezm some 2,500 years ago, the Caspian connection was broken and the Aral formed. And, these events (rerouting the river, cutting off the Caspian, and creating the Aral Sea) were acts of the Khorezmians, whose capital Khiva is the subject of Chapter 19 in this volume.

This change in the river flow allowed the population of Khiva, Urgench, and other Khorezm cities to grow well beyond what their original oases would have supported, amassing wealth and power as they grew. The Khorezmians went on to become one of the greatest and most powerful civilizations in the Silk Road period (Golden, 2011; Liu, 2010;

Tolstov, 2005). Once the Amu Darya was diverted, it was a constant battle to keep it that way. Tolstov cites Kos regarding the Khorezmians' active interventions to keep the waters from returning to their old path.

> … a change of the current towards … Uzboi would bring the river flow to the lower delta leaving all the Khivian oasis, with its large population engaged in centuries-old irrigated agriculture, without water and without a source of substance. In connection with this a person uses artificial measures to salvage this unstable balance of nature and, thus perhaps, gives no opportunity for the river Uzboi to appear again.

To this Tolstov (p. 295) added a few words of further explanation.

> These artificial measures' consisted first of all in the system of artificial irrigation which from the start became an essential regulator in the history of this mercurial river which almost seems the work of humans … . Creation of the irrigational network was included in the national memory as the reason for the drying up of the Uzboi. There are reasons to believe that the huge consumption of water for irrigation led to the termination of the flow to Sarykamysh and the resultant drying up which stopped the flow of water to the Uzboi, which never renewed its flow subsequently.

This addresses the question of how the Amu Darya was severed from the Caspian Sea and drained instead to the Aral Sea. The river was serving the interests of the Khorezmians. Note that both the Uzboi and Lake Sarykamysh, its source of water, desiccated as the result of the diversion although today Sarykamysh apparently receives irrigation drainage waters.

But, how did the Caspian connection return for a relatively brief period in the middle ages and why were explorers along the Silk Road at that time oblivious to the Aral Sea, as observed by Wood? Tolstov figured this out, too. The Aral Sea was created and maintained by the Khorezm civilization. In the mid-13th century, the Mongol Empire swept out of western China, passing through Central Asia then Persia all the way to Moscow and beyond. Tolstov reasoned that, when Khorezmia fell to the Mongolian invasion, the conquerors destroyed its cities and water works, freeing the Amu Darya, or at least a major branch of it, to refill Lake Sarykamysh, which in turn fed the Uzboi channel, providing enough water to reconnect the Caspian Sea (Tolstov, 2005, p. 295).

> Now it is possible to mark the contours of this history. The Mongolian invasion destroyed the irrigation facilities of Khorezm and led to the destruction of the way of life which was established over two millennia on the Lower Amu Darya. Excessive waters from the Amu Darya had broken through to the west via the channel of the Darylak to Lake Sarykamysh.

What then accounts for the disappearing Aral Sea? In his study of the area, Wood had reached this general conclusion about the body of water (1876, pp. 114–115).

> ... the lake has degenerated from an extensive and tolerably deep body of water, possessed of a copious overflow, to a shallow pool, which has at certain historical epochs been shallower still, or has even perhaps dried up sufficiently to have caused its practical disappearance from the map.

Presumably, in the process of losing the input from the Amu Darya that was diverted to the Caspian, enough water was diverted from the Aral Sea to cause it to at least partially drain, making it invisible to the considerable number of European explorers, including Marco Polo, who came through during that time.

The Mongols maintained a continuing occupation of Russia and Central Asia for a century. They were adept at integrating the best practices they encountered. Tolstov (2005) claims that the Mongols had great respect for Khorezm, adopted the culture, and rebuilt trade along the Silk Road. The Aral Region was on the boundary between the three principle Mongol hordes, the Jochi or Golden Horde, the Ilkhanids, and the Chagatai (Golden, 2011). One branch, the Golden Horde, made its capital at Tsaritzyn, near the current city of Volgograd on the lower Volga River, necessitating supply lines through the Aral Region. As to the return of the Aral and the disconnection from the Caspian, we can surmise that eventually enough waterworks were repaired to restore the Khorezm irrigation system and re-divert the Amu Darya to the Aral Sea, again cutting off the outlet into the Caspian.

ANCIENT ECHOES IN MODERN DESIGN: NATURE AS MAN'S SERVANT

The Aral Sea disaster is thus hardly the first narrative written for the "Blue Sea" or the "Sea of Islands." Whether this account captures the exact details of events not well recorded in history is immaterial to the importance of the information. The point is that the Amu Darya was reinvented by the different civilizations that flourished at various points along its changing course. Waterways and lakes ebbed and flowed. The microclimate fluctuated. And the idea of remaking a lake or river was clearly a rather plastic concept, not at all considered in the "fixed feature" frame with which the average person of today views forms of nature. Even before modern technology made the reshaping of nature relatively easy, Central Asians were molding the Amu Darya as they wished. They did not stop at one river. The Syr Darya was also modified, as Wood (1876, pp. 98–99) comments:

> ... generally speaking, it may be said that the flow, or the cessation of flow, of the waters of the Syr Darya, by its different branches, are under the control of artificial means, and that the intelligent application of a very moderate amount of labor would turn the river into any direction across the Kizilkoom desert, short of one actually running uphill. The operation is facilitated by the circumstance that during more than one half of the year, the volume of water flowing in the channel is at least from three to four times greater than that of the remainder of the year. At the epoch of minimum flow, the Syr darya is therefore a shallow stream, with a low velocity, and a canal can then be excavated in such a position as would ensure a large body of water passing into it on the advent of the succeeding floods. The set of the river having thus been influenced, no long time would elapse before the whole volume of the stream would change its direction into the artificial course, and flow to a new outlet.

By way of explanation for such manipulation of nature, the very prescient Wood (1876, p. 100) made a profound observation that the harsh desert environment had itself demanded adaptive ingenuity that invited people to attempt to tame their surroundings actively.

> ... it may be that the desolation of these deserts is a relic of the battle of the infant human intellect with the as yet uncomprehended phenomena of nature; of a struggle which has indeed left ruin behind it, but in which were nevertheless gleaned the first rudiments of that knowledge which has produced the modern art of irrigation, and which may once more restore culture and productiveness to the wastes of Turkestan.

The result was that the Amu Darya and Syr Darya were as much under the whip of humans as the whim of nature.[3] Recall Tolstov's description that it was a "mercurial river which almost seems the work of humans." That the Russians and then the Soviets were also of the plastic rather than fixed perspective is centrally important in our Aral Sea story. By the time the Russians arrived as conquerors and the Soviets as colonial masters, they were armed with concepts of scale, modern technology, and modes of organization, which would enhance their powers and the resulting destruction exponentially.

ENTER THE SOVIETS: DESIGN OF THE ARAL ECOCIDE

The Aral Story is very much about the legacy of the Soviet occupation of Central Asia. At that time, the history was being written within the Soviet paradigm of thought and action.

The Soviet Paradigm of "Eco-Perestroika"

Marx's view of social progress adopted capitalism's ability to free people from dependence on the earth, rendering nature merely a utilitarian object to be obliterated by owners and workers alike (Worster, 1985). Thus, both of the dramatically divergent paths taken by the two major contesting systems of the 20th century shared this fundamental unifying characteristic. Along with the domination of women, minorities and native peoples, both the capitalist and communist systems fully embodied the exploitation of nature (see, e.g., Bookchin, 2005).[3]

On the Soviets' part, the idea of taking from nature for the good of society and state was central to the operative paradigm of thought. It was evidenced, for example, in 1953, by one scientist's enthusiasm for the earliest push to transform deserts into farm fields (Glantz et al., 1993, p. 175).

> ... the Party ... and the Soviet government are doing everything possible to transform nature ... the grand projects outlined by Stalin's genius ... will make it possible to master the forces of nature in the USSR.

The Soviet Union would go on to commit ecocide on its lands and people (Feshbach & Friendly, 1992). Its "culture of contamination" would leave a horrific inheritance for future generations while subjecting its current citizens to extraordinary hazards (Edelstein, Tysiachniouk, & Smirnova, 2007). As will be seen, much of this damage lay unseen, unidentified, unrecognized, or unacknowledged until the late 1980s, when the period of *glasnost* (openness) and *perestroika* (restructuring) occurred under Mikhail Gorbachev. Open disclosure and discussions could finally occur, acknowledging and exposing the underlying paradigm that set in motion the Aral Sea disaster. Playing off the word for "restructuring" in droll Russian humor, an emergent term for this paradigm was "*eco-perestroika*." Here it was nature, not society that was restructured (Glantz et al., 1993, p. 175). Soviet power was enhanced

> ... by restructuring the relationship between society and the natural world. Soviet leaders believed that, armed with Soviet engineering know-how, rivers could be diverted, precipitation increased at will, deserts made green, and so forth.

Illustrative of this exercise of human power over nature was the Karakum Canal, built in the 1950s, which diverted 15% of the Amu Darya's flow into Turkmenistan.[4] The Aral Disaster must be considered in light of the cumulative consequences of the restructuring of nature characterized succinctly by Glantz et al. (1993, p. 175).

> Soviet Communist rule has bequeathed a legacy of unparalleled widespread environmental degradation, impoverishment and mismanagement. This may prove to be one of history's bitterest condemnations of Soviet Socialism – its wanton treatment of nature … .

It may have been the worst case, but it was certainly not the exception to the rule.

The Soviet period was also one of mass mobilization of labor and resources for meeting state goals. Sometimes directives were explicitly dictated by powerful leaders such as Stalin, and other times they emerged from unnamed bureaucracies in order to accomplish some stated objective of the leadership. From the early Five Year Plan that unleashed the collectivization of peasants and the war on the kulaks, the pattern was clear. Central planning was capable of forcing social restructuring for the good of the state, whether it was ordering hundreds of thousands of people to the fields to pick a crop or exiling them permanently. Rivers were dammed or diverted as a matter of routine. Such edicts came without public discussion or consent. And resistance was violently punished. People learned quickly not to question, at least publicly. Those who did were simply exiled, sent to work camps, or killed. The others learned from these misfortunes (see Figes, 2007). The mold was set for later decisions, including Khrushchev's Virgin Lands Initiative. The Aral region was at the remotest fringe of the Soviet empire. It could be exploited only if labor was somehow deployed for the task.

Beginning in the 1950s, as a function of the Virgin Lands Initiative, people were relocated en masse to marginal areas designated in state plans for new food production. The Aral Region was a major resettlement area. Once they had combined fertile land with sufficient farm labor, supplying the water was the third requisite link. The Virgin Lands Initiative would therefore provide pretense for those wishing to exploit the waters then flowing into the Aral Sea, setting the stage for the Aral Sea disaster.

KHRUSHCHEV AND THE ASSAULT ON VIRGIN LANDS

Nikita Khrushchev made his name advocating massive improvements in the Soviet Union's ability to feed its people. His succession of Stalin was in no small part due to the success of his plans for agricultural expansion intended to address famine and escape dependency on international food assistance. The 1954 policy to expand agriculture onto virgin and idle lands epitomized the Soviet view of nature as resource. It bore the slogan "We cannot wait for favors from nature, our goal is to take them from it," (Zonn, Glantz, & Rubinstein, 1994, p. 143; also Glantz et al., 1993, p. 175).

The Aral Sea disaster grew out of the centrality of food in both foreign policy and domestic policy under Khrushchev. In the west, the open display of this focus was Khrushchev's shoe pounding visit to the United States. It was as if global supremacy would not be won alone in secret military science cities or in space, but also in the race for agricultural productivity. And the Soviet Union had a lot of catching up to do. Khrushchev's visit to the U.S. Corn Belt showed him the superiority of the American reliance on private farmers with huge farms in a subsidized free market pushed to monocropped or monocultural agriculture using ever advancing mechanical and chemical approaches for expansion into otherwise marginal lands, subsidized, of course, with irrigation water. Soviets had collectivized farms, creating a mechanized social structure but a primitive technological approach to agriculture. The virgin lands scheme would expand the reach of agriculture using some elements of the American approach, the monocropping, chemicals, and water. But it would rely heavily on human labor as a means of offsetting the investment in advanced mechanization. The added benefit was that the use of human labor offered a pretext to further redistribute and mix the population.

The program also appears to have been modeled to some degree after the contemporary American initiated "Green Revolution," developed after World War II, to help highly populated nations expand their capacity to feed growing populations. The Green Revolution had the side benefit of expanding American influence and finding new markets for emerging products. In fact, it created the model for international agribusiness as practiced today. The modus of the Green Revolution was increased use of industrial farming, employing large holdings, mechanization, expansion onto marginal lands often requiring intensive irrigation, mono-cropping for efficiency, use of hybrid rather than native seed, and pest and weed control through pesticides and herbicide application. The program assumed plentiful oil and water and access to advanced chemical poisons produced originally for chemical warfare, something the Soviets had in abundance.

The virgin lands scheme would push agriculture into uncultivated areas so that wheat and other crops would be grown in the arid and semiarid steppes of western Siberia and Kazakhstan (Glantz et al., 1993). Khrushchev's image of farmland was drawn from the fertile Ukraine, but the virgin lands in Central Asia were very different. To set the program in motion, 650,000 new settlers were sent to Kazakhstan, including the Aral region, to the horror of Kazak pastoralists. At first the imported agriculture succeeded, but then the cycle of drought and other endemic problems of the area interceded. Compounded by the damage caused by plowing up the fragile soils, widespread degradation and desertification occurred, hampering even a return to grazing. This virgin lands expedition was a foundation for the eventual Aral Sea debacle. It should also have been a warning.

The decision to maximize cotton production demanded diversion of water to irrigation, but it also brought about the problems of highly intensive irrigation through distribution systems that were ancient. Used modestly, the result might not be too damaging. Used intensively in modern agriculture, it not only drained the Sea but also caused the second prong of the Aral disaster. Not only was the water of the Aral Seas' two tributaries cut off, but it was then overconcentrated in arid but arable lands, setting the stage for soil water logging, salinization, loss of soil fertility, and surface ponding. The decision to intensively grow mono-crop cotton then required massive inputs of fertilizer and pesticide, adding another dimension to water, soil and eventually air pollution. Like layers of phyllo dough, each contributed an integral layer in building the final dish. The decision to add each ingredient was always part of the bigger plan, yet required design and implementation of its own allowing the experts involved to be distracted by their part but never seeing the whole. Even if the best ingredients had been used, the outcome was unpalatable.

THE INGREDIENTS OF DELIBERATE DESIGN

The demise of the Aral Sea was no accident. The case for the Aral tragedy being a disaster by design rests on several components of its history. First, there was the paradigm of the Soviet system for controlling nature and society discussed earlier. The predisposition for disaster was thus hardwired into the very process of thought emerging from this time. Second, there was the "Soviet system" and its planned economy, dictated from the top but instigated from a cadre of bureaucrats and technocrats under pressure to perform. Experts suffered from overspecialization, interdisciplinary disconnect, a dearth of holistic, biocentric and bioregional thinking, and a philosophy of taking from nature. Third, deliberate policy decisions and concrete implementations advanced the Aral Disaster forward during the Soviet period, beginning with the virgin lands policy, discussed earlier. Those fulfilling central plans often understood the consequences, making the Aral's death premeditated, in the sense that the death of the sea was foreseen and accepted. The outcomes of the policies implemented were crystal clear and well understood. More than this, the destruction of the Aral Sea was perhaps even desired. Rather than an unintended consequence of policies pursued, its demise may have been a secondary benefit from the perspective of the policy setters. Fourth, the Aral region was, from a Soviet perspective, a marginal place inhabited by marginal people. It was far away from the center; the expectation was for the place to meet colonial needs. Finally, the Soviet system had an authoritarian structure capable of forcing its agenda on others. These five points are explored in the remainder of the chapter.

It is possible to assemble a reasonably clear picture of how the design for the Aral's demise emerged and was implemented, when the implications were understood by government and scientists, by the public in the USSR, and by the people in the Aral region and finally, how these respective actors responded. The evidence was skillfully put together in a 1993 article by Mickey Glantz, Alvin Rubinstein, and Igor Zonn, and in Murray Feshbach and Alfred Friendly's remarkable chapter "Sea of Troubles" in *Ecocide in the USSR: Health and Nature Under Siege*. The strength of these two accounts is in no small part that the authors were direct observers of at least some facets of the unfolding Aral story.

After an exhaustive examination of every issue of contamination and degradation in the USSR, Feshbach and Friendly (1992, pp. 73–75) were in a position to weigh the relative devastation of the Soviet system and discern the cause. Their conclusion, the

> ... worst case of ecocide in the Soviet Union [was the] the murder of the Aral Sea and the contamination of the cotton fields that have swallowed the rivers that once fed it ... . The sea turning to desert was a symbol of a sixty-year pattern of ecocide by deliberate design.

Who perpetrated such ecocide?

The Soviet System

Yanitsky offers a clear description of the "Soviet system," detailing its core and peripheral organizational structure with which it maintained an effective grip on the vast Soviet Union. Members of the system nucleus included "All the groups which held key positions in the sphere of the planning and distribution of all types of society's resources ..." Included in the nucleus, accordingly, were "ruling elites" of ministries and departments, "local organs of power," the "areopagus of state science," the party "central apparatus," Komsomol (the young Communists), trade unions, and all planning bodies.

In order for the nucleus to reach across the land, it required a broad peripheral apparatus as well. This function was fulfilled, first off, by those who served the System zealously in return for rich material rewards as well as opportunities for upward mobility. The second peripheral group might not be so zealous, as their services were compelled, and they lacked any choice. These were rank and file soldiers or members of the army of migratory workers, a million people driven from their rural homes, enslaved by the System and forced to live as migrant laborers in temporary barracks near whatever task they had been assigned to complete (Yanitsky, 1993, pp. 95–96). Yanitsky (1993, p. 92) decries how the System exploited and changed people to serve its ends, commenting that

> Dozens of 'Great Works of Communism' exploited the natural human desires of the masses ... then leaving them to 'the mercy of fate' ... . Perhaps the most serious legacy of the rule of the System was the creation of a 'new' type of personality: an individual without roots, a hired hand, an exploiter and quick profit seeker.

Drawing from its nucleus and periphery, as well as the planned economy and the bureaucratic structure, the System was designed for the ultimate exploitation found at the core of Soviet values. It was, in Yanitsky's words (p. 92),

> ... a powerful infrastructure for exploiting the country's natural and human resources. In effect, everything outside the limits of the System was regarded simply as a resource to be used to guarantee the System's survival. Hence the growing pollution of the natural environment, the deterioration of our social surroundings, and the complete disregard for national cultures, unique natural landscapes, historical landmarks and local traditions.

In fact, it was just the organizational machine capable of destroying the Aral Sea and region.

The Soviet Planned Economy

The Soviet economy was spurred by successive Five Year Plans, each with far-reaching goals. The Soviets liked to think big and dictated policy in sweeping concepts. New policies were matched with very detailed outcomes expected. In perhaps a good dress rehearsal for the current obsession with "outcomes assessment," the Soviets who carried out the plans were under strict pressure to meet quotas and satisfy the plan. When taken literally, such pressure too often resulted in the social sanction of aberrant and destructive behavior. When not taken literally, or to cover unintended failure, the system invited falsehoods, inflating achievements so that the plan could rhetorically be fulfilled even if, in reality, Soviet society was engaged in an elaborate ruse to fool itself. Both conditions of excess and disguised underachievement were common.

A number of key elements of this system are worthy of note: its bureaucratized nature, the reliance on crude cost–benefit analysis, the ability to ignore warnings about adverse impact and impending failure, and the ability to invent alternative explanations to support the policy, even engaging "objective" scientists in propaganda generation.

Bureaucracy

Like famous parodies of bureaucracy, Yanitsky chronicles the foundation of State socialism as

> a powerful stratum of 'Soviet clerks' ... people with average education and very modest incomes, who are employed to perform low-skilled routine paperwork. These office workers have always been "cogs" in the System; they were created by it, materially and spiritually, and they are totally dependent on it. Without them, however, it would not last a single day. This stratum is one of the main advocates of the values created by the System.

Such office workers, Yanitsky suggests, lived collectively in a work group of some 40 individuals who acted like a large family. Rather than work, however, the "priorities became obedience to the System and obeying the laws of the 'large family.'" As a result of collectivism, they further enjoyed "personal freedom from responsibility." It was virtually impossible to be fired. Finally, these office workers were "alienated from the process and results of their labour." Yanitsky pointed to Soviet planning organizations as the exemplar (pp. 95–96).

> In several centres in the capital, hundreds of thousands of people used to plan population centres, that is the standard living environments for tens of millions of people whom they did not know, in places they had never seen.

Implied here is that they did a miserable job of it.

Minvodkhoz, the Soviet Ministry of Land Reclamation and Water Resources, the planning bureaucracy central to the making of the Aral Sea disaster, was similarly alienated from the process and results of its labor. As a huge organization, it employed one and half million people, about 1% of the entire Soviet labor force. Lacking any culture of questioning, policy revision or social learning, its task was to set goals to meet policy dictates and then to assure that these planned goals were met. Thus, in the Virgin Lands Initiative, unrealistically high farm production targets were set that were impossible given the natural limitations of the lands to be farmed. To increase the challenge, it was forbidden to admit or acknowledge that such limits existed. Heavy doses of pesticides and fertilizers were employed to drive up agricultural output and artificially overpower these limiting factors. By meeting output targets, the agency could not only ignore natural limits but also maintain its totally unrealistic planning process. In other words, as Feshbach and Friendly (1992, p. 64) put it, the approach provided just the

> ... chemical miracle [needed] to propel agricultural output high enough to compensate for the basic structural weaknesses Soviet leaders either would not or could not address.

Protection of "the plan," not agricultural productivity, took precedence.

The obsession with protecting the plan became even more perverse. Unable to easily enforce targets for agricultural output, the bureaucracy substituted planned targets for pesticide application, instead. It then shifted emphasis from outputs to inputs. As a result, the ineffectual overuse of pesticides became a core component in bureaucratized Soviet farming. Whether or not the outcome made any sense, plan fulfillment was assured. Prominent ecologist Alexei Yablokov, at the time of Perestroika, the leader of Parliamentary environmentalists, explained this convoluted logic to Feshbach and Friendly (1992, p. 66):

> An "expenditure psychology" prevails in the USSR's agroindustrial complex; the basic growth measurements are not bigger harvests or lower production costs but the total volume of outlays for pesticides, fertilizer, capital investments, etc. The work indices for crop-protection units are tied directly to fulfilling the plan for chemical treatment of the crops regardless of the actual situation in a given month, season, economic zone or region. Thus in the race to put the highest figures in the ledger books, pesticides are sometimes simply carried out to the fields and dumped alongside them.

The same level of absurd reasoning propelled the decision to destroy the Aral Sea. Bureaucracies not only protect but also seek to perpetuate their missions. The design of the Aral Disaster, viewed in this light, becomes less of a grand plan for cotton production than a grand plan for perpetuating the mission of the Soviet agency Minvodkhoz, which made the primary decisions regarding irrigation expansion in Central Asia and was, as a result, poised to direct the rescue of the Aral Sea now required. Thus, critics charged Minvodkhoz (Feshbach and Friendly, 1992, p. 83) with benefitting

> ... from constantly digging canals and obtaining billions of rubles from the state budget for this purpose. To insure its operational territory for years to come, it set out to exhaust all the water resources of Central Asia and then to increase the area under

> irrigation. The resulting water shortage would be the pretext for promoting the "Project of the Century" – the diversion of Siberian rivers to expand irrigated acreage even further.

The decision-making calculus of bureaucracy is the cost–benefit analysis, a tool that readily allows political decisions to appear objective.

Premeditation by Cost–Benefit Analysis

The design of the Aral disaster embodies a judgment dating from before the Communist Revolution that the sea had little value. At the turn of the 1900s, a Russian scientist, Voyeykov, first questioned the Aral's importance and proposed that the waters would be put to better use were the lake desiccated and the waters diverted for irrigation of crops. Such reasoning is later mirrored in these words of a Soviet official (Glantz et al., 1993, p. 187):

> It is necessary to drain the Aral Sea in the future … . The time has come for the rational use of the water resources of the Aral basin for the benefit of the Soviet people building communism.

Later, Stalin is alleged to have stated, "Any drop of water flowing down the Syr and Amu Darya, that reaches the Aral Sea, is a drop of wasted water" (Wehrheim & Martius, 2008, p. 1).

Such underlying valuation clearly influenced subsequent "objective" analysis. In particular, a cost–benefit analysis of the inherent trade-offs served as the basis for and justification of the decision of Minvodhkhoz's officials in the 1960s to starve the Aral Sea for irrigation. Calculation of these costs and benefits, however, was a crude exercise in distorted thinking. First, it embodied the prevailing thought that the Aral Sea had no value. Then, it employed greatly exaggerated forecasts to conclude that doubled rice and cotton harvests were possible. The scale was quite tilted by the time the head of the agency compared the benefits, calculated to be 11.2 billion rubles in cotton earnings, with the cost of sacrificing Aral fishing, sea transport, and the coastal economy, estimated at only one-twelfth that value (Feshbach & Friendly, 1992). Glantz et al. (1993, p. 187) summarize the calculation in simple terms:

> The perceived value of water in the Aral Sea was compared at least in a qualitative way to the expected value of water for other uses. For example, scientists have suggested that the value of water used to irrigate cotton fields would be several times higher than if the same amount of water were allowed to flow into the Sea.

Not only was the destruction of the regional fishing industry discounted, but also water was valued as "free" (Glantz et al., 1993). The calculations also omitted impacts to health, to ecology, to regional climate, and of "fields covered by poison rain" (Feshbach & Friendly, 1992, p. 87), as well as secondary, social, and long-term consequences (Glantz et al., 1993). Capitalism, apparently, was not the only system to leave out externalities. The same kind of tradeoff analysis was made with regard to the construction of the Turkmen Karakum Canal, which drained off 15% of the Amu Darya before it even entered Uzbekistan. Dissenters from these valuations were ignored (Glantz et al., 1993).

CREATING THE DISASTER: DELIBERATE AND PLANNED POLICY

Central Asia came under Soviet control in the 1920s and remained there until the end of the USSR in 1991. Arbitrary boundaries were drawn for the Central Asian republics, influencing the eventual independent nations of the region. Following the World War II, planners targeted this region for maximum cotton production as part of Moscow's calculated decision to strive for "cotton independence." The USSR would achieve self-sufficiency in production of cotton fiber and in textile manufacture as well. Factories would be built outside of Central Asia, where the cotton fiber would be produced. There was an abundance of fertile but arid land; irrigation was required. Virtually all of the water in the Aral Sea Basin would be needed to contribute to the effort (Glantz et al., 1993; Lipovsky, 1995).

Agricultural policy began to push cotton monocropping heavily. In the 25 years leading up to the end of the USSR, cotton production went from comprising half the area under cultivation in Uzbekistan to two-thirds. The result of this shift was greater reliance on pesticides, fertilizers, and irrigation water. Eventually, a steady decline in soil productivity occurred. As greater amounts of water were diverted from the Syr Darya and Amu Darya to allow expansion of cotton farming, by the end of the 1970s, the Aral Sea was being fed by an ever decreasing water supply from the Amu Darya alone. The gambit worked for a while. Cotton production doubled from the 1960s through the mid-1970s before rates began to fall (Glantz et al., 1993).

The monocropped fields of cotton were extremely vulnerable to attack by boll weevil, requiring intensive pesticide application of 54 pounds per acre, 180 times the amount used in cotton cultivation elsewhere. Herbicides were also utilized, including dioxins, which although banned in 1971, continued to be used secretly in the 1980s. Unprotected farm workers were routinely doused. Rates of childhood illness went through the roof; nearly half the region's total population was ill (Feshbach & Friendly, 1992).

During the annual harvest or when extra labor was needed, schoolchildren from every school were ordered to work in the cotton fields for a few weeks at a time. This action placed children directly into the highly poisoned setting. Breathing problems were documented among schoolchildren after herbicide-use beginning in the 1970s. Feshbach and Friendly (1992, p. 80) directly observed such reckless behavior when they witnessed a farm field supervisor so much under pressure to meet his quotas that he kept sixth graders working in the fields even when defoliants were being sprayed. When the authors protested, he explained that "*he had to fulfill his plan.*"

IGNORING (OR MAYBE WELCOMING) WARNINGS OF THE ARAL'S DEMISE

From the onset of intensive cotton irrigation, Soviet scientists predicted that the Aral Sea would decline. This was beyond question. For example, in 1968, the head of the Soviet Hydro-Technical Institute commented ... "it is obvious to everyone that evaporation of the Aral Sea is inevitable" (Feshbach & Friendly, 1992, p. 88).

Scientists also understood the other problems that would emerge. Thus the fate of the Aral watershed, including salinization and other problems, was presumably appreciated by those, in the 1960s, who made the decision to divert water for intensive cotton production. Mitigations for these impacts were also understood. But the consequences were callously disregarded in the name of economic development even though it was clear to any informed observer that any benefits of the program could not long be sustained (Glantz et al., 1993).

It was not common for such doubts to be expressed or the truth to be laid out. The truth might not be tolerated even if one survived sharing the secret. Even, later on, during the perestroika period, it was risky to speak the truth, although the nature of the risk had dramatically changed. When the USSR's top hydraulic engineer, P.A. Polad-Zade, lost his patience with Save the Sea advocates whose cause he viewed as already lost, he publicly declared (Feshbach & Friendly, 1992, p. 88):

> It is time for all the wailing to stop. The case is closed, and the people here will have to learn to live without the sea.

The statement led to such an outcry that he retracted it the next day. Feshbach and Friendly comment that Polad-Zade had made a rare error of that time, he had been too honest.

SCIENTIFIC RATIONALIZATIONS IN FAVOR OF ARAL'S DEMISE

In the Soviet system, advancement as a scientist or academician was greatly aided by party membership. And party membership was placed in jeopardy by unsanctioned behavior. There were strong incentives for producing work that advanced "the party line" and strong disincentives for crossing it. Some scientists may have been guided by self-interest, fear for their families, or specific coercion. Others may have carried out genuine work that was misdirected or wrong. And others may have just been plain old hacks (apparatchiks, or functionaries of the party). In any of these cases, what is commonly found is scientific legitimization of political ends.

While a number of warnings about impending disaster evidenced some independent thinking among scientists, most scientists appear to, at least publicly, have supported the official line on the Aral Sea. The cost and benefit analyses discussed earlier were one approach by scientists to justify these actions. Another involved the production of "expert facts" that assuaged or minimized public concern. This work could not be publicly challenged at the time for what it was.

During *glasnost*, Andrei Monin, a corresponding member of the USSR Academy of Sciences, made a blistering public attack against three scientists who helped to justify government policy to destroy the Aral Sea (Glantz et al., 1993, pp. 177–178). First, Monin's earth scientist colleague at the Academy of Sciences, Yevgeny Federov, was publically lambasted for allowing his "branch of science [to be] assigned the role of servant to certain powers that be."

Second, Monin deconstructed the work of another scientist who repeated overstated predictions from the 1960s for doubling both the yields of crops per unit of land in the Aral region and doubling the amount of irrigated land. The actual outcomes were far lower (e.g., irrigated land had increased by half). Moreover, Monin took the opportunity to note that per capita production in "prosperous" Central Asia was well below norms for meat (26%), milk (42%), and fruits and vegetables (53%). If crop production had not increased, what then, he asked, was the effect of the policies to expand agriculture (Glantz et al., 1993, pp. 177–178)?

> What had increased? The use of chemicals. Cotton fields are inundated with ten (!) times more pesticides than on average elsewhere in the Soviet Union or in the USA. What else has increased? Consumption of irrigation water.

Finally, Monin attacked the "scientific phraseology" used by officials to disguise the rising loss of irrigated land to salinization that rendered the view along a route from Khiva to Nukus "… as white as a snow-covered steppe, a lifeless plain from horizon to horizon" (Glantz et al., 1993, pp. 177–178). In sum, he posited that scientists and their organizations at best failed to challenge actions they knew to be destructive; at worst they gave flawed policy justification.

Monin attacked scientists and theories with equal vengeance. Some of the theories he found to be dubious existed to explain away the impact of the declining sea (Glantz et al., 1993). These included the notion that evaporation and transpiration of reed beds in wetlands along the Aral shore robbed the sea of water; therefore, wetlands should be drained. In another theory, there was no real worry of desiccation because the Aral Sea could be sustained on as little as one-tenth of its original stream flow; 90% of its water could be safely diverted. Then there were the imagined benefit theories. One posited that drainage water from the irrigated fields should be returned to the Aral Sea because the drained water was "rich in biogens." Of similar benefit, wind erosion would spread beneficial excess agricultural chemicals improving the region's soils. And, a final contested theory, should desiccation of the Aral Sea occur, it would not affect the region's climate because the lost evaporation and transpiration would be offset by losses from the required expansion of irrigation ditches.

The lies did not stop with the circulation of spurious theories. Peer response very likely upheld these theories and findings at the time they were proposed. They would have been published in journals, been cited and positively reviewed, and otherwise (wrongly) recognized for their nonexistent truth value. In addition, conduits of such useful information likely advanced in their careers, received good work assignments, had favored access to housing and food, received passes for the Black Sea beaches or desirable sanatoria (rest spas), and generally were accepted as highly productive members of the academy and society.

Finally, a debate apparently arose among Soviet geographers about whether the Aral region had previously desiccated when the region cycled through a dry period, with the possible implication that, if true, the Aral could dry up naturally, again. In contrast, Tolstov had demonstrated Aral desiccation had previously occurred due to human activities.

In fact, both causal factors were operative in the Aral Sea disaster. Although entirely caused by human actions, natural cycles influenced how the Aral Sea disaster developed. The 1950s was a very wet decade and irrigation had no noticeable impact on the Sea. In contrast, in the 1960s, low rainfall in the mountains caused dramatic reductions in river flow into the Aral Sea, resulting in major irrigation effects on sea levels. This difference, however, was at first misinterpreted because the Caspian Sea level had also fallen, suggesting a natural regional climatic effect. Thus, appreciation of the full gravity of irrigation from its tributaries for the Aral Sea was not fully understood until the 1970s, when Aral water levels dropped precipitously but the Caspian levels did not (Glantz et al., 1993).

Lipovsky (1995, pp. 6–7) reports that such nuances could not surface easily during Soviet times because experts were influenced by the Marx/Lenin doctrine that

> … the dominant factor is not nature's influence on man but rather man's influence on nature. All theories regarding the desiccation of Central Asia were declared reactionary and distortive of reality.

In sum, the focus was on the exploitation of the Sea, not over possible resulting impacts and their relative causes.

CORRUPTION

Corruption enforced by fear was another significant contribution to the Aral Sea disaster. It is important to remember that Gorbachev undertook restructuring (perestroika) to rid the Communist party and the Soviet government of corruption that interfered with their ability to create and implement rational policies. The problem was certainly evident with Aral Sea cotton. Endemic corruption had long been a major factor in the exploitation of the Aral region all the way across the chain of control. First, in Central Asia, local cotton lords arose who paid off authorities to control the inflow of funds for irrigation projects, which they then embezzled. They also controlled the outflow of profits, inflating harvest claims to earn bonuses for exceeding the plan. Cotton lords also earned money by charging protection from black marketers and drug dealers. These cotton lords maintained control using what Feshbach and Friendly (1992, p. 80) termed a "feudal-style control over subjects."[5] Meanwhile, cotton profits that made it back to Moscow from the periphery were skimmed off by Moscow politicians up the line until (p. 74), "The corrupt Party elite waxed rich off mythical cotton." Ironically, the only return to the region was funding for additional irrigation canals to expand cotton farming further. In this sense, Feshbach and Friendly (1992, p. 74) concluded, "the real crop was ravaging the land."

COMPLICITY AT THE SYSTEM'S PERIPHERY

During our trip across Uzbekistan, I had occasion to spend considerable time with a former ranking Soviet official who was integral to carrying out the irrigation policy that led to the destruction of the Aral Sea. I asked him to allow an interview that specifically explored his own complicity in causing the disaster. He agreed but then asked not to be identified.

The interview established that he and others knew full well the implications of the actions that they undertook. They did not approve. However, they went along. At the time, there appeared to be no other alternative. If Moscow ordered more cotton and more irrigation, then it was the role of local officials in Central Asia to fulfill their orders, not to question them. Although he did not elaborate, it was implied that the consequences of disobedience would end the career and possibly the life of the objector. What would not be accomplished, he made clear, was ending the policy implementation. Somebody else would just be ordered to carry it out. There was no point to objecting and little recourse but to aid in destroying the Aral Sea.

It was evident that the informant cared very much about the Aral Sea and regretted his role. But as to the question of whether he had a choice, there were no doubts.

THE DISASTER OCCURS

The Aral Sea disaster started off with baby steps. The first irrigation canal was hand dug from 1939 to 1941 in eastern Uzbekistan and diverted water from the Syr Darya. Before long, the 215-mile long "Great Fergana Canal" resulted in water contamination by pesticides. Such projects proliferated after World War II (Feshbach & Friendly, 1992). As the network of current irrigation canals seen in the maps at the beginning of this volume took form, the principle channels through which Amu Darya and Syr Darya water was sent to irrigate fields came to function as artificial rivers. A case in point was the 1,200 km Karakum Canal. Large drainage lakes, such as Lake Sarykamysh, were also part of the network. This overall reshaping of the natural waterways of the region amounted to a major transformation of nature, as Glantz et al. (1993, p. 185),

> Diverting Amudarya and Syrdarya water to irrigate desert sands can be viewed as an attempt to transform one natural ecosystem, the desert, into a manmade one, a 'garden', 'a grandiose project that was to transform the region from an arid wasteland into a fertile oasis.'

The pace of this transformation was rapid. Overall, irrigation grew at a rate of 2.3% annually between 1965 and 1986. Eventually, the drainage network stretched over 100,000 mi and included 8,000 vertical drains (see maps). Across a wide area, water tables rose from 22 to 32 ft up to within a foot of the surface, resulting in the inundation of 985 settlements in the Aral basin. Contaminated water was pumped off into freshwater ponds used for drinking water. Soils became waterlogged; 3 million acres became so saline that it could no longer be farmed. Another 8.6 million acres could not be farmed because it all had become swamps (Feshbach & Friendly, 1992).

Such ecological damage had repercussions for whether harvests met targets dictated in the Five Year Plans. As salinization cut harvests in half by 1987, Uzbekistan fell 1.5 million tons behind its plan commitment. To make up for losses, peasants were ordered to relinquish the small private plots where they grew fruits and vegetables. Popular protests with the slogan "*Down with the cotton, long live the orchards*" failed to reverse the decision. The result of losing their private plots was that farm families were forced to subsist on a poor diet; by 1987 Uzbekistan's people were getting less than half of the milk and a quarter of the meat required according to Soviet standards. Acute child malnutrition became prevalent throughout the region. Starving children in the Aral region were even documented in the Soviet press (Feshbach & Friendly, 1992).

By the early 1980s, Uzbek novelist Sagdulla Karamatov wrote (Feshbach & Friendly, 1992, p. 78) "There is no water for the people, where are we to find water for the trees?"

As if taking the cue from Karamatov's comment, the government determined to fell trees across the region in the errant belief that transpiration depleted water resources. By 1987 tree cover was reduced from 15% to 1% (Feshbach & Friendly, 1992), affecting shade, water temperature, and wildlife habitat. Watering trees was no longer an issue.

It was evident that an ecological disaster was looming. Throughout the 1960s and 1970s, awareness of Sea level drop was widespread and impacts for ecosystems and economy were predicted by scientists in their publications and meetings. Degradation of wetlands and deltas was noted in the mid-1960s but recommendations for limiting artificial irrigation of fertile soils were ignored. By the next decade, Soviet cosmonauts circling the earth spotted huge dust storms. There was an ecological collapse around the sea, evident in the absence of wildlife and declining forests. Uzbek scientists began to report findings that fish catch had crashed. Because spawning grounds were being degraded, they feared a total collapse, an outcome also predicted by the Soviet scientist Borovsky. In the mid-1960s the signs of this collapse were so clear that a resolution was passed by the Council of Ministers of the USSR calling for preservation of the fisheries (Glantz et al., 1993).

As the result of the combined alterations to the Aral Sea, its function as a "huge heat pump and cooling system" ceased moderating regional climate. As a consequence, the region began to suffer drought, summers were hotter, and winters came earlier and became more severe (Feshbach & Friendly, 1992, p. 78).

By the early 1990s, salinity in the Aral Sea had tripled from the pre-1960 levels, nearly matching ocean water. The sea bottom was desertifying and wind erosion was producing major sand storms big enough, as noted, to be visible from space. By the mid-1970s, evidence on wind erosion convinced Borovsky the scientist, that saline dusts blown from the Aral sea bed would threaten farmlands far to the east (Glantz et al., 1993).

Meanwhile, Muynak, the Aral shoreline city that in the late 1950s processed over 100 million pounds of fish from the sea, found itself 44 mi inland by 1990. A desperate attempt to dig a canal was made to keep it connected to the receding sea. Refrigerated fish were being brought in by rail and truck from the ocean 2,000 mi away. Writing at that time, Feshbach and Friendly (1992, p. 79) term this fish import "... a surreal example of uneconomic improvisation."

One Aral region official described this period as "the triumphant march backwards." (Feshbach and Friendly, 1992, p. 74). He continued,

> The Aral Sea represents our anguish ... but not just the Aral Sea. Violations of crop rotation, overapplication of toxic chemicals, the mindless opening up of more and more new lands without a strictly scientific approach and without accurate calculations – all this led to a situation whereby our most fertile soil ceased to be so, and dozens of diseases proliferated.

As opportunity after opportunity was lost to stop the disaster and warnings went unheeded, Feshbach and Friendly (1992, p. 88) watched helplessly as the Aral disappeared. They bemoaned that "The greatest single, man-made ecological catastrophe in history advanced unhindered."

THE DISASTER COMES TO LIGHT

Mickey Glantz (1999) has described the Aral Sea disaster as a prime example of a "creeping environmental problem." As my own work on environmental contamination showed, it is often the case that problems "incubate" for a long time before they are recognized and understood. Their recognition by experts and government does not necessarily translate into awareness on the part of the public. Even for people right in the midst of the event, the fact that usual expectations are not fulfilled or that unexpected occurrences are experienced does not at first suggest a pattern that is permanent. Such events proceed inexorably forward at a pace almost below the human threshold of perception and certainly below the threshold of attention and response. Attentive locals can nevertheless recognize changed patterns. However, the Soviet Union presented special problems to moving from recognition to public awareness. Indeed, the Aral disaster was happening for a long time before it became a matter of public discourse.

The Period of Transition: Glasnost and Perestroika

Widespread public awareness to the Aral disaster coincided with the period of *perestroika* (restructuring) instituted by Soviet leader Mikhail Gorbachev[6] beginning in 1985 and of *glasnost* (openness) triggered by the April 1986 Chernobyl disaster. In the latter instance, the entire world knew about the horrific nuclear disaster for 2 days before the Soviet people, including the populations most affected by released radiation, were informed. The revelation forced the Soviet Union to flip abruptly from a secretive society to one of open disclosure and transparency. The move toward democracy indicated by both of these reforms hit a crescendo in the period of 1987–1988. This was a heady period in which transition was in the air. The breakup of the Soviet Union in 1991 signaled the failure to address change quickly enough. Understanding this period is important to the Aral Sea issue because it was the one point where the issue was publicly recognized and might have been addressed in time to make a difference.

The time of *glasnost* and *perestroika* was an amazing but brief period of open discourse, civil action, civil society development, and democratization. An important insiders' view of the period is given by environmental sociologist Oleg Yanitsky (1993), who also conducted numerous interviews with environmental leaders and public surveys. He describes the period as a faceoff between two forces. First, there was the existing closed system of government, operating by centralization and use of directives and armed force, which initially resisted change. And, second, there was an emergent civil society and environmental mass movement that sought reform according to a very divergent set of values: nature protection, altruism, innovation, global thinking, and practice of universalist ethics.

As Yanitsky describes it, stepping into the role of nature defender at the beginning of the era was replete with danger. The environmental movement had emerged quickly after *glasnost*. It grew rapidly despite the obstacles. As nature protection organizations popped up across the rivers of the USSR, each was named after young volunteers' shot confronting poachers or other resource abusers. Vocal leaders of the movement faced a different threat, the high social cost of activism (Yanitsky, 1993, p. 51).

> Only yesterday you were a normal engineer, a parent, a car enthusiast, a fisherman, a trade union member. But around you today there is a "dead space," because you took the risk ... publicly – I stress, not in a tea-break conversation, but in front of a group of people ... . All of the links which sustained your accustomed life are now broken. No-one will help you, because you have become the sworn enemy of a powerful department and of the no less powerful local leadership. And what about your family, career and place in the housing queue? Think about it; would your spiritual resources have been sufficient?

But by the heydey of perestroika, from 1988 to 1989, the situation had changed. Everywhere in the USSR, an "extremely simple form of popular self-organization," the environmental club, cropped up to combine traditional modes of association with "a new embryonic form of collective action ... . All of Russia was a ferment of meetings." And half the meetings addressed the environment. When discussion clubs proved too informal to confront bureaucrats of the Soviet party and state, the local clubs took the leap from talking to activism, engaging in picketing and collecting petition signatures. Deputations were sent "up above" to the level where decisions were made, the Central Committee of the Communist party and the Supreme Soviet of the USSR (Yanitsky, 1993, pp. 64–65).

The press played a central role in erasing the isolation and vulnerability of the activist and elevating the issues, as Yanitsky (1993, pp. 64–65) notes:

> The great majority of environmental clubs arose with the support or patronage of the local press. In 1989 and 1990 local papers helped scattered groups to unite, provided them with information on the local environmental situation, supported and popularized their activity through sympathetic coverage, and provided finance for expeditions ... . Their main contribution was to act as powerful legal protectors.

The press played a broader role as well. As the repressive lid was lifted off of Soviet society, many of the long hidden ills of the Soviet period came to light as, everywhere, ugly revelations were made. The full effect of repression by Stalin and others was opened to public view. A plethora of environmental ills was also revealed across the Soviet Union.

The Aral disaster was big news during this period, driven by this newly unleashed free press. Articles published nationwide announced the latest revelations about the receding sea and tens of thousands of victimized Karakalpaks. These stories repeatedly hit the theme that this was a caused disaster, brought on by poor planning and water management, and blamed "barbaric" Soviet officials for "scandalously" sacrificing the sea for their cotton quotas. These revelations had a big impact in Central Asia, where for the first time people learned the full truth about the Aral disaster that engulfed them (Glantz et al., 1993).

As information continued to mobilize people, a further evolution in organization became possible. Parliamentary elections in Autumn 1988 provided two firsts in Soviet history, the first "more or less democratic elections" and the first time "that social movements made themselves felt as a political force." Then in 1989, an electoral Green movement was launched and 300 environmentally oriented deputies were elected to the union parliament, 12% of the body (Yanitsky, 1993, pp. 64, 65, 90, 123).

The next stage saw the development of broad campaigns on issues relating to environmental health (Yanitsky, 1993). Although addressing a range of major problems around the vast nation, these included the campaign to save the Aral Sea discussed by Kamalov in Chapter 7.

A final stage of civil society involvement arose to offer an innovative way to respond to the emerging impediments to addressing all of these new environmental problems that had taken national stage. These included a lack of funding, few trained personnel, and a need to deal with new problems being discovered constantly. Komsomol, the young Communist organization, invited activists to become entrepreneurs setting up businesses through regional "environmental centers." These were an important reform, constituting the first public groups allowed to exist independent of the state politically and economically (Yanitsky, 1993, p. 68).

Awakening the Region

As the full extent of the Aral debacle came to light, the people closest to and most dependent on the Aral Sea faced a multiple whammy. Their home environment was profoundly degraded. Their livelihood was removed, plunging them into economic hardship. And, they faced an epidemic of disease that affected them at the personal and family levels even as it became a cumulative blow to their communities and region. These outcomes emerged over the period of the Aral's demise from the 1960s, and were suffered by individuals, families, and communities especially in the most rural and remote areas. But these tragedies were happening behind a public veil.

It is interesting that, in the region where the Aral disaster was unfolding, it was not a significant topic of public discussion before *glasnost*. The extent of the problems and their potential permanence was not widely understood. Central Asians had been largely omitted from any national discourse on the topic. Little coverage initially occurred in the media and the public was ill informed. The changes stirring in Moscow were initially absent from Central Asia. When issues were brought forward, there was no way of translating talk into action. Officials were passive and continued to follow the latest economic Five Year Plan as usual (Glantz et al., 1993).

With *glasnost*, information began to reach the people of Central Asia and the Aral region. Research documenting the full scope of the health disaster was before then restricted from the public, as Kuznetsov (cited by Glantz et al., 1993, p. 189) described:

> Unfortunately, secrecy over an entire series of research results in the 1970s, especially medical-epidemiological data, precluded their publication at that time and the predictions associated with them did not become available to the public in time.

But with *glasnost*, the full dimensions of this health disaster began to emerge as public information became available.

Now it came to light just how poor the health of the Aral region's people was. For example, in the late 1980s and early 1990s, half of the children who died in the USSR before their first birthday were Central Asians. Moreover, health care there was abysmal, suffering from a combination of poor medical treatment, few doctors, poor training, little supply, and few facilities. Feshbach and Friendly (1992, pp. 82–83) reasoned that such treatment had to create local resentment.

> The poverty of public health in Central Asia, like the ecological damage done by the cotton monoculture, could be seen as a dramatic enlargement and intensification of the pillage and neglect inflicted on most of the Soviet countryside. Magnified in the Aral Sea region and the Amu Darya basin, agricultural policy, however, looked to growing numbers of Central Asia like part of a historic conspiracy by Russians to exploit their Muslim subjects.

The fact that the Soviet Union brought about the Aral disaster in Central Asia cannot, in the end, be separated from the facts that the region's population is Muslim, that the region had been forced to become subservient in the 1920s, and the Aral desiccation was implemented as a policy predicated on their remaining passive. Writing of all Central Asia and not just the Aral region itself, observers Feshbach and Friendly (1992, p. 75) wrote that response to the Aral threat posed

> ... the first major political challenge to Moscow since Bolsheviks violently suppressed independence movements in Central Asia in the 1920s ... . As a symbol of environmental abuse, the Aral heightened public outrage over what outsiders had done to the Central Asian homeland.

Indeed, from the onset of *glasnost*, as the Aral Sea became a public issue in the region, the fate of the Sea became linked in political discourse to the fate of the region's population, including emerging urges for nationalism. In November 1988, public protests broke out headed by Birlik, a student group rejecting the Russian language and promoting the use of the Uzbek language. A year later, in an unauthorized rally in Tashkent's Lenin Square, Birlik had broadened its agenda to include ecological and economic issues (i.e., restoration of a feudal economy) and calling for Uzbeks to become masters of their own land. Feshbach and Friendly (1992, p. 86) (see also Yanitsky, 1993, p. 225) were in the audience. They quote a young woman speaking to the crowd beneath a towering statue of Lenin.

> The Communist Party Central Committee has wasted all of our treasures and keeps taking away the fruits of our land. A future generation is going to ask us: "What were you doing at that time? What were you thinking about? How did you get us into this situation? Where's our wealth? And what will we answer them? How could we look them in the eye?"

A layer of animosity between the Soviet Union and Central Asia was reframed by a regional writer, Naser, as a problem of colonialism (Glantz et al., 1993, p. 197).

> No writer has publicly acknowledged that Central Asia's problem, symbolized by the water shortage and the Aral problem, is fundamentally a colonial one ... . Unless systematic changes are undertaken to break the traditional colonial relationship between Moscow and Central Asia, I cannot see how the core problems, including the Aral's disposition, can be solved.

In a real sense, the Aral disaster had come to be viewed as a symbol of the region's treatment by the USSR.

Beyond the health crisis, two further aspects of the Aral crisis were cited in emerging demands for sovereignty. First, destruction of Uzbek ecology through cotton production and, second, the failure to compensate Uzbekistan for that destruction (Critchlow, 1991; Feshbach & Friendly, 1992). Local patterns of "silence, collaboration and apathy" gave way to "impatience and resentment" (Feshbach & Friendly, 1992, p. 83).

As a visible case in point, Karakalpak leader Tulepbergen Kaipbergenov complained publicly in 1989 about the negative changes he saw around him (Feshbach & Friendly, 1992, p. 73).[7]

> Our earth no longer smells like soil but like chemicals ... the level of chemical residues washed back from irrigated lands is so high that the fish die ... two out of every three people examined in public health dispensaries are ill – mainly with typhoid, cancer of the esophagus and hepatitis ... . Worst of all, most of the sick are children; [some] doctors recommend against breast feeding because the mothers' milk is toxic ... .

Kaipbergenov made it clear whom he blamed for these conditions (Feshbach and Friendly, 1992, p.74).

> We did not will these things on ourselves. They came to us thanks to the tender mercies of the central planning bodies, essentially a consequence of an utterly criminal economic policy.

Around the same time, Oktir Hashimov, editor of literary magazine *Sharq Yulduzi* asked a question that should have been posed much earlier (Feshbach & Friendly, 1992, p. 80).

> Who can guarantee that a gigantic tragedy will not come to pass in Central Asia and that the land will not become uninhabitable?

Much as revelations about Chernobyl led indeed to radical restructuring of the USSR, the Aral Sea inflamed political discourse in Central Asia. If Chernobyl demonstrated any one thing, it was that the Soviet Union could be neither trusted to prevent problems nor counted on to address them once unleashed. The Aral Sea confirmed these conclusions.

THE EROSION OF CIVIL SOCIETY AND FAILURE OF THE SOVIET UNION

The lack of action to address the Aral disaster was a microcosm of events across the Soviet Union. Not only were effective actions lacking to address ecological crises, but also the gains made under *glasnost* and *perestroika* began to erode. This was evident both at the national level and for the Aral region.

Komsomol closed the successful experiment with environmental centers by early 1991 (Yanitsky, 1993, p. 68). Around the same time, other disappointments began to take the steam out of the environmental movement. The strong Green legislative faction headed by ecologist Alexei Yablokov succeeded in getting a resolution adopted in 1990 entitled "On Urgent Measures to Improve the State of the Environment." The Aral disaster was part of its focus. For the deputies, this was a way to empower people across the USSR to address problems in their home regions that they could never change through the legislative process. But the resolution actually caused a rift in the ecological movement because major environmental organizations viewed these problems as too big for people to solve themselves and the idea of working as part of a government initiative as a loss of independence. They responded to the Green deputies in anger. The people had elected the Green deputies to power to assure that government addressed big problems such as Chernobyl, the Volga River, and the Aral Sea, as well as thousands of smaller issues. These could not just be thrown back on civil society to deal with. Not only did such squabbling divide the ecological movement, but also when the resolution, like most, amounted to talk absent action to improve the ecological situation, it discouraged enthusiasm for the newly won legislative democracy as an effective replacement for an ineffectual Soviet system (Yanitsky, 1993).

A similar chain of failure occurred for the Aral Sea. A high-powered Commission on the Aral Sea appointed in 1987 brought its recommendations to the highest levels of government, resulting in the January 1988 resolution "On the Radical Restructuring of Environmental Protection Work in the Country." In response, Andrei A. Gromyko, President of the Soviet Union during the first part of Gorbachev's leadership,[8] published an article on the country's ecological problems in the journal *Kommunist* shortly before his retirement in 1988. Along with official failures and "a certain deformation of public awareness," Gromyko blamed what he termed "the cost-is-no object" principle of economic management for protecting the environment "in an unsatisfactory manner" (Glantz et al., 1993, p. 179).

But these were just words. A government fact finding scientific expedition was organized in 1988 that traveled through Central Asia, concluding that the Aral was a "symbol of the calamity" befalling nature across the USSR. However, effective response to the acute environmental distress observed was hampered by officials still in place within the Gorbachev administration who had once caused the problems to begin with. Large sums of money were spent by the Ministry of Water Resources to "solve" the Aral problem by building yet more of the canals that drained water away from the sea (Glantz et al., 1993).

On September 30, 1988 a resolution was passed by the Central Committee and the Council of Ministers entitled "On Measures for a Fundamental Improvement of the Ecological and Sanitary Conditions" in the Aral Sea region, and for increased effectiveness in the utilization, strengthening, and safeguarding of water and land resources in this basin. But the subsequent economic crisis leading up to the fall of the Soviet Union prevented action.

The same fate met a joint resolution of Central Asian leaders of June 23, 1990 asking Moscow to "declare the Aral region one of national calamity and to provide real help." Much discussion had now occurred. But all of the public discourse about the need to do something never lead to an agreement about what to do; as a result, little happened before the end of the USSR late in 1991 (Glantz et al., 1993).

THE END OF THE USSR

The Gorbachev administration's ineffectuality at responding to the magnitude of environmental problems eroded public confidence. Over time, the administration itself began to implode as party reforms failed and power struggles prevented business from being done and the economy went into a tailspin (Glantz et al., 1993).

Other factors in the breakup of the USSR related to the Aral region. The rise of political militancy in Central Asia scared Russians living in the region, leading to a major brain drain as hundreds of thousands fled. Ironically, it was the departure of Russian skilled labor coinciding with the collapse of Soviet funding that destroyed any chance of rapid ecological recovery (Feshbach & Friendly, 1992).

A creeping demographic trend helped to force the issue of Uzbek independence. Uzbekistan saw 23% population growth between 1979 and 1989. By the year 2000, it was projected to account for one-tenth of the population of the entire Soviet Union. It had few skilled workers, a surplus of labor, high unemployment, and only 6.5 acres of arable land per person. Its young men threatened to overwhelm the Russian draft with a projected one in three conscripts, challenging the hegemony of ethnic Russians there. It was time for Russia to give in to Central Asian nationalism. By Fall 1991, Central Asian republics declared independence (Feshbach & Friendly, 1992).

For the Aral Sea, the political tipping point, not to mention the ecological one, had been missed. The ball was now tossed to new states facing major immediate challenges with little funding, lots of competing demands, and fundamental divisions within the region. They would not do any better than the USSR despite recognizing the urgency of the situation. Meanwhile, there were effective limits to just how much independence was achieved.

Central Asian leaders independently came to the same conclusion, namely that their best course was to remain close to the Russian Federation. One reason was the need to address the Aral Sea disaster. Both before and after independence, many in the region demanded that the Soviets deliver Siberian rivers to replenish the Aral Sea and its tributaries (Feshbach & Friendly, 1992).

Three additional factors add irony to a postmortem on perestroika. The environmental movement was severely weakened after the transition, lacking the Soviet system as a common and easy target. Without achievements in hand, they faced a loss of public support. The amazing civil society so evident during *glasnost* and *perestroika* largely vanished.

A second irony was that the transition to a new system protected the wealth of the most privileged under the Soviet system. As leadership emerged for the new states, too often people discovered familiar old faces in the positions of responsibility. Even with the move to privatize modes of production in the new states, what was called "nomenklatura privatization" occurred. This meant that former party and management elite became factory owners.

Finally, in sharp contrast, the economic downturn after the fall of the USSR made most others worse off. In 1991, Yanitsky asked survey respondents "if you had known in 1985 what the changes begun in the country would lead to, would you have supported them?" More than half of those polled, particularly older respondents, said "no" (Yanitsky, 1993, p. 100). Presumably the nomenklatura were not included in the survey.

In looking, then, at the entirety of perestroika and *glasnost*, Yanitsky (1993, p. 50) observes a succession of three states of mind that applied nationally and most likely in the Aral region as well.

> At the onset of perestroika, there was apathy and a lack of belief in the possibility of change. Then came a psychological upsurge, as a new confidence arose in the possibility of rapid and painless changes for the better. Now, when this confidence is again practically exhausted, we are observing even deeper psychological despair, social apathy and escapism.

As a final disappointment, the experiment with democracy, at least in the Russian Federation, ended to no small degree on New Year's Eve 1999, when the federation's first President, Boris Yeltsin, resigned and named

Vladimir Putin his successor, shifting society back toward militarism, secrecy, and sharp reductions in free speech and public involvement. Once again it became dangerous to be a journalist or environmentalist. The question of whether the nature-exploitive paradigm shifted back to where it was under the Soviet system is a matter of some current uncertainty.[9] Similar concerns can be raised for former Soviet Union states not in the Russian Federation.

The question of mitigation for the Aral Sea disaster began long before the breakup of the Soviet Union. And it still continues, long afterward.

MITIGATION

In environmental impact assessment, the concept of mitigation implies changing the project, the way the project is being done, or the conditions within which the project occurs in order to avoid one or more adverse consequences. Undesired impacts, at least in theory, are thus prevented from even occurring. Once the worst possible outcome has been achieved, however, the term mitigation takes on a different connotation. Now it entails compensating for the negative outcomes through efforts to restore or repair or remediate the damage done or to restore or replace the functions or qualities lost by the project outcomes. Some mitigation involves monetary compensation for losses but not restoration. Affected people may feel they were made whole even if damage to the environment persists.

Up the Sea without a Paddle

The Aral Sea disaster is a story about the path not taken, of alternatives, mitigating strategies and precautionary decisions not made. All or part of the disaster might have been avoided had demands for cotton been moderated according to water availability beyond that needed by the Aral Sea. Cotton yield could then have been enhanced through efficiency, delivering river water using lined channels, and drip irrigation to avoid wasting water. A return to traditional sustainable approaches to farming – long practiced in Central Asia – would have maintained soil quality through crop rotation and alfalfa plantings. Such approaches would have worked better than the massive water redistribution scheme instituted to support irrigation (Feshbach & Friendly, 1992).

In the late 1980s and early 1990s, the Soviet government seemed ready to act on the Aral Sea. Doctors were sent to the Aral region to stem international criticism about the health crisis there. Other "actions" were really thin veils of disguise for inaction. A special study commission was formed. In October 1988, a decree was promulgated pledging to improve the efficiency of irrigation and boost the flow to the Aral in both the Amu Darya and Syr Darya by increasing increments between 1990 and 2005. While these were earnest sounding but ineffectual efforts, my favorite inaction was the announcement of a public contest "to develop concepts for restoring the Aral Sea" (Feshbach & Friendly, 1992, p. 87). The competition yielded over 200 proposals, of which seven were selected for further consideration (Glantz et al., 1993). I found no evidence that any proposal was finally selected and clearly no miracle cures emerged. Overall, during this period, 5 billion rubles was appropriated to the Aral Sea effort, less than half the funds required to reverse the escalating desiccation. Ecologists termed these steps as "bureaucratic blather" and "a mirage" (Feshbach & Friendly, 1992, p. 87).

Meanwhile, a very different response was taking shape in the Aral Region. In a 1990s' meeting in Nukus, Glantz witnessed a play and song lamenting the loss of the Aral Sea that featured an old man sitting in a rowboat with no bottom. He observed that programs such as this were designed to help prepare people to accept the loss of the sea (Glantz et al., 1993). Adapting to the worst scenario was an effort to help Aral region residents face their new reality. Ironically, those having designed the Aral's demise never seem to have had anyone to help them face reality.

Schemes to Save the Aral Sea

Over time, many schemes emerged for saving the Aral Sea (summarized in Table 1), including the use of sustainable alternatives. What is instructive is that, given diverse options, the proposed solutions have tended to favor the large-scale big impact alternatives over the local, small-scale sustainable options. These large-scale big impact alternatives have proven too expensive and infeasible to ever be practical. And yet, the fascination and promise of big alternatives is what has captured the Central Asian imagination.

THE PROJECT OF THE CENTURY

Earlier, reference was made to the Soviet agency Minvodkhoz, who, having made the primary decisions to divert water away from the Aral Sea, next hoped to promote the "Project of the Century" in an effort to reverse the

Table 1. Schemes for Saving the Aral Sea.

- Allowing numerous small saline lakes to form in place of the Sea
- Preservation focused only on the deltas
- Reducing irrigation diversions or increasing efficiency
- Piping in water from the Caspian Sea
- Planting less water intensive crops
- Directing polluted waste water into the sea
- Deliberately melting Pamir glaciers
- Diverting Siberian rivers to the Aral

Note: Compiled by the author from Glantz, Rubinstein, and Zonn (1993).
Source: Author.

damage it had caused by diverting Siberian rivers to the Aral sea. The major prize was not diverting water from the sea, but rather managing one of the biggest water projects ever contemplated. This far-fetched idea even gained popularity among the local population for a while. In the period of *glasnost* and *perestroika*, the idea of the Siberian diversion became the only hope for Aral region residents that something could and would be done.

Central Asian activists pushing for the Siberian river diversion made a utilitarian and self-serving argument that the waters were not being utilized as they flowed to the Artic, whereas their own need was great.[10] These leaders formed an alliance with Minvodkhoz to dig a 1,470 mi long canal over 40 years to deliver 27,000 km^3/year from the River Ob to the Aral Basin to replace losses of 44,000 km^3/year in the Amu and Syr (Feshbach & Friendly, 1992).

Such mitigation schemes were hatched long after the Aral Sea was heading to its demise and continued even after the Aral passed its final tipping point and accelerated its decline. The Aral Sea had slipped beyond the ability to slow, stop, or reverse its positive feedback loop of desiccation. This was a big sea, a huge disaster, and it invited solutions of a gargantuan scope. Mirroring similar thinking in the Americas, where a major river was reversed for hydroelectric production at James Bay, diversions of water from one river to another represented an old Soviet large-scale conception of controlling nature for human ends.

During the 1990s, there were numerous discussions about the fate of the Aral Sea, with proposals that ranged from "let it fail" to full restoration. There were also a number of partial restoration stages in between, including partial refilling, the two Aral solution (separating the Big from Little Aral) and promoting small disconnected lakes within the region fed by the rivers.

There was a question about whether entirely refilling the Aral was necessary to its health and functionality, perhaps an Aral half full might be preferable to an Aral totally empty. Scientists at the time sought to define what ecological services had been lost as a way to determine how far restoration needed to go (Glantz et al., 1993). The Siberian River diversion was a later version of a late 1960s' Minvodkhoz plan to divert the Siberian waters to the Volga and Don Rivers (Feshbach & Friendly, 1992). In 1967, when research was undertaken to prepare for water transfer into the Volga basin, scientists demanded that further irrigation diversions of the Syr Darya and Amu Darya be predicated on interbasin transfer to the Aral Sea as well. The result was the even bigger 1,600 mi "Sibaral canal" proposal that would have connected the Aral Sea to two Siberian Rivers, the Ob and Irtysh (Glantz et al., 1993).

FALLACIES OF THE PROJECT OF THE CENTURY

However, Glantz was told point blank in the 1980s by Soviet scientist Michael Borovsky that if the Ob was indeed diverted south "... not one drop of transferred Siberian water would go to save the Aral Sea" (Glantz et al., 1993, p. 173)

In fact, rather than saving the Aral Sea, the river diversion project hinged on the Aral drying up. The desiccated sea bed would provide a vast new area for cotton cultivation. The diverted Siberian river water would then be used to irrigate this cotton. Alternatively, a 1981 map showed the seabed turned into rice paddies expected to produce 4,400 pounds per acre. Through 1987, backers of these desiccation schemes called the Aral "nature's error" and sought to see it "die in a beautiful manner" (Feshbach & Friendly, 1992, p. 83).

For others not desiring the Aral's demise, existence of the Siberian diversion plan played a major role in rationalizing continued irrigation diversions. It was like going into debt. One might spend the Aral's waters today knowing that a windfall of Siberian river water would be coming soon. It would be large enough to both pay back one's debt to the Aral Sea and still allow expanded cotton cultivation (Glantz et al., 1993).

Then there was the matter of public support. The interbasin transfers proved to be the most contentious of the strategies floated to save the Aral. In Russia, a major defense of the Siberia Rivers emerged, lead by environmentalists allied with those concerned about the project's high cost. In 1987, one of the new emergent Moscow NGOs Green World was founded to research the most ecological dangerous projects, choosing as one of its first major topics for independent research "the project of the century" (Yanitsky, 1993, p. 228). Besides cost, it was argued that the project would destroy the fisheries and wildlife of these rivers, cause flood damage, damage flood plains, inundate historical sites, degrade water quality, affect spring flows to agricultural fields and diminish navigability (Glantz et al., 1993). As Gorbachev caved to public pressure, the scope of the project was at first limited to replenishment of the Volga and then shut down entirely. Loss of the Siberian diversion was not taken lightly by Central Asians, who used it to fuel conspiracy theories about colonial attitudes in Russia toward them (Feshbach & Friendly, 1992, p. 183).

Too Little Too Late: Reaching and Missing the Tipping Point

With large-scale rescue schemes proving infeasible late in the waning days of the USSR, sustainable alternatives finally came to the fore. However, these were also the waning days of the Aral Sea. In 1991, a last ditch attempt was made to institute cooperative water basin management on the Amu Darya in order to reduce water loss and contamination. The effort failed because Moscow had abandoned its central control over land reclamation and no funds were made available for water conservation (Feshbach & Friendly, 1992).

Continuing Dreams of an Aral Rescue

Even though the prospect of Siberian water had been snatched from their drying Sea, belief that this diversion project might someday revive and save the Aral persisted, resurfacing as the long awaited salvation of the region over the ensuing decades. The "Sibaral diversion" was viewed as the Aral's "phoenix rising." It would simultaneously restore the lost Sea while spurring economic development in the investment-starved Aral region (Feshbach & Friendly, 1992, p. 83).

I have personally witnessed continuing evidence of this belief. When I spoke at Tashkent's Financial University in February 2012, the only Karakalpak person in the room was a dean and professor. His question about the Siberian diversion implied his continued belief that the sea could still be saved in this manner.

But the Aral has growing competition for Siberian water, should it ever arrive. There are many other uses of additional water. Besides continued or expanded cotton farming, there is now rapid population growth, promised an increase in water demand for supply and sewerage as well as for industrial activities. The water might come, possibly through a smaller project, but the Sea might be left still low and dry (Glantz et al., 1993).

ENVIRONMENTAL STIGMA AND ENVIRONMENTAL INJUSTICE

The Aral diversion program emerged from Moscow planners as the result of decision-making from the top echelons of government much the way that the biological weapons effort did. The former was not couched in secrecy, but the decision was above public discussion. As in the case of Vozrezdheniya (Edelstein, Chapter 4), the attack on the Aral Sea underscores the willingness of the Soviet system to carry out programs that risked civilian populations and their home environments.

In both cases, poisons were field released. Ironically, the research on the bioweapons was ruthlessly scientific and systematic. There was a need to know how tortured a death in animal proxies a given agent might produce under different variables of application. Hundreds of medical scientists were engaged in the closed city effort to perfect biological weapons, while nearby human populations were medically underserved. There was no parallel measurement of pesticide exposures or other impacts of the Aral diversions and no medical monitoring to assess health consequences. Other than measurements of cotton and grain production per unit, there was no comparable concern with dose and response. Poor plant outcomes were subjected to intense scrutiny but productive plantings were not. From the central planners to the regional administrators down to the farm hands working on the *kolkhoz* (collective farm), just as with the parallel organizational systems conducting bioweapon testing, many people were coerced into going along with steps they may not have agreed with. Others complied without coercion, doing what was expected of them out of passivity, patriotism or economic necessity.

The principle that emerges from the parallel events on the killing fields of Rebirth Island and on the pesticide intensive farms of the Aral Sea region was that this remote wonderful place had been selected in an inverted contest for triage. The Aral Sea, its region, and its people would be sacrificed for what were considered higher goods. To explore the lessons of this sad legacy, I will next discuss two issues, environmental stigma and injustice and the authoritarian structures of decision-making. Both of these elements fill in further explanation for how the Aral Sea was allowed to meet a designed and premeditated demise.

The Theory of Environmental Stigma

This author has often written about environmental stigma (Edelstein, 1991, 1993, 2001, 2004), the changed perception of environments that are known to be contaminated. The stigma tends to perpetuate itself in public knowledge as people begin to avoid such places and, as a result, their economic value plummets.

How does contamination affect future use of a site? Environmental stigma is accompanied by the notion of perpetual jeopardy. Once an environment is stigmatized, activities are more likely to occur there that intensify the stigma and invite further insult. Thus, not only the stigma but also stigmatizing activities persist. Environments selected for contaminating activities are often chosen because they carry some prior stigma. Already spoiled, there is certain logic to continuing to despoil. At the point that the environment is considered to be dangerous, stigma is maximized. It takes a major "clean-up" of the contamination in some form to put stigma in abeyance, although it may never be erased.

How are people affected by environmental stigma? Environmental stigma becomes social stigma when people become associated with stigmatized environments. People viewed as living in degraded environments are looked down upon. Sometimes their exposure to contamination is viewed as if it were contagious. If it is further known that people have been exposed to environmental contaminants, the stigma is intensified. There is a tendency to assume that people somehow deserve the injustices they experience and, association with a contaminated place is one aspect of this rationalization.

Social contamination may also precede and guide environmental contamination. The initial selection of an environment for stigmatized activity may come from its association with stigmatized people as well as the attributes or history of the place. Thus, siting of hazardous facilities and other degraded and contaminated activities often reflects social prejudice about who is marginal and expendable. It is this relationship that gave birth to the entire field of environmental justice (Bullard, 2000).

Environmental Stigma and Injustice in the Aral Region

The Case of Biological Weapons Testing

The Soviets viewed what is now western Kazakhstan and Uzbekistan as highly expendable places. The locations were far from Moscow and desolate except for local villages and nomadic peoples. The Aral Sea, its fish, its attributes as a natural setting, its role as home and place of livelihood, all were callously sacrificed for the Soviet "greater good." Pesticides were applied at levels difficult to justify and progressive salinization was tolerated. Nuclear weapons were tested in western Kazakhstan. Biological weapons were manufactured in Kazakhstan. The bioweapons' testing ground was sited at Vozrezdheniye because it was an isolated and remote island from the standpoint of defense and secrecy. But it was not equally remote from the standpoint of collateral damage. The act of placement made a statement that the environment and people of the region were expendable. Risk tolerance was set high until downwind populations were exposed to pathogens. Monkeys, apes, and other animals were the official test subjects. But a little inadvertent exposure to human subjects was also in order and scientifically valuable even if not specifically planned for.

Such thinking was not out of keeping with the Soviet approach. Take the case of Muslumovo, in the Urals. In the aftermath of the Kyustym-57 nuclear disaster, while most people in the heavy fallout zone were evacuated, some populations were actually kept in the vicinity (Edelstein & Tysiachniouk, 2007). Such was the situation for Muslumovo's population living along the highly radioactive Techa River. People were not allowed to leave, and were carefully studied to identify the consequences of their radioactive exposures. One of the many activist groups formed by residents explained their situation by its name, "White Mice." The victims in that situation were Bashkir and Tartar people. Like the Karakalpaks and Kazakhs, they were Muslim minorities in the Soviet Union, a category deemed marginal by the larger society (Edelstein & Tyschiachniouk, 2007).

Cotton Production and Loss of the Aral Sea

Selection of the Aral region for bioweapons' testing is reminiscent of its prior selection under the virgin lands policy as a place for relocating people to marginal lands that would be farmed. The denigration of the Aral Sea doomed those whose life and livelihood were connected there. The control of health information and tolerance of poor health outcomes in the face of poor health care services all indicate a low regard for the people of the region. They were last in services because they were considered least. The tolerance for high levels of pesticide exposure similarly suggests a disdain for the well-being of these people. And, as remote residents of an autonomous republic, Karakalpaks may not be viewed much better by other Central Asians than they were by the Soviets. As with indigenous peoples in many parts of the globe, it is hard to tease out environmental stigma from social stigma. Both are indicators of vulnerability to ready sacrifice.

Perpetual jeopardy not only reflects social prejudice, but also permanent changes introduced into a region's climate and ecology, as is evidenced in the larger Aral region. When severe drought occurred in Uzbekistan early in the millennium, with only 15% of expected rainfall, Karakalpakstan and Khorezm were particularly impacted. The drought was explained away by some observers as a "man-made drought" because Karakalpaks just happened to be last in line along a long river. Rather, the human cause was that continued cotton cultivation and water diversion had altered the local climate and placed the region at ongoing risk (Wegerich, 2002).

The persistent neglect suffered by this region suggests that stigma persists even though many people have relocated (see Cerny, Chapter 14). Karakalpakstan is known for its poor health care and material well-being (see Lall, Chapter 16). It is far from the cultural and economic centers of Uzbekistan and particularly from the capital Tashkent. When I lectured at the Financial Institute in Tashkent about the Aral disaster in Winter 2012, I was surprised that no one in the audience had ever been in the Autonomous Republic of Karakalpakstan except one Karakalpak professor. Other than him, no one knew about the Aral Sea disaster in any detail and overall knowledge was not much better than I find with my students in the United States. Thus, at a minimum, if not stigma, there is little common ground and basis for appreciating the situation of people in the far western Aral region. That distance of experience may be as significant as the distance in kilometers (Edelstein, 2000). And the formation of national identity around perceived concepts of cultural pride may be just as divergent (see Alaya, Chapter 20).

AUTHORITARIAN SYSTEMS

When we contemplate what has befallen the Aral region over the past century, it is interesting to look for an intersecting line that connects the story of Vozrezdheniye (Edelstein, Chapter 4) and the larger desiccation of the Aral Sea and contamination of the Aral region. That connector is the role of authoritarian organizational structures in creating this integrated disaster. I end this chapter with a few thoughts on the role of authoritarianism in creating the Aral disaster.

Authoritarian structures arise to maintain social order, provide security, and ensure the rationalization of organizational truths. They combine top-down authoritarian decisions with means of delivery that derive legitimacy from that authority. Two events in 19th century American history are instructive. The first was the government's war against "varmints" in the United States on behalf of ranchers, a government initiative that led to the systematic removal of wolves from the lower 48 states. Meanwhile, in the west, the near eradication of the American Buffalo by highly efficient means of mass murder was tolerated as part of the U.S. policy to control a perceived human varmint, the Native Americans, subjugating and ensuring their removal to "civilizing" reservations.

In the United States, by the 1920s, several threads merged to create the modern authoritarian organization. Ford's success with the Model T assembly line production inspired a new theory of scientific management developed by Frederick Taylor. Then there was the emergence of the technocrats, who advocated the objective hand of the engineer as the rational alternative to the corruption of the politicians (Akin, 1977). Engineers would proceed to reshape American nature according to a gospel of efficiency. The problem was, of course, that you cannot get rid of fundamental questions of value in choosing what ends are pursued by "efficient means." A gospel of efficiency without a clear consideration of endpoints is a false gospel indeed. And efficiency in the pursuit of evil ends creates aberrations, as evidenced by the Nazis.

As Richard Rubenstein wrote of the Holocaust in *The Cunning of History* (1975), while mass murder of innocents was hardly a new phenomenon, what made the Nazi Genocide of Jews and other target populations distinctive was the German's application of systematic methods of planning and use of especially developed technology to commit their crimes. The method of murder belayed a level of deliberation and intent that moved murder to an even deeper level of depravity. The Germans had done for murder what Henry Ford did for automobile manufacture, even employing his techniques along the way. They exalted not only in bloodletting, but also in maximizing their efficiency while doing so. Their acts, thus, united the most banal expression of anti-Semitism and the mythology of otherness with the unemotional, indifferent and impersonal technocratic aspects of the modern corporation. Killing was now merely an industrial process.

These themes appear as well in the evolution of the Soviet system, with its emphasis on Five Year Plans, rule by fear, massive relocations of humanity, and an engineering approach intent on reshaping the earth.[11] Mixing the view that nature is merely a reservoir of resources for human use with efficiency minded elements of the modern organization – the planning and the engineering – exploitation of nature and humans was pushed to new heights. From the top-down, exploitation was built into the operating system without checks and balances.

Advances in authoritarian social structure were demanded by the modern police state. Well before the advent of terrorism and Homeland Security, Mumford (1964) warned of the authoritarian nature of certain technics that render democratic society impossible. The nuclear age particularly triggered Mumford's concern, with its demands for control and secrecy. Other weapon systems, including bioweapons and chemical weapons, demanded authoritarian police control as well.

The Soviets' militarization of food production and the diversion of chemical weapons to pesticides converted agriculture into an authoritarian technic as well. And in a world where the state and large corporations contend for hegemony, such corporations are increasingly the proxy of the authoritarian state, increasingly even fighting its wars and serving as crucial outsources. Today, farming in an open society readily pits the small local farmer against the interests of large multinational corporations – the agro-industrial system – where something as small as the seed saved from last harvest becomes a point of contention.

In the post-9/11 period, Mumford's warnings resound. Nothing demands more state power than security. Security demands secrecy and legitimates abuse of limits set by civilian society. The case study of Vozrezdheniye illustrates the organization of scientists, engineers, and the military that Eisenhower (1960) famously warned about in the heart of the Cold War period. In this regard, the enrollment of scientists in the creation of weapons of mass destruction diminishes every shred of objectivity's virtue. The military industrial complex that develops becomes a controlling institution of its own with the power to pursue an agenda quite divergent from even the most democratic of societies. We see from the Aral disaster that much the same might be said about the agro-industrial complex.

Authoritarian technics produced parallels between west and east, both "cultures of contamination." Take the case of the secret Soviet nuclear complex Mayak in the Ural Mountainsand the U.S. Hanford Nuclear Reservation it was modeled after. Remarkable similarities are found with regard to secret culture, cover up of accidents, treatment of the local population, and contamination of the environment. The major difference was that Mayak caused the world's first major nuclear disaster, "Kyshtym-57," which was covered up until the *glasnost* period of the late 1980s and 1990s. Numerous radioactive releases from Hanford were also covered up, the secrets held even longer by the US than by the Soviets (Edelstein, 2007). It was almost as if the respective playbooks had been written by the same hand.

Then there is the Aral debacle itself, disaster by design, carried out by an authoritarian society with control over both nature and people.[12] Enforcement in a top-down planned society succeeds in getting people to suspend their own judgment in favor of carrying out the orders of the state.

In the end, what we learn from the Aral debacle about authoritarian systems is why they fail. Rather than "learning organizations," they are cut off from unfiltered and uncensored input. Guided from within, they close up their public information, make puppets of objective scientists, use fear and other threats to limit independent thought, proceed on outdated paradigms, and distance themselves from their victims (natural and human) and the consequences of their actions.

The funny thing about disaster by design is its most serious flaw. A design is only an idea. And it is only as good as the pre-and post-implementation steps one takes to critically examine its viability and implication. The Soviet approach made all the wrong moves. It eschewed the need to take precaution, to really examine need, to assess impacts and alternatives before making decisions, to implement or enforce mitigations, to monitor outcomes, and to apply correctives. As a result, it proceeded based on the imposition of a black box design concept on the real world. Particularly in the case of the Aral Sea disaster, where the problem emerged in increments and change was highly feasible at any point, we see the self-reinforcing power of authoritarian systems that operate as if they were closed systems but in reality are not. This systematic cutting off of learning guaranteed failure in a situation where correction would have been easy and effective until the end.

NOTES

1. Writing about Khorezm, the southern Aral region (Tolstov, p. 303). See also Lance Jennott (2001). It is interesting to note that Jenkinson's original map of his travels is displayed by Jennott and does NOT show the Aral Sea.
2. Further cited by Cerny in Chapter 14.
3. In fact early civilizations were "hyrdaulic societies" developed in irrigated dry landscapes (Worster, 1985).
4. Certainly Americans had parallel technocratic predilections as well, using engineers to reshape the land, particularly in the American West (see Horowitz, Chapter 8; also Edelstein et al., 2007).

5. Feshbach and Friendly (1992, p. 80) share a journalist's report of how one Uzbek cotton lord dealt with subjects who offended him. "*He cast them into an underground prison, tortured them with a red-hot iron and beat pregnant women with a whip before the eyes of their husbands.*"

6. Gorbachev actually held three different titles during his leadership of the Soviet Union, the last being entitled formally, the "President." From 1985 until 1988 he was secretary of the Communist Party. When the comparatively symbolic President (head of the Presidium) Andre Gromyko retired in 1988, Gorbachev took that title, as well, then combining them in a more definitive title of President. He did not have too long to enjoy the new title before the end of the USSR in 1991.

7. Feshbach and Friendly took Kaipbergenov's quotes from a transcript of the proceeding of the Congress of People's Deputies, May 30, 1989 and *Izvestiya*, June 1, 1989, p. 5.

8. See note 6.

9. At the time this is written, Russia is implementing rules that would label as "foreign agents" most environmental activists in the country because their work is heavily funded by grants from other nations. These agents would then be subject to special rules, restrictions, and perhaps sanctions. Other limitations on environmental activism are underway, including limits on foreign funding (Herszenhorn & Barry, 2012).

10. The argument is reminiscent of the early claims that the Aral was not productive compared to cotton production.

11. Stalin's mass murders lacked the Nazi's level of efficiency, even if they shared the banality.

12. Again, there are parallels in water diversions in the United States (see Horowitz, Chapter 8), in the secret weapons development program (see Edelstein 2007) and in the imposition of energy projects following the Cheney Energy Plan with its exemptions of corporations from compliance with U.S. laws and public involvement.

REFERENCES

Akin, W. E. (1977). *Technocracy and the American dream.* Berkeley, CA: University of California Press.

Bookchin, M. (2005). *The ecology of freedom: The emergence and dissolution of hierarchy.* Oakland, CA: AK Press.

Bullard, R. (2000). *Dumping in dixie: Race, class, and environmental quality.* Boulder, CO: Westview.

Critchlow, J. (1991). *Nationalism in Uzbekistan: A Soviet Republic's road to sovereignty.* Boulder, CO: Westview Press.

Edelstein, M. R. (1991). Ecological threats and spoiled identities: Radon gas and environmental stigma. In S. Couch & J. Stephen Kroll-Smith (Eds.), *Communities at risk: Community responses to technological hazards* (pp. 205–226). Worster: Peter Lang.

Edelstein, M. R. (1993). When the honeymoon is over: Environmental stigma and distrust in the siting of a hazardous waste disposal facility in Niagara Falls, New York. In W. Freudenburg & T. Youn (Eds.), *Research in social problems and public policy* (Vol. 5, pp. 75–96). Greenwich, CT: JAI Press Inc.

Edelstein, M. R. (2000). 'Outsiders Just Don't Understand': Personalization of risk and the boundary between modernity and postmodernity. In M. Cohen (Ed.), *Risk in the modern age: Social theory, science and environmental decision-making* (pp. 123–142). New York, NY: St. Martin's Press.

Edelstein, M. R. (2001). Crying over spoiled milk: Contamination, visibility and expectation in environmental stigma. In J. Flynn, P. Slovic & H. Kunreuther (Eds.), *Risk, media, and stigma* (pp. 41–68). London: EarthScan.

Edelstein, M. R. (2004). *Contaminated communities: Coping with residential toxic exposure* (2nd ed.). Boulder, CO: Westview Press.

Edelstein, M. R. (2007). Hanford: The closed city and its downwind victims. In M. R. Edelstein, M. Tysiachniouk & L. V. Smirnova (Eds.), *Cultures of contamination: Legacies of pollution in Russia and the U.S. Research in social problems and public policy* (Vol. 14, pp. 253–307). New York, NY: Elsevier.

Edelstein, M. R., & Tyschiachniouk, M. (2007). Psycho-social consequences due to radioactive contamination in the Techa River Region of Russia. In M. R. Edelstein, M. Tysiachniouk, & L. V. Smirnova (Eds.), *Cultures of contamination: Legacies of pollution in Russia and the U.S.* Vol. 14 of Research in Social Problems and Public Policy (pp. 185–204). Elsevier.

Edelstein, M. R., Tysiachniouk, M., & Smirnova, L. V. (Eds.). (2007). *Cultures of contamination: Legacies of pollution in Russia and the U.S.* (Vol. 14). Research in Social Problems and Public Policy. New York, NY: Elsevier.

Eisenhower, D. D. (1960). *Military-Industrial Complex Speech, Dwight D. Eisenhower, 1961.* Public Papers of the Presidents, pp. 1035–1040. Retrieved from http://coursesa.matrix.msu.edu/~hst306/documents/indust.html. Accessed on July 27, 2012.

Feshbach, M., & Friendly, A., Jr. (1992). *Ecocide in the USSR: Health and nature under Siege.* New York, NY: Basic Books.

Figes, O. (2007). *The whisperers: Private lives in Stalin's Russia.* New York, NY: Picador.

Glantz, M., Rubinstein, A., & Zonn, I. (1993). Tragedy in the Aral Sea Basin: Looking back to look ahead? *Global Environmental Change, 3*(2, June), 174–198.

Glantz, M. H. (Ed.). (1999). *Creeping environmental problems and sustainable development in the Aral Sea basin.* New York, NY: Cambridge University Press.

Golden, P. (2011). *Central Asia in world's history.* New York, NY: Oxford University Press.

Hershenhorn, D., & Barry, E. (2012). Russia halting Group's access to U.S. money. *The New York Times*, September 19.

Jennott, L. (2001). *Anthony Jenkinson's Explorations on the Land Route to China*, 1558–1560. Retrieved from http://depts.washington.edu/silkroad/texts/jenkinson/pics/q4.jpg. Accessed on July 22, 2012.

Lipovsky, I. (1995). The deterioration of the ecological situation in Central Asia: Causes and possible consequences. *Europe-Asia Studies, 47*(7), 1109–1122.

Liu, X. (2010). *The silk road in word history.* New York, NY: Oxford University Press.

Mumford, L. (1964). Authoritarian and democratic technics. *Technology and Culture, 5*(1), 1–8.

Rubenstein, R. (1975). *The cunning of history.* New York, NY: Harper and Row.

Tolstov, S. P. (2005). Following the tracks of ancient Khorezmian civilization. *Central Asia and the Caucasus, 2*(14), 154–162.

Wegerich, K. (2002). *Natural drought or human made water scarcity in Uzbekistan?*

Wehrheim, P., & Martius, C. (2008). Chapter 1: Farmers, cotton, water, and models – Introduction and Overview. In P. Wehrheim, A. Schoeller-Schletter & C. Martius (Eds.),

Continuity and change: Land and water use reforms in rural Uzbekistan—Socio-economic Analyses from the Region Khorezm. Halle (pp. 1–15). Leibnitz, Germany: Leibniz Institute of Agricultural Development in Central and Eastern Europe (IAMO).

Wood, H. (1876). *The shores of Lake Aral.* London: Smith, Holder & Co.

Worster, D. (1985). *Rivers of empire: Water, aridity , and the growth of the American west.* New York, NY: Oxford University Press.

Yanitsky, O. (1993). *Russian environmentalism: Leading figures, facts, opinions.* Moscow: Mezhdunarodnyje Otnoshenija Publishing House.

Zonn, I., Glantz, M. H., & Rubinstein, A. (1994). The virgin lands scheme in the former Soviet Union. In M. Glantz (Ed.), *Drought follows the plow: Cultivating marginal areas* (pp. 135–150). New York, NY: Cambridge University Press.

SECTION THREE
COTTON, COTTON EVERYWHERE, BUT NOT A DROP TO DRINK: AGRICULTURE AS THE VILLAIN

Because we tend to make a close association between farming and the land, agriculture is not widely recognized for its destructive impacts to the environment and health. The shape of agricultural production is very place specific in many ways. The Soviets collectivized farms, farmers and machinery, deliberately destroying the preexisting peasant culture. The rural small farmer worldwide has been dramatically impacted by neocolonialism and globalization, with a massive out migration to cities resulting.

The United States tradition of the small farmer was transformed in the 20th century into a new form of agribusiness, operating on larger and larger scales. After World War II, tractor factories retooled from making tanks. Chemical weapons were redirected to the control of pests. The model of the "green revolution" was promoted as a means to increase worldwide food production to feed fast rising populations. Farmers everywhere branched out onto marginal lands using inappropriate techniques not likely to work there. Irrigation was increasingly required as dry lands were tilled. Increasingly agriculture was founded on oil and water inputs. Local markets were eclipsed by global markets and food became the major world traveler. The model globalized an approach to farming that meshes with the multinational corporate interests that have developed to exploit crop production everywhere, including an increasing emphasis on plants grown for products and now for energy, rather than food. If from the point the plow first broke the earth and introduced anthropocentric erosion on a large scale, agriculture has had a major environmental impact, made worse by the current mix of fertilization, pesticides, mechanization. And now genetic engineering and industrialized agriculture have magnified these impacts to an extraordinary degree. In short, the proposition that the design of our food systems is a threat to sustainability is not a hard reach. Agriculture, as practiced today, is a major form of human-caused environmental disaster.

In this third section of *Disaster by Design*, we examine the disastrous agricultural policies begun in Soviet times and which, to a lesser degree, still characterize Uzbekistan and Central Asia. In the design of our disaster, we find all of the elements of an unsustainable system. The outline is almost dumbfounding in its simplicity. To meet demands for Soviet "cotton independence" and, more recently, Uzbek export, agriculture is dramatically expanded onto marginal soils in an arid climate, thus requiring extensive irrigation and overuse of water with minimal efficiency. A monocultured crop that is water needy is employed. Water is diverted way beyond the capacity of the overall system. Not one grain of ecological thought is applied. And, ditch-by-ditch, plot-by-plot, a disaster of the magnitude of the Aral Sea is created.

Although the push for Soviet cotton production was the specific cause of the Aral Sea disaster, other crops played a significant role over time, notably rice and wheat, that added to the problem. Over time, since independence, Uzbekistan has acted to reduce the percentage of cotton grown, and it has dropped to about a quarter of the total crop. The key point here, however, is the one made by Kamalov in section two. It was never the cotton, per se, that caused the disaster. It was the poor water management that accompanied the push for agricultural productivity. In this section, however, we see that the approach to agriculture itself is no small part of the other linked Aral Sea disasters we are concerned about, the salinization and contamination of soil, water and air.

When we look at Uzbekistan through the lens of agriculture, it is useful to remind ourselves that agriculture is not only the production sector which puts food on our tables from farms far away, but that for many countries and parts of the world, agriculture is the main livelihood of the population – and that present and future stability of this livelihood is an important concern. Unfortunately, the global agricultural sector rarely still operates according to the bucolic images we may remember from yesteryear, but rather, is practiced with input and output demands that encroach on nature's limits, especially in local contexts. Agricultural machinery, initially invented strictly as labor saving devices for individual farmers, has become co-opted as a means to maximize output at all costs, except where it is too expensive for farmers to keep up with modernization. Agriculture in Uzbekistan is an example of a way of life that has become unbalanced, not to say heavily influenced – by politics, economics and most recently, by market forces. The by-products of globalization that give us access to all kinds of goods including tropical hardwoods and exotic fruits, imported meats, inexpensive clothing and strawberries in winter include monocropping, crop yield maximization and long production chains. The resulting unbalance has become a cause of greatest concern in countries where strict environmental limitations were ignored in order to produce crops that are in high global demand, or where politics consistently interfere with sound science. Cotton is such a crop. The four selections in this section offer considerable insight into its impact and the alternative course of action that should be taken.

Michael Wilson, in Chapter 10 "What Went Wrong: The Case of Un-Ecological Agriculture," provides a thorough review of agriculture's role in the design of disaster. Wilson, a horticulturalist, was a member of the project exchange team. As Wilson lays it out, we begin to see how growing crops becomes a form of ecocide, assaulting the environment on every front. This situation has certainly been true for the cotton-dependent economy that was introduced to the Central Asian steppe by Russia in the 1800s, and which grew to the scale of obsession in the Soviet days that destroyed the Aral Sea. The conventional approach to agriculture, used in modern factory farming, has been employed to extract as much cotton as possible from a given unit of soil and water. The fact that energy inputs also soared, in the form of mechanization, pesticides and herbicides and fertilization, or that ground water, soil, animals and people were poisoned, made little difference because these factors were not taken into account in the Soviet (nor for that matter capitalist) systems. Wilson's detailed analysis covers every aspect of Uzbek agriculture, including the delivery of irrigation waters and the selection of plant varieties. He also comments on landscaping practices. And, if the methodology of plant production in Uzbekistan has been designed for disaster, Wilson gives a detailed blueprint for how, instead, to redesign it for sustainability.

Chapter 11, "The Nonarable Aral: Loss of Productivity in Uzbek Agriculture," was contributed by Farhod Ahrorov and Ilkhomjon Niyazov, both part of our exchange's Uzbek team. The analysis presented allows for a comparison of the Aral region in Western and central Uzbekistan with other areas of the nation, providing an opportunity to see some of the effects of introducing widespread farming in the west that contributed to the Aral Sea disaster. And, indeed, the data makes clear, the farming in the Aral region occurs on poorer soils, is more dependent on irrigation and uses more water than agriculture elsewhere in Uzbekistan. It is never clarified whether these results reflect the consequences of the Aral Sea disaster as much as the cause. What is clear is that there is an intended loss of productivity involved.

In Chapter 12, "Cotton in Our Ears: Water, Agriculture and Climatic Change in the Post Aral Context," Shavkat Hasanov and his colleagues take the issues raised by Ahrorov and Niyazov one step further. Focusing on the Zarafshan Valley in Central Uzbekistan, stretching from Bukhara in the Aral region in the west to the central area around Samarkand, these authors show some telling differences. While increased use of water in agriculture indicated overall a loss of efficiency at any location, these inefficiencies were heightened in the Bukhara region. The western section of the valley not only had the poorest water quality, but also the most intensive irrigation levels. In a second study, the authors prepared a computer model utilizing an index of water overuse, yet again finding a significant inverse relationship between overuse of water and farmland productivity.

Climate change is a variable that might affect these relationships. Warmer temperatures increase the stress on plants and may increase their water demand. Evaporation increases in warmer weather, meaning that even more water is lost. To gauge whether climate is a factor, the authors included in their modeling study an assessment of the effects of temperature rise on agricultural productivity. They found a tentative relationship showing that even a 1% increase in temperature produces measurable losses of productivity. If confirmed, the study indicates further complications for the relationship of irrigation to water overuse and declining productivity in the western Aral regions.

So, how then do we untangle the consequences of water inputs and what to do about them? This precisely is the challenge undertaken by Inna Rudenko and her colleagues in Chapter 13, "Water Footprints: Integrated Water Resource Management to the Rescue in the Aral Sea Basin," the fourth and final chapter in this section. The authors provide an overview of water usage in Uzbekistan, the regional stresses on the province of Khorezm as a cotton producer, and evaluate the impacts of policy, technology and public awareness. By focusing on Khorezm, they present a sophisticated analysis for how to identify sustainable alternative courses of action, measuring water inputs, on one hand, and comparing the production of agricultural crops (i.e., cotton) and "value added" finished items (i.e., jeans production) on the other. By calculating the "virtual water use" embedded in both agriculture and industry, the ecological costs of both are identified and compared. The two are then integrated through the concept of value added product production. The level of production sought is gauged by the tradeoff point between economic gain and loss of sustainability. The trick to finding that point is improving economic profit and, simultaneously, minimizing water use. It is the ability to pinpoint the "win/win" moment when these two values are simultaneously maximized without conflicting that is the beauty of the method. And the outcome is a detailed prescription for sustainable development in Uzbekistan and Central Asia.

In fact, drawing cumulatively from these four contributions, one not only understands the cotton monocrop for its role in causing the Aral Sea disaster, but one also appreciates Kamalov's point from the earlier section. It is not cotton, per se, that is the problem, it is the overall failure to do water planning that allows for inefficient unproductive and environmentally harmful choices to persist.

Astrid Cerny and Michael R. Edelstein
Editors

CHAPTER 10

WHAT WENT WRONG: THE CASE OF UN-ECOLOGICAL AGRICULTURE

Michael Wilson

ABSTRACT

There is a connection between cotton production and the Aral Sea disaster in Uzbekistan. Large-scale cotton production utilizes the practices of conventional agriculture and has severe environmental consequences in arid regions. Some of these problems, such as salinization, currently exist in Uzbekistan as a result of cotton production and these conventional farming practices. This chapter is a review of cotton production, the environmental consequences of conventional agriculture, and its relationship to the Aral Sea Disaster. Storm water management with biofiltration, sustainable farming practices, efficient irrigation, ecological horticultural practices, and a water conservation program are remedies that can help to reduce the environmental degradation caused by cotton production and restore some of the water resources in Uzbekistan.

Disaster by Design: The Aral Sea and its Lessons for Sustainability
Research in Social Problems and Public Policy, Volume 20, 159–174

ISSN: 0196-1152/doi:10.1108/S0196-1152(2012)0000020020

INTRODUCTION

The Aral Sea disaster and its relationship to the cultivation of cotton are classic examples of the results of conventional agricultural practices. The consequences of monoculture, excessive tillage, improper irrigation, synthetic fertilizer, pesticides and herbicides, plant gene manipulation, and animal feeding operations all contribute to environmental degradation. The cultivation of cotton in Uzbekistan relies on most of these practices.

Russia began the cultivation of cotton in Central Asia as early as 1860 when the Civil War in the United States threatened its supply. At that time, Central Asia supplied less than 10% of Russia's cotton. After the 1930s the Soviet Union became self-sufficient in cotton and in the 1950s even began to export it. Between 1913 and 1940, the area devoted to cotton in the Uzbek republic of the Soviet Union increased from 441,660 to 1,022,600 ha (Djanibekov, Rudenko, Lamers, & Bobojonov, 2010). To increase yields even further, the Soviet government depended on the intensive use of machinery, fertilizers, and pesticides. In the 1960s there was an expansion in cotton production with the creation of a specially designed irrigation program. The result was an increase by 23% in area devoted to cotton growing in the 1960s and 1970s, reaching 2,000,000 ha. After 1960, 61% of the arable land in Uzbekistan was devoted to cotton production (Djanibekov et al., 2010).

By the mid-1960s, Soviet scientists began to notice a drop in the water levels in the Aral Sea and an increase in salinity, but did nothing to improve the faulty irrigation system. The irrigation system was expanded as the hectares devoted to growing cotton increased, which also demanded an increase in the use of pesticides. As the agricultural sector of the Soviet Union in Central Asia (mainly Uzbekistan and Turkmenistan) increased with cotton and wheat production, the Aral Sea slowly shrank in size. Currently, the sea is divided into a northern sea in Kazakhstan and a remnant southern sea in Uzbekistan. Although the portion in Kazakhstan is becoming stabilized, the Aral Sea in Uzbekistan continues to dry up. With water levels still diminishing, the former lakebed is becoming a desert.

Uzbekistan today has over 1.47 million hectares under cotton cultivation that consumes 20 km^3 of water every year. The delivery of this water continues to be handled via an old and inefficient irrigation system of open, earth-lined canals. Open canals lead to considerable loss through evaporation and seepage, estimated at a total loss of about two-thirds of the water withdrawn from the rivers (Rudenko & Lamers, 2010).

Irrigation leads to one form of salinization in the fields. Because of the earth-lined irrigation canals, the system loses water into the soil from capillary action, which pulls the water into the porous material. The seepage of water into the soil is creating a high groundwater profile, causing waterlogged soils, and pushing salt upward to the top layers of the soil profile. Increasing salinization in turn leads to a decline in crop yields because the cotton plant cannot perform well with excessive, saline soils.

Agriculture is the largest consumer of water in Uzbekistan with about 90% of the water available used for irrigation. There is competition for the remaining water resources between industry, horticulture, domestic use, rural, urban, and municipal use. Agriculture is a major consumer of water while being a direct contributor to Uzbekistan's food security and income for the large rural population, but leaves little water for other sectors, like industrial and municipal use (Roll et al., 2003).

CONVENTIONAL AGRICULTURE AND ENVIRONMENTAL DESTRUCTION

Cotton is grown in most of the world by utilizing the methods of conventional agriculture and this is certainly the case in Uzbekistan. This industrial form of agriculture or factory farming is responsible for a large amount of environmental damage, particularly to water resources. Conventional agriculture also entails significant energy consumption, thus contributing to global warming, as does deforestation and desertification. The common practices of conventional agriculture that are most evidently in use for cotton production in Uzbekistan include intensive tillage, monoculture planting, the use of pesticides, irrigation, and synthetic fertilizer.

Deep, intensive tillage is a common practice with the cultivation of cotton. The deep turning over of soil on a regular schedule is environmentally damaging water resources. Excessive tillage is a primary cause of sedimentation in waterways from erosion. In addition, residual fertilizer and pesticides are easily leached out of loose, tilled soils and get into the waterways through inadequate drainage. The conventional agricultural methods used in the cultivation of cotton recommend that soils be tilled prior to planting and after harvest. A common practice with conventional agriculture is to till soils while the fields are fallow to control weed growth. In addition, it is common to cultivate between the plants as they are growing to control weeds. All of this adds to sediment runoff (Gliessman, 2007).

Currently, Uzbekistan is growing two species of cotton, which are *Gossypium hirsutum* and *G. barbadense*. This selection is not the problem. The problem is that it is grown in monoculture in an arid climate. This monoculture approach is another damaging practice of conventional agriculture. When a crop is grown in monoculture, it requires strict monitoring and the use of pesticides if a disease or insect begins to infect the crop. Even with pesticide treatments there is a chance of losing the crop, and with a monoculture the whole crop is lost. Cotton production in most global systems and certainly in Uzbekistan is a monoculture crop.

The use of synthetic fertilizer contributes to eutrophication in waterways. In coordination with excessive tillage, synthetic fertilizer will easily leach out of soil. This in turn contributes to excessive nutrients entering waterways and creates an accelerated rate of growth of algae, which depletes the oxygen levels in the water. Also, as the fertilizer breaks down, the residues are minerals that easily leach out of the soil with irrigation. Proper soil fertility with good water-holding capacity is preferred for growing cotton. Chemical fertilizers compromise soil fertility through continuous use of nitrogen and the resulting accelerated decomposition of organic material developed from crop residue (Eyhorn, Ratter, & Ramakrishnan, 2005).

Chemical pest and weed control is a standard in conventional agriculture and is a common practice in cotton growing. Cotton is susceptible to a number of diseases and insects, so a large number of chemicals are needed to produce a high-yield crop. In time, many of the insects become resistant to the pesticides and stronger doses of pesticide or new, different pesticides are needed to combat the pest problem. It has been estimated that 11% of the world's pesticides are used on 2.4% of arable land globally for the cultivation of cotton (Kooistra, Pyburn, & Termorshuizen, 2006). In addition, cotton production consumes roughly 50% of pesticides used in developing countries. In Uzbekistan, defoliants with mostly sodium chlorate, which is a source of concentrated salt in the soil, are applied on 75% of the surface; pesticide use is as high as 54.6 kg/ha in some areas (Kooistra et al., 2006). In the past, under the Soviet Union, as the yield of cotton decreased, large amounts of agro-chemicals were applied to increase yields. These chemicals partially leach out, but they also partially remain in the soil. While chemicals for cotton production are a globally known problem, the practice that has affected the Aral Sea the most is irrigation. Irrigation and improper drainage are the driving forces that make agro-chemicals leach into adjacent waterways. Irrigation conducted on a large scale is capable of depleting water resources and this is the case with the Aral Sea disaster.

Uzbekistan utilizes an antiquated method of flood or furrow irrigation. With this method, water is diverted from rivers into irrigation channels that are directed to the fields. The water efficiency of this type of system is only about 40% due to evaporation, seepage, and mismanagement (Kooistra et al., 2006). The extensive network of irrigation channels that are connected to the Syr Darya and the Amu Darya are 70% unlined, and provide for a high loss of water due to capillary movement into the surrounding soils. In Uzbekistan, only 12% of these irrigation canals are waterproofed and the rest of the irrigation system is unlined, with much of the water seeping out and adding to the groundwater stock. The efficiency of the total system in the Aral Sea Basin remains low with only about one-third of the water withdrawn that is actually reaching the farmer's fields (Rudenko & Lamers, 2010). Open channels are an inefficient way to irrigate and further require that fields are leveled on a yearly schedule to keep the water evenly distributed over the field. Uneven fields create inefficiency and a greater amount of water is needed to properly water the whole crop. Incredibly, the inefficiency and the need for level fields are both compounded in recent years by the fact that the fields are usually flooded one or two times prior to planting in an attempt to flush out the accumulated salts (from previous inefficiency) to help with germination and growth in the early stages (Fig. 1).

So far we have seen that technical standards and inputs are inefficient and detrimental to short-term and long-term agricultural productivity. There are secondary implications from these practices. Unlined channel irrigation results in a lot of water loss through evaporation and capillary action. Salinization is a widespread result stemming directly from current irrigation practices – with the frightening reality that almost half of the land in Uzbekistan is now bordering on the unproductive. Saline soils are created when the irrigated water dissolves calcium carbonate and soluble salts in the soil (Kooistra et al., 2006). The agricultural system is thus a vicious circle of reinforcing negative consequences as the saline fields need to be flushed, requiring more water, which stresses the irrigation system and the water management, resulting in poor yields, which prompts more flushing, and so on.

On a smaller but equally widespread scale, ornamental horticulture and home gardens are other common uses of water resources. The common irrigation techniques utilized here are also the furrow and flood methods. Vegetable gardens are furrowed in the same manner as agricultural crops. Ornamental gardens and cottage gardens tend to be watered through irrigation channels.

Fig. 1. Typical Irrigation Channel in Uzbekistan: Note that the Darker Soil Shows Waterlogging on Both Sides of the Irrigation Channel. Photo: Michael Wilson.

Interestingly, in the urban areas there is a tendency to use a western approach to ornamental gardening. Turf grass is used in all the parks and public spaces, but often watered from the same faulty irrigation systems. Turf grass is not a native grass and is known for its extreme water-use requirements to thrive in arid areas. Furthermore, the use of typical landscape annuals like marigolds, petunia, and snapdragons, which are tropical in nature, will also necessitate copious amounts of water for such non-native plants to perform well in arid climates (Fig. 2).

All of these practices working in conjunction with each other cause environmental degradation. The pesticide residues clinging everywhere in the arid climate have created health and quality-of-life issues. Clean water is rare across much of western Uzbekistan. In particular, the inefficient

Fig. 2. Cottage Garden at a Restaurant in Bukhara Utilizing Flood and Furrow Irrigation. Photo: Michael Wilson.

irrigation system has disturbed the hydrological balance of inflow from the two rivers that normally would supply the Aral Sea and keep the salinity of the water at a stable level to support wildlife and a fishery.

WORKING TOWARD A SUSTAINABLE AGRO-SYSTEM

It is unlikely that the Aral Sea will ever be restored back to its original surface area and evidence suggests that there is an interest in gas and oil exploration in the Aralkum Desert.[1] Given the world's demand for energy, this interest in exploring for oil and gas is understandable, but the attention would be better placed on water. It is more important in the short term that Uzbekistan protects its rapidly diminishing freshwater supply. In order to support agriculture, on which much of Uzbekistan's economy is based, and have adequate water supplies for non-agricultural industry and domestic use, Uzbekistan needs to plan for sustainable water availability.

Such planning will necessitate many changes, careful planning, and adaptation of technologies. The agricultural sector needs to diversify not only to improve environmental conditions but also to support a growing population. Several actions for water conservation should be considered that include a sustainable agro-system, storm water management, riparian restoration, ecological horticultural practices, and water conservation policies.

The agricultural sector is the largest consumer of water resources and the one that is in greatest need of change to reduce environmental degradation directly resulting from current practices. Uzbekistan currently has the opportunity to become one of the leading supplier nations of organic cotton. This can only occur with a transition period from existing chemical intensive practices, to new approaches leading to a sustainable agricultural system that provides higher incomes and overall food security. A diversified agricultural system with crop rotations will help provide food security, along with income, and move Uzbekistan toward reducing the negative consequences resulting from cultivating cotton.

Methods for sustainable agriculture have been in practice to develop soil fertility, water-holding capacity, natural pest management, and efficient irrigation. These include methods that work in arid environments where water management is at a premium. Integrated pest management (IPM) and conservation farming practices are suitable methods to work toward a sustainable system. IPM is a plan that reduces the amount of pesticides needed to control diseases and insects. Conservation farming is simply conserving the current inputs and resources utilized in an attempt to reduce the amount needed. Gliessman (2007) considers this level one of four as a conversion to agro-ecology or a sustainable farm. Levels 2 through 4 include conversion to organic agriculture, a redesign of the agro-system, and a connection of farmers to markets. Conservation farming and IPM, which conserve pesticide use, are a good first step for Uzbekistan to work toward sustainable agriculture because this practice is easy to implement immediately.

Conservation tillage and no-till systems can be good methods to overcome the environmental destruction brought on by deep intensive tillage. This method was developed in 1943 by Edward Faulkner, an agronomist, who questioned the use of the plow after witnessing the Dust Bowl era in the United States. Over the past few years, farm technology has developed no-till seeders where it is not necessary to break up the ground so deeply, which in turn minimizes soil erosion. The stubble from the previous year is left behind and is cut after harvest to provide mulch. In 2006, experiments on soil moisture in Khorezm Province showed that soil on a no-till site had the highest amount of soil moisture and a loamy soil due to organic residues. This site was not flooded or irrigated. The study concluded that a reason for the higher soil moisture was the effect of the organic residue at the soil surface, which prevented rapid water evaporation. The organic residue also provided maintenance of higher water-holding potential by avoiding tillage (Massucati, 2006). One drawback though is that no-till seeding does require a greater amount of pesticides and herbicides according to some reports. It is a common practice with these systems to use herbicides previous to planting to remove any plant growth, which helps to warm the ground by allowing sunlight to reach the soil surface for better germination.

Currently, cotton is crop rotated with winter wheat in Uzbekistan after cotton is harvested. This is an advantage in diversification for soil health and food security. However, a more diversified agro-system will provide more food products, reducing the need for imported food and provide soil fertility that is more conducive to growing cotton. Crop rotations are the simplest ways to diversify the agro-system and improve soil fertility. Rotations can be made with nitrogen-fixing crops, soybeans, and other green manures, reducing the need for synthetic fertilizers. In addition, pest and disease problems will be reduced and if there were to be a breakout of disease, the whole crop would not be lost because the farmer would have several different crops growing simultaneously in adjoining fields.

Soil fertility is important for water-holding capacity and it can be built up with the use of organic material added to the soil. Organic farming recommendations for crop rotation are not to grow cotton on the same field for more than 2 years in a row. The following year another crop can be grown that can be a green manure cover crop, another market crop, or both. Cotton that has been grown after pulses, like soybeans, chickpeas, or pigeon peas, has produced good yields and farmers should include them in their crop rotation program to take advantage of nitrogen fixation. Other good crops to use in rotation are chilies, vegetables, sugarcane, and wheat. Green manure crops can also be sown in-between cotton rows and cut down before or at flowering to be used as mulch or incorporated into the soil. Intercropping is a technique where a row of cotton is replaced every few meters with a green manure or a trap plant, like sunflower, as some pests for cotton prefer these plants and will spare the cotton (Eyhorn et al., 2005).

The uses of manure, compost, cover crops, and green manures are all good additions to provide a healthy, living soil and eliminate the need for synthetic fertilizers. In terms of manure collection, it is difficult to do on a large scale in Uzbekistan due to the common practice of range feeding livestock. However, nighttime confinement does occur in some localities and that provides a source. Composting of vegetative matter is more feasible and may be easier to accomplish. In time, compost can create alkaline conditions and must be mitigated with the use of nitrogen-fixing crops, a proper crop rotation program, and additional manure.

DISCUSSION

Uzbekistan has a history of high pesticide use for the cultivation of cotton that resulted from the use of conventional agriculture and this can be mitigated with sustainable methods. The use of pesticides can be reduced with a proper IPM plan and the use of organic pesticides approved by organic farming associations. Instead of spraying on a regular schedule, insect populations are monitored and are treated when the population is the most vulnerable to the pesticide to prevent additional breeding. The IPM approach works well with conservation farming, and also can be a transition to organic and biological pesticides that are approved by organic farming standards.

A number of natural pesticides have been developed in the United States and India for cotton that can be applied in Uzbekistan. Some of these work well in combination with beneficial insects and intercropping. These methods are being researched currently in India with successful yields of cotton. Habitat for beneficial insects can be achieved in conjunction with a biofiltered storm water system that collects agricultural runoff by utilizing host plants for these predator insects. Also, by providing habitat for insects, there is a food source that will attract avian species, which also will help with insect control.

Excessive pesticide use and inefficient irrigation systems have made the greatest environmental impact to the Aral Sea Basin, both of them are indicative of old technology and ideas. One newer approach to more efficient water management is drip irrigation. Drip irrigation is a good alternative to the current system, though one downside is that it does require infrastructure investment and incurs high maintenance costs. One important decision point is whether to install subsurface or surface drip irrigation systems. New subsurface drip irrigation systems are being developed that work well with no-till seeding equipment because the irrigation lines do not have to be removed to operate the seeding equipment. The irrigation system is installed about 18 in. below the soil surface and is unaffected by the seeding equipment. Another common method is surface irrigation heads that are only 6 in. above the surface and water low to the ground with a heavy droplet that is less prone to evaporation, as opposed to a mist style irrigation head, which loses more water in evaporation. In the hot and sunny Aral Sea Basin, improved irrigation would introduce significant water savings by any method of water conserving and efficient irrigation over the current channel style.

Several possibilities exist for Uzbekistan to increase cotton yields and improve its agricultural sector. The examples of drip irrigation, improved fertilizer efficiency with sub-surface irrigation, the use of improved varieties of cotton that are drought and salt tolerant, and appropriate crop rotations for soil fertility are all possibilities. The introduction of conservation farming practices, which can lead to organic farming, can show some immediate benefits. It may be too soon to suggest a conversion to completely organic farming and its application may be limited with low availability of organic fertilizers, like manure. Organic farming recommendations for the cultivation of cotton do not accept genetically modified (GM) cotton despite the advantage of improved yields but there is a selection of natural varieties and hybrids which are recommended (Eyhorn et al., 2005). Currently, GM cotton, which may improve yields, is not cultivated in Uzbekistan due to the government's concern about long-term effects on human and animal health (Djanibekov et al., 2010). Conservation farming, which reduces the cost of external resources, is the most cost effective way to work toward a sustainable agro-system and goals can be made to develop into other methods of organic farming.

CLEANING WATER THROUGH HABITAT

Irrigated agriculture in Central Asia began about 3,000 years ago to meet the needs of an expanding population. When the Russians began to expand their reach into the Central Asian lands in the 1800s, they significantly expanded an extensive, unlined canal system for irrigation that had been developed by local people over time for more modest agricultural needs (Rudenko & Lamers, 2010). Later, Soviet engineers expanded the system to its fullest extent, and this is still the current, active irrigation system. It is in use for cotton production, wheat, corn and other crops, and in rural locations also for home gardens. As illustrated earlier, this vast network consists mostly of earthen conduits of various widths and dimensions. One modernization technique, which is totally feasible, is to line the canals with cement. This will significantly reduce the movement of water into surrounding soils, and begin to reduce the high water table and waterlogged conditions that had developed. It has been estimated that renovating the irrigation system and lining the canals with concrete can save 10–22 km^3 of water per year (Rudenko & Lamers, 2010). Where the irrigation canals begin to get narrow along the secondary, tertiary, and final tributaries there is an added benefit in laying pipes. Enclosing the channels as pipes then prevents evaporation as well as well as seepage. Converting canals to pipelines has an added advantage. It becomes easier to locally store water in tanks or cisterns that can be used with pumping systems for modern irrigation systems, aiding the metering of water during wetter and drier seasons. Then the amount of water can be regulated according to demand.

The effects of agricultural runoff and urban storm water can be mitigated with best management practices (BMP). BMP is the practice of treating storm water at its source and includes techniques like biofiltered detention basins and retention ponds. These detention basins utilize plant material to biologically filter pollutants, such that agricultural runoff can be channeled through these systems to remove pollutants and provide cleaner effluent. There is already a large drainage system in Uzbekistan that is part of the massive irrigation system. Biofiltered detention systems can be attached to drainage networks, with the additional benefit of providing habitat. These systems utilize floodplain plants that are used to a cycle of flooding and then undergoing dry periods.

Biofiltered detention basins are the equivalent of constructed wetlands. These have been studied since the 1960s as a way to mitigate the effects of storm water runoff and wastewater. They are often used due to their cost effectiveness and the environmental benefits that include habitat creation. Once the plant materials used in these systems are established, there is a small maintenance cost compared to chemical and mechanical methods of water purification. *Typha* species in particular have been shown to remove not only excess nitrogen and phosphorous from water but also heavy metals that include cadmium, copper, lead, and zinc (Coon, Bernard, & Seischab, 2000). There is also a positive effect on suspended solid removal and control of water temperature (Coon et al., 2000). The plant materials chosen must be adaptable and native to the region, like *Typha minima* and *Phragmites australis*, which are indigenous to central Asia. *Tamarix ramosissima* and *Elaegnus angustifolia* (Russian Olive), which are also indigenous, can be utilized with these systems, as these plants thrive under saline conditions, allowing for phytoremediation that can help to remove salts that may otherwise enter waterways. *E. angustifolia* has the added benefit of fixing atmospheric nitrogen that makes it amendable to poor soils (Dirr, 1983) and may be a good species to help repair poor soils, allowing other native species to establish themselves. The use of native plants in these systems will also provide wildlife benefits by creating habitat, providing cover and food source for both avian species and insects, and aiding a sustainable agro-system.

In Uzbekistan almost everyone has a home garden with some ornamental and edible plants, so it is worth mentioning ornamental horticulture and its water footprint. Unfortunately, inefficient irrigation and tropical plants both use large amounts of water to perform well. Although the amount of water utilized for ornamental horticulture is minimal compared to agriculture, sustainable landscaping practices in selecting ornamentals and cultivating the home garden can help to conserve water resources.

Most of the plants used for ornamental horticulture currently in use in Uzbekistan are not native to central Asia. *Campis radicans*, *Yucca filamentosa*, and *Agave* species, which are indigenous to North America, are commonly used ornamental plants. Although these plants are drought tolerant and may grow well in Uzbekistan, they are not native, with *C. radicans* having potential to become an invasive species. Native plants not only perform well in their intended range due to adaption to the environment, but also provide a food source for insect and avian species. For example, *Phragmites australis* has been shown to support over 170 species here in its native range (Tallamy, 2009).

It is important to bear in mind that tropical annuals and turf grass, once introduced, require a large amount of water to thrive in an arid environment and will only perform well with regular fertilization. Given Uzbekistan's current water management predicament, these are not good choices and should be eliminated. Instead, *Rosa* and *Althea* are commonly used in Uzbekistan and perform well in arid conditions. These plants can be utilized as native alternatives to water hungry tropical plants. The use of native plants can help to reduce water consumption for ornamental plants and has the benefit of helping to preserve species that may become lost over time.

Effective water management, as we have seen, will entail several layers of intervention. Some are largely technological, while others involve planting vegetation on and off the fields for water conservation. One additional method that has been overlooked so far, but which could be highly effective, is riparian restoration. This restoration will help to slow down the effects of erosion by reducing suspended solids in the water. This is a problem particularly in the drainage network and in natural waterways that no longer have healthy riparian buffers. A healthy riparian zone will protect the edges of the rivers and irrigation canals, preventing washouts and soil loss.

The plant material used for restoration will again provide the added benefit of phytoremediation by absorbing chemicals that may be carried with runoff. A riparian buffer with tall trees will provide shade, helping to keep the water cooler and slowing down evaporation. In the drainage network of the irrigation system, biofiltered storm water systems can be coordinated and combined with planting riparian buffers to increase the amount of created habitat. Riparian restoration of the rivers and lining of the irrigation system will have the greatest impact on preserving water resources for Uzbekistan.

A PLEA FOR CONSERVATION

A comprehensive conservation plan, not limited to agriculture and horticulture, can tie all of these concepts together. This plan can also raise public awareness through education about water conservation and storm water runoff, while putting an emphasis on the benefits of clean, potable water. The government of Uzbekistan is in a position to take the lead by developing policies and regulations for conservation, along with monitoring industrial and agricultural point pollution to reinforce the conservation plans.

The advantage of conservation is that the process can begin immediately with little investment. For example, with conservation farming practices and IPM approach to insect control, there is an immediate cost reduction because the amount of pesticides needed is not as great. This is true with other agricultural inputs as well. Conservation in urban areas and for domestic use will show quick results for the water users with a reduction in expenses associated with water.

Conservation is a good first step since it can begin quickly. However a long-term plan for implementing conservation farming, restructuring of the agricultural practices to include organic farming, changing home gardening practices, and developing storm water management needs to be developed by the government of Uzbekistan. This should be a long-range plan that addresses not only the most severely damaged areas, but also those that are still in good shape, like the Fergana Valley, to preserve them from further environmental degradation. This plan can work on a region-to-region schedule. All restoration plans take time to implement and do not show immediate results. It took over 50 years of poor agricultural habits to create the Aral Sea disaster and it could take just as long to notice the full results of good agricultural practices. However, given the fact that some towns have run dry and many fields are saline, there is no time like the present to get started.

If all of these recommendations or even some of these are adopted, they will provide a holistic approach toward reducing the effects of the current conventional agriculture practices. Lining of the irrigation canals and efficient irrigation methods will reduce water loss and provide water for other sectors of Uzbek society. Many of these approaches need to work in conjunction with each other to become successful. For example, good soil fertility will increase the water-holding capacity and reduce the need for irrigation. An IPM approach that treats only targeted pests helps to protect beneficial insects that are harbored in newly created habitat. Riparian restoration will create habitat as it prevents erosion. Natural pesticides and beneficial insects will reduce the environmental conditions that are normally created with chemical pesticides. Diversity will occur with crop rotations that are also part of the soil fertility program by not exhausting soil conditions, which is common with a monoculture. A diversified, sustainable agro-system may not restore the Aral Sea, but will provide stable clean water resources for the future.

NOTE

1. The Aralkum refers to the desert in the previous lakebed.

REFERENCES

Coon, W. F., Bernard, J. M., & Seischab, F. K. (2000). *Effects of a cattail wetland on water quality of irondequoit creek near rochester, New York.* Water-Resources Investigations Report 00-4032, US Geological Survey, Ithaca, New York. Retrieved from http://ny.water.usgs.gov/pubs/wri/wri004032/index.html

Dirr, M. (1983). *Manual of woody landscape plants* (3rd ed.). Champaign, IL: Stripes Publishing Co.

Djanibekov, N., Rudenko, I., Lamers, J. P., & Bobojonov, A. (2010). Pros and cons of cotton production in Uzbekistan. *Pinstrup-Anderson, Per: Case Study # 7-9 of the Program: "Food Policy for Developing Countries: The Role of Government in the Global Food System".* Ithaca, NY: Cornell University. Retrieved from http://cip.cornell.edu/dns.gfs/1279121771

Eyhorn, F., Ratter, S., & Ramakrishnan, M. (2005). *Organic cotton crop guide* (1st ed.). Frick: Research Institute of Organic Agriculture FiBL. ISBN 3-906081-67-2. Retrieved from http://www.fibl.org/fileadmin/documents/en/development-cooperation/production-systems/cotton-guide-large.pdf

Gliessman, S. R. (2007). *Agroecology: The ecology of sustainable food systems.* Boca Raton, FL: CRC Press Taylor and Francis Group.

Kooistra, K. J., Pyburn, R., & Termorshiuzen, A. J. (2006). *The sustainability of cotton. Consequences for man and environment*, Science Shop Wageningen University & Research Centre. Report 223. ISBN: 90-6754-90-8585-000-2.

Massucati, L. F. P. (2006). *Monitoring of soil macrofauna and soil moisture in a cotton field: an assessment of the ecological potential in irrigated agriculture in Central Asia (Khorezm province, Uzbekistan)*. PhD dissertation, University of Rheinischen Friedrich-Wilhelms at Bonn. Retrieved from http://www.zef.de/.../downloads/Publications/MasterTheses/Massucati-MSc.pdf

Roll, G., Alexeeva, N., Aladin, N., Plotnikov, I., Sokolov, V., Sarsembekov, T., & Micklin, P. (2003). *The Aral Sea management experience and lessons learned.* Lake Basin Management Initiative Regional Workshop for Europe, June 18–21. Central Asia and the Americas, Saint Michaels College in Vermont. Retrieved from http://www.worldlakes.org/uploads/Aral_Sea_15Jun03.pdf

Rudenko, I., & Lamers, J. P. A. (2010). The Aral Sea: An ecological disaster. *In Pinstrup-Anderson, Per: Case Study # 7-9 of the Program: "Food Policy for Developing Countries: The Role of Government in the Global Food System".* Ithaca, NY: Cornell University. Retrieved from http://www.khorezm.zef.de/publications

Tallamy, D. W. (2009). *Bringing nature home: How you can sustain wildlife with native plants.* Portland, OR: Timber Press.

CHAPTER 11

THE NONARABLE ARAL: LOSS OF PRODUCTIVITY IN UZBEK AGRICULTURE

Farhod Ahrorov and Ilkhomjon Niyazov

ABSTRACT

Agriculture accounts for most of the land use in Uzbekistan, and is consequently the main determinant of environmental quality. As a result, due to the excessive use of production inputs and irrigation, land is widely degraded in Uzbekistan. This chapter explores characteristics and changes in land use, agricultural production, soil ecosystems, and water availability, and their effects on both soil degradation and limits on sustainable development in different agro-ecological zones of Uzbekistan. Secondary data from various sources was collected and processed, and descriptive statistical methods were used to analyze trends, causes, and effects of land degradation. Zones with lower land quality have higher water-use indicators. There is a positive correlation between the land grade and fertilizer application coefficients. Agriculture from the east zone showed sustainable water consumption, the highest average land-quality grade, and has the highest fertilizer-use coefficient. The down-stream west zone has the highest water-use and the lowest fertilizer-use indicators. The analysis revealed a great potential for water saving in the west and southeast agricultural zones of Uzbekistan.

Disaster by Design: The Aral Sea and its Lessons for Sustainability
Research in Social Problems and Public Policy, Volume 20, 175–186

ISSN: 0196-1152/doi:10.1108/S0196-1152(2012)0000020021

INTRODUCTION

Soil quality is crucial for agriculture's sustainable development. Indeed, it determines the productivity of agricultural production, which is the basis for food safety and the livelihood of rural people. The continuous decline of soil quality is observed in the irrigated lands of Uzbekistan, where the excessive exploitation of natural ecosystems under agricultural production to follow state orders caused severe environmental degradation.

The policy issues of agricultural and rural development in Uzbekistan, as well as the environmental and soil problems in Central Asia, have been both studied and documented. The most current study on the effects of policy on agricultural development is provided by Lerman, Csaki, and Feder (2004). Considerable research, focusing on the consequences of agricultural activities on soil quality and other environmental factors, addresses mainly the Aral Sea region (see Lal, Suleimenov, & Steward, 2007). But there are glaring lapses in the literature. Mikhalev and Reimov (2008) highlight land degradation problems among Central Asian countries but without a thorough explanation of its causes and effects. Such causes and consequences include soil degradation (loss of fertility, salinization, water-logging), degradation of pastures, degradation of forests, as well as erosion, landslides, and mudflows.

There is a total lack of research comparing countries according to the specific aspects of agri-environmental issues limiting the development of agriculture. An explanation of soil degradation problems and their impacts on agricultural development within each country will help to understand specific circumstances of soil degradation processes. To contribute toward this eventual outcome, the objectives of this chapter are to explore characteristics and changes in land use, agricultural production, soil ecosystems, water availability, and their effects on soil degradation and the limits on sustainable development in various agro-ecological zones of Uzbekistan.

DATA SOURCE AND METHODOLOGY

Published data on population, land use practices, agricultural production, and soil degradation was collected for analysis.[1] As a result of this review, Uzbekistan was found to have five distinguishable agri-ecological zones with different natural and economic characteristics (UZGI, 2001):

- *East (Andijan, Ferghana, Namangan)*: plains and mountain zones with well-drained, good soils, irrigated and some rain-fed.
- *Central*: poorly drained, saline soils, irrigated, low yields.
- *Central-East (Syrdarya, Tashkent, Jizzak, Samarkand, Bukhara and Navoiy)*: plains and mountain zones with well-drained, good soils, irrigated and some rain-fed; plain zones with poorly drained, saline soils, irrigated.
- *Southeast (Kashgadarya, Surkhandarya)*: plains and mountain zones with well-drained, good soils, rain-fed and irrigated; plain zones with poorly drained, saline soils, irrigated.
- *West (Khorezm, Karakalpakstan)*: poorly drained, irrigated, saline soils, low yields.

RESULTS AND DISCUSSION

In 1992, 65.2% of the land in Uzbekistan was used for agriculture and 10.5% of agricultural land was irrigated. The figure hardly changed between 1992 and 2007 (FAOSTAT, 2009). The agricultural sector significantly contributes to the economy of the country. Uzbekistan is the world's fifth largest cotton producer and has become self-sufficient in grain production in the post-Soviet period. Uzbekistan is pursuing a gradual strategy of partial changes in the area of agrarian reform (Spoor, Ten propositions on rural poverty and agrarian transition in Central Eurasia, 2007). The reforms in agriculture of Uzbekistan are guided by the principle that "private farming" uses natural resources more efficiently, and thus contribute less to environmental degradation. The state fixes the area that farmers may cultivate to produce two strategic crops: cotton and wheat. Cotton is the main export crop, contributing approximately 25% of foreign exchange revenues, and a significant source of tax revenue. Wheat is considered essential to achieving food security, and it also offers an easy rotation with cotton, although cotton–wheat is not necessarily the best rotation for maintaining soil fertility. The analysis of land use changes showed that despite an announced withdrawal from cotton monoculture policy, that crop nevertheless makes up the largest portion of cropland (Table 1). The share of cropland under cotton cultivation decreased slightly in the east and central-east zones, while the area under wheat cultivation has shown an increase in all zones. The yield of cotton in the east zone decreased to 0.4 ton per hectare. In other zones, yield per hectare increased for both crops.

Table 1. Arable Land Use Dynamics.

Region	Crops	Arable Land (%)			Yield (Mg/ha)		
		2001	2009	Change	2001	2009	Change
East	Cotton	35.4	33.1	−2.3	2.9	2.7	−0.2
	Wheat	27.3	32.7	5.4	4.3	5.4	1.1
Central-east	Cotton	34.3	31.9	−2.4	2.1	2.4	0.3
	Wheat	28	40.2	12.2	2.9	4.6	1.7
Southeast	Cotton	34.3	35.1	0.8	2.3	2.5	0.2
	Wheat	26.9	39.7	12.8	2.7	4.7	2.0
West	Cotton	22.6	25.8	3.2	2.0	2.2	0.2
	Wheat	7.5	10.8	3.4	2.2	3.9	1.7

Source: MAWR (2007) and State Committee of Republic of Uzbekistan on statistics (2010).

Table 2. Irrigation and Water Consumption.

Regions	Water Use, m^3/mg	Water-Use, Coefficient (Cotton)	Water Use, m^3/mg	Water-Use, Coefficient
	Cotton	Cotton	Wheat	Wheat
East	4,240	0.83	1,970	0.65
Central-east	4,720	0.92	3,040	1.01
Southeast	4,330	0.84	2,970	0.98
West	7,970	1.55	4,420	1.47

Source: MAWR (2007).
Note: The coefficient of water use is a metric that allows comparison to the national average water use, set here at 1.

The first factor for analysis is reliance on artificial irrigation, a widely developed approach in Uzbekistan. Irrigation is of the surface flow type, including furrow irrigation (63.9%), strip (31.6%), and basin (4.5%) irrigation. Earth gravitation is used in furrow-type irrigation, where water runs through artificial furrows between plants. With strip irrigation technology, fields are flooded with water for a short period. Basin-type irrigation is mostly used to produce rice. Sprinklers are not used. Drip irrigation, the most water-efficient approach to irrigation, is only employed in several small pilot projects. Table 2 shows water consumption by main cash crops for the five agricultural zones.

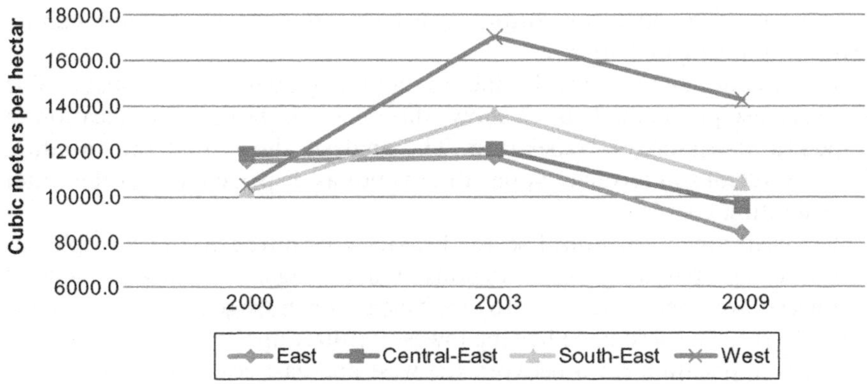

Fig. 1. Temporal Changes In Irrigation Water Use (m^3/Ha). *Source*: MAWR (2007) and State Committee of Republic of Uzbekistan on statistics (2010).

Table 2 shows that water use between two crops and across zones varies considerably. The east zone has the lowest water-use coefficient, with 0.83 and 0.65 scales for both crops. The farmers from the west zone used 1.55 times more water to produce 1 ton of cotton than the country average, and 1.47 times more for wheat. This is explained by the availability of water for irrigation purposes. The west zone is located in the downstream of two main rivers, the Amu Darya and Syr Darya.

But does the greater availability of irrigation water in the west imply that water there is more sustainable? This question is explored in Fig. 1, which reveals that the standard deviation of the mean of per hectare water use from 2000–2009 is equal to 292 m^3 in the east zone, 2494 m^3 in the central-east zone, 1610 m^3 in the southeast zone, and 4457 m^3 in the west zone.

This suggests that the east and southeast zones have sustainable water availability and consumption patterns. In contrast, west zone agriculture shows high volatility of water use, while precipitation dynamics is similar for the whole region. Volatility and high deviation of water usage indicate extensive dependence on water availability for west zone agriculture. Accordingly, we may conclude that agricultural technologies and practices of the region are less sustainable.

Despite the benefits of irrigation, which include increasing agricultural productivity and improving rural welfare, flood irrigation has negative impacts. In addition to high water use and low efficiency, the environmental problems are subjects for concern. These include excessive water-logging, soil salinization, water depletion, and water quality degradation. Yield reductions of 20–30% for cotton have been observed at medium salinity levels in the irrigated soils.

The second factor contributing to land degradation is unsustainable fertilizer-use practices. Table 3 shows differences in fertilizer application in agricultural regions of Uzbekistan. The analysis shows that the variation between regions in fertilizer application is not as dispersed as was the water consumption coefficient.

The east zone agricultural sector heavily uses nitrogen and phosphorus fertilizers compared to other regions. On average, 0.2 tons of mineral fertilizers are used, with a 1.16 coefficient for nitrogen and a 1.46 for phosphorus. The west zone has the lowest fertilizer application. The average fertilizer application ratio between the west and east zones is 2/3.

The soils in agricultural lands are classified by the State Committee on Land Resources, Geodesy, Cartography, and State Cadastre. A land productivity classification system, termed the *Bonitet*, is used by land resource managers and Government Officials to classify land into classes based on their potential productivity. The average value of the *Bonitet* for Uzbekistan is 53 (MAWR, 2004, 2007). Soils with a *Bonitet* range of 41–60 are regarded as average. By stratifying *Bonitet* values into classes, a semiquantitative evaluation of the soil resources is achieved. What are the results?

Only 25.3% of the land is classified as "good" and of a "high" fertility (Table 4). Compared to the beginning of the 1960s, the land class of irrigated soils dropped all over Uzbekistan by 10 units (UNDP, Environmental profile of Uzbekistan based on indicators, 2008a, 2008b) and 5 units

Table 3. Mean Fertilizer Use across Regions of Uzbekistan (1996–2008).

Region	Fertilizers Consumption, '000 ton				Fertilizers Consumption, t/ha			Fertilizer Application Coefficient		
	Total	*N*	P_2O_5	K_2O	*N*	P_2O_5	K_2O	*N*	P_2O_5	K_2O
East	173.6	136.1	34.3	3.2	0.150	0.038	0.004	1.16	1.46	0.98
Central-east	276.2	223.5	46.2	6.5	0.127	0.026	0.004	0.99	1.02	1.03
Southeast	121.3	100.5	18.0	2.8	0.121	0.022	0.003	0.94	0.84	0.94
West	104.6	90.8	12.1	1.7	0.117	0.016	0.002	0.91	0.60	0.61
Uzbekistan	677	551	110.6	15.4	0.129	0.026	0.004	1.00	1.00	1.00

Source: Authors' estimation based on Ergashev (1999), UNDP, Environmental statistics: agricultural use of fertilizers by regions (2009).

Table 4. Soil Quality Distribution in Uzbekistan on Scale of I (poorest) to X (best).

Soil Class	I (0–10)	II (11–20)	III (21–30)	IV (31–40)	V (41–50)	VI (51–60)	VII (61–70)	VIII (71–80)	IX (81–90)	X (91–100)	Unrated	Total, 1,000 ha	2004 Average Land Grade, 1991	2006 Average Land Grade, 2004	% Change
East	0.0	0.3	4.1	18.1	18.9	18.6	19.9	13.8	3.9	0.2	2.1	755.7	64	57	–7
Central-east	0.0	0.3	2.3	14.3	29.8	20.0	18.7	9.2	2.5	0.0	2.9	1,520.8	60	54	–6
Southeast	0.0	0.3	1.2	16.4	34.1	20.5	13.8	8.3	3.7	0.0	1.7	724.2	59	54	–5
West	0.0	0.3	6.5	38.6	16.7	13.3	10.7	3.6	0.6	0.1	9.6	697.6	47	45	–2
Uzbekistan	0.0	0.3	3.2	20.1	26.0	18.5	16.5	8.9	2.7	0.1	3.8	3,698.3	58	53	–5

Source: MAWR (2004, 2007).

compared to 1991. The highest decrease of soil quality was observed in the east and central-east zones. The condition of irrigated lands in Uzbekistan is in most cases classified as "medium" or "low" and in the two western regions – Khorezm and Karakalpakstan – as the "lowest."

Alarmingly, across Uzbekistan, the humus in the soil, which is the basis of its fertility, has decreased by 30–40% in the last 50 years. A continuation of excess irrigation has led to a loss of organic matter and thus to decreasing soil fertility. The soils with humus content of $\leq 1\%$ equals 65.4% of all irrigated lands. An indicator of the extent of the problem is that the average humus content for European agricultural soils is 3%. Yet 28.1% of irrigated Uzbek agricultural lands have between 1% and 3% of humus and 65.4% fall at or below 1% humus content (Fahrenhorst, Haubrok, & Sydow, 1990).

The *Bonitet* system was developed for cotton monoculture during the Soviet period (Ul Hassan, Noble, & Kazbekov, 2005). Table 5 shows correlations between the *Bonitet* score and various land-use indicators. Analysis of correlation coefficients shows a strong relationship between average land Bonitet grades and cotton yields, suggesting that cotton yields are higher on better soils. Wheat yields are not shown to have the same dependence on quality soils. Moreover, for both cotton and wheat, poorer soils are correlated with higher levels of irrigation. The rate of fertilization has the highest correlation to the best soils. Studies prove that over-irrigation leads to high yields in the short run and water erosion of soils over long time periods (Soil_institute, 2001). Farmers in the east zone and partly in the central-east zone use more intensive technologies, with a

Table 5. Correlation Between Bonitet Land Grade And Land Use Indicators.

	Bonitet Grade	Cotton Yield, t/ha	Wheat Yield, t/ha	Water-Use Coefficient of Cotton	Water-Use Coefficient of Wheat	Fertilizer-Use Average Coefficient
East	57	2.63	4.57	0.83	0.65	1.2
Central-east	54	2.43	3.14	0.92	1.01	1.01
Southeast	54	2.4	2.17	0.84	0.98	0.91
West	45	2.1	3.94	1.55	1.46	0.71
r	1	0.704**	0.242	−0.752**	−0.772**	0.527*
Sig. (two-tailed)	0	0.002	0.349	0.001	0	0.03

*Significant at 0.1 the level; **Significant at 0.05 the level.
Source: Authors.

higher cost–benefit ratio. This region is a traditionally settled area with a long history of irrigated agriculture. In contrast, other parts of the country, especially the west zone, have relatively newly developed lands (Spoor, Agricultural Restructuring and Trends in Rural Inequalities in Central Asia, 2004).

CONCLUSIONS

Agricultural policy in Uzbekistanhas a dual purpose; calling for both wheat self-sufficiency, and sustainable currency flow from cotton exports. The policy is hardly sustainable. The area under cotton and wheat cultivation makes up two-thirds of total arable land, and the cotton–wheat rotation does not significantly contribute to soil quality. The obsession with growing cotton impeded the use of crop rotation, which is the basis for soil conservation. Balanced land use would have developed fodder crops and animal husbandry, which produce the organic fertilizers necessary for soil fertility. Instead, large-scale and unbalanced application of mineral fertilizers and pesticides has undermined the natural biological processes and degraded soil quality.

Uzbekistan's excessive reliance on agriculture results in intensive land use and excessive use of chemicals, which is detrimental to soil quality. This short-term policy to achieve high productivity levels using chemicals and irrigation contradicts the long-term goal of sustainability. One can easily see how the system is self-defeating. The government procures the bulk of the cotton harvest from producers at the state procurement price, which is calculated every year by subtracting official costs from export revenues. However, due to overstated costs, the price received by farmers typically lies considerably below export parity under market conditions. This inflated cost represents the main source of tax revenue from agriculture. High taxes in turn imply reduced profitability of cotton production and the need for state orders to ensure that cotton production levels are maintained. Reflecting adverse production incentives, deteriorating land quality consequently results in gradually declining cotton yields.

The infrastructure for irrigation and drainage has been used for more than 30 years without modernization and rehabilitation. Due to operational difficulties, many existing drainage systems are malfunctioning or out of order. The average efficiency of irrigation practice is 0.66–0.68 overall of Uzbekistan, and the operational efficiency of irrigation systems is even lower (0.58–0.60). With on the order of 40% of irrigation diversions being lost, the strong correlation between sustainable water availability and water-use efficiency is revealed. The employment of high leakage irrigation systems dramatically reduces efficiencies. In contrast, zones with low temporal dispersion of available irrigation water show higher efficiency of water consumption. A high variation of water-use in downstream regions of the west and central-east zones is explained by a high sensitivity of agricultural practices to long-term irrigation water availability.

Trends in fertilizer consumption show a significant bias for nitrogen fertilizer use, increasing yields in the short-term at the expense of long-term soil productivity. Cotton monoculture requires the large-scale use of chemical fertilizers and pesticides. Between 20 and 90 kg of pesticides and 300 and 500 kg of mineral fertilizers per hectare/year were used in the past. Total fertilizer use in Uzbekistan did not change significantly from 1996 to 2009. The share of nitrogen fertilizers used in agriculture increased from 56.2% to 77.8%. A decrease in the consumption of phosphorus and potassium was replaced by nitrogen use, which leads to the long-term reduction of overall soil productivity.

Agriculture in the east zone showed sustainable water consumption, and highest average land-quality grade. The east zone has the highest fertilizer-use coefficient. The downstream west zone has the poorest soils, the highest water use, and the lowest fertilizer use. The analysis revealed the high potential of water saving in the west and southeast zones.

NOTE

1. These sources were used for the collection of baseline data: Websites: http://www.isric.org; http://www.fao.org; http://faostat.fao.org; and http://enrin.grida.no/aral/aralsea; http://www.statistics.uz. Official statistics: (SLCR, 2002), (MAWR, 2007) Previous studies: (Samylov et al., 2006), (Ul Hassan et al., 2005), (Gleick, 2000). (Klotzli, 1994), (Spoor, Agricultural Restructuring and Trends in Rural Inequalities in Central Asia, 2004), (Guadagni, Raiser, Crole-Rees, & Khidirov, 2005), (Lerman et al., 2004), (Wehrheim, Schoeller-Schletter, & Martius, 2008), (Cai, McKinney, & Lasdon, 2003).

REFERENCES

Cai, X., McKinney, D., & Lasdon, L. (2003). Integrated hydrologic-agronomic-economic model for river basin management. *Journal of Water Resources Planning and Management*, *129*(1), 4–17.

Ergashev, A.-K. (1999). Ecological and socio-economic aspects of the use of mineral fertilizers in Aral Sea Basin countries. *67th IFA Annual Conference*. IFA, Manila.

Fahrenhorst, C., Haubrok, A., & Sydow, M. (1990). Übernahme der Bodengesellschaftskarte Berlin in das Umweltinformationssystem Berlin und Zuordnung von Bodeninformati onen. Im Auftrag der Senatsverwaltung für Stadtentwicklung und Umweltschutz.

FAOSTAT. (2009). *FAOSTAT* (Food and agricultural organization). Retrieved from http://faostat.fao.org/site/377/default.aspx#ancor. Accessed on April 8, 2008.

Gleick, P. H. (2000). The changing water paradigm: A look at twenty first century water resources development. *Water International*, *1*(25), 127–138.

Guadagni, M., Raiser, M., Crole-Rees, A., & Khidirov, D. (2005). *Cotton taxation in Uzbekistan. Opportunities for reform*. ECSSD Working Paper No. 41. The World Bank, Washington, DC.

Klotzli, S. (1994). T*he water and soil crisis in Central Asia. A source for future conflict?* ENCOP Occasional Paper No. 11. Center for Security Policy and Conflict Research Zurich/Swiss Peace Foundation, Berne.

Lal, R., Suleimenov, M., & Steward, B. A. (2007). *Climate change and terrestrial carbon sequestration in Central Asia*. London: Routledge.

Lerman, Z., Csaki, C., & Feder, G. (2004). *Agriculture in transition: Land policies and evolving farm structures in post-Soviet countries*. Lanham: Lexington books.

MAWR. (2004). *Distribution of agricultural area of Uzbekistan by grounds*. Tashkent: Ministry of Agriculture and Water resources of Uzbekistan.

MAWR. (2007). *National report of MAWR*. Tashkent: Ministry of Agriculture and Water resources of Uzbekistan.

Mikhalev, V., & Reimov, A. (2008, April). Land Degradation in Central Asia. *Development and transition*, *2008*(09), Retrieved 05/2009 from: http://www.developmentandtransition.net/index.cfm?module = ActiveWeb&page = WebPage&DocumentID = 673#.

Samylov, S. V., Abdujalilov, U. A., Ahmadjanov, B., Bekmirzaev, I., Grigoryants, A., Dotsenko, N., et al. (2006). *The national report environment and natural resources use*. Tashkent: Chinor ENK.

SLCR. (2002). *National report*. State Committee on Land Resources, Geodesy, Cartography and State Cadastre of Uzbekistan, Tashkent.

Soil_institute. (2001). *Briefly report. Investigation of the meliorative conditions of irrigation soils in Uzbekistan*. Genesis and Reclamation of Salt Affected Soils Department. Tashkent: 19. Soil Institute, Briefly report. Investigation of the meliorative conditions of irrigation soils in UzbekistanScientific Soil and Agro-Chemistry Institute.

Spoor, M. (2004). Agricultural restructuring and trends in rural inequalities in Central Asia. *Civil* Society and Social Movements Programme Paper No. 13, United Nations Research Institute for Social Development, Geneva.

Spoor, M. (2007). *Ten propositions on rural poverty and agrarian transition in Central Eurasia*. Idea Working Papers, IBEI, Barcelona.

State Committee of Republic of Uzbekistan on statistics. (2010). *Agriculture of Uzbekistan 2010*. Tashkent: ООП ГКС.

Ul Hassan, M., Noble, A., & Kazbekov, J. (2005). *Bright spots in Uzbekistan, reversing land and water degradation while improving livelihoods: Key developments and sustaining ingredients for transition economies of former Soviet Union. International Water Management Institute*. Colombo: IWMI.

UNDP. (2008a). *Environmental atlas of Uzbekistan*. Tashkent: Mega Basim.

UNDP. (2008). *Environmental profile of Uzbekistan based on indicators*. Tashkent: Mega Basim.

UNDP. (2009). *Environmetal statistics: Agricultural use of fertilizers by regions*. (UNDP CO in Uzbekistan). Retrieved from http://www.statistics.uz/map/1988/

Uzgi. (2001). *Preliminary report of Uzbekistan National Team*. Subcomponent A1 of the Water and Environmental Management, GEF project, Tashkent.

Wehrheim, P., Schoeller-Schletter, A., & Martius, C. (2008). *Continuity and change land and water use reforms in rural Uzbekistan. Studies on the Agricultural and Food Sector in Central and Eastern Europe* (Vol. 43). Halle: Leibniz-Institut für Agrarentwicklung in Mittel- und Osteuropa (IAMO).

CHAPTER 12

COTTON IN OUR EARS: WATER, AGRICULTURE, AND CLIMATIC CHANGE IN THE POST ARAL CONTEXT

Shavkat Hasanov, Mirza Nomman Ahmed, Ibragim Ganiev and Olim Murtazaev

ABSTRACT

When agricultural productivity was studied in areas of the Zarafshan valley incorporating 26 rayons *or counties, a clear relationship was found between increased use of water and decreased efficiency. Moreover, inefficiencies were most evident in the Bukhara region, where water quality is the poorest. There, an overall slide in productivity was evident. A shift to sustainable water use and farming practices was called for. Convergent findings were found in a second, computer modeling study. On an index of water overuse, a significant relationship between water overuse and a decline in agricultural production was again demonstrated. The potential impacts of climate change on agricutural productivity were also modeled, revealing a tentative relationship between a decline of productivity and a 1% increase in temperature.*

Disaster by Design: The Aral Sea and its Lessons for Sustainability
Research in Social Problems and Public Policy, Volume 20, 187–195

ISSN: 0196-1152/doi:10.1108/S0196-1152(2012)0000020022

INTRODUCTION

Modern Uzbekistan inherited from Soviet times a strong economic dependence upon cash export crops, mostly cotton. Despite its thirsty nature, cotton grows well in the dry climate of the Uzbek steppe. The collectivized farming of the Soviet period has given way to private farms, but the dependence on cotton has not been significantly changed. The interesting issue that arises for the agricultural economist is whether increased efficiency from private farming is offset by the continued reliance on an energy intensive crop. The focus on cotton as the cause of the Aral and broader disasters of the West needs to be broadened to include additional factors that also affect farm productivity to address this issue.

The agricultural sector still remains the most important sector of the economy, responsible for almost 35–50% of the livelihood of the population, directly or indirectly linked to the sector. Uzbekistan's agriculture has undergone gradual structural changes. After three rounds of farm restructuring, there has been a net decrease in the number of farms in the country to 81,300 while the average size of a farm increased by 60.7 hectares. Where the main crops are wheat and cotton, the average farm plot area increased 92 hectares (MAWR, 2011). Also, agricultural commodities in the country today are being produced mostly by private farms and households ("dehqon"). The principal agricultural products, cotton and grain, have fixed state procurement prices and are produced by private farms. Households, in the meantime, play a significant role in the production of food crops, for example, supplying 84% of potatoes to market, 67% of vegetables, 52% of fruits, and nearly the total amount of meat and milk in the country (SCS Uz., 2009).

There are a number of factors making Uzbekistan's agricultural systems unsustainable. These include water scarcity and land degradation responsible for rapidly increasing costs of input resources in recent years and likely to cause continued increases in the near future. Irrigation plays an essential role in agriculture, and almost all crops produced in the country are taken from irrigated lands. However, the main sources of water for irrigation come from neighboring countries, resulting in scarcity pressure for most farms. Finally, the fast population growth in the country results in ever increasing demand for food and other commodities. These factors make farming an increasingly difficult endeavor.

Promising to complicate the situation further are the impacts of global warming and its specific implications for the agricultural sector in Central

Asia. The Central Asian countries climate ranges from arid to semi-arid with variability on both temporal and spatial scales. Climate change in Central Asian countries manifests itself in many ways, including temperature increases, rainfall reduction in the arid plains, and increases in the mountain areas and last but not the least, accelerated glacial melt. Climate change thus joins other threats to agricultural productivity as a challenge to food security.

This chapter examines two convergent methods for addressing factors affecting farm productivity. In the first, an analysis is made of farm efficiency in Samarkand, Navoiy, and Bukhara regions, all located in the Zarafshan valley of Uzbekistan, in the center of the country and to the east of the Aral region. The western parts of the region, downstream in Bukhara, are considered part of the Aral impact region for this volume. *Rayon-* (i.e., county-) level statistics were used to conduct a "Data Envelopment Analysis (DEA)" in order to investigate trends in the farm sector. The data was then used to conduct a systems analysis of inputs (i.e., scarce resources) and outputs (i.e., agricultural production) to determine the economic value of agriculture using the approach of management sciences and economics. In the second study, a computer modeling project, an additional factor is added, the effects of climate change on agricultural productivity, controlling for such factors as fertilization, water use, land characteristics, labor and technology.

STUDY ONE: METHODOLOGY AND FINDINGS

Efficient farming balances inputs and outputs to maximize profit. A number of Uzbek studies have previously examined the efficiency of the entire irrigated agriculture system, in and of itself, and with regard to special inputs or crops during the Pre-independence period of the country (Abduganiev, 1983). Such studies lacked systematic and robust data, relying on simplistic statistics designed to demonstrate causality but lacking the requisite information.

In this study, a modern non-parametric approach was used that flexibly allows for measuring efficiency in complex systems from empirical observation where there are multiple outputs and inputs.[1] Distances between two study areas (e.g., Samarkand versus Bukhara) were measured against a common technology (e.g., irrigation), along with issues of technical efficiency, scale efficiency, and technical change.

Study Site and Data Collection

The study area covered three regions of the country, all located in the Zarafshan valley of Uzbekistan. The valley is one of the biggest inter-mountain areas of Central Asia and is situated between the mountain ridges of Turkistan and Zarafshan. Historically, the Zarafshan River was a center of irrigated agriculture. The valley occupies around one fifth of the country's irrigated land and produces one third of the total agricultural commodities of Uzbekistan.

We compared agricultural data for yield per hectare of cotton for 2007 through 2009 as related to three input variables: water usage per hectare of cotton grown, related labor costs, and capital costs, including the expense of seed, tractor and machinery use, and fertilizer. Cotton yields varied among different farms in the sample, largely due to accessibility to water. Irrigated farms produce yields between 1.5 tons per hectare to 4.5 tons per hectare. These yield numbers are lower than the average yields in major cotton producing countries such as China and the United Status (Gemma, 2003). Descriptive statistics are presented in Table 1.

Results

The analysis of farm performance is carried out using the gross yield per hectare of cotton as an indicator. Fig. 1 shows the technical efficiency of cotton production over 3 years for 26 *rayons* (counties) included in the sample. The findings clearly indicate a wide variation of efficiency across *rayon* and over the 3 years examined.

Rayons were then grouped according to inefficiency based upon performance in 2009. The four least efficient received technical efficiency (TE) scores below 70% (Olot, 69.4%, Shofirkon 61.8%, Karaulbazar 46.3%, and Bukhara city 63.0%). They are all located in the Bukhara region, downstream of the Zarafshan river valley, where water is a very sensitive resource. The second least efficient group was almost entirely from Bukhara region.

Then there is the issue of whether farmers in a given *rayon* showed net improvement over the three study years or lost ground. In Fig. 3, TFP (total factor productivity) is averaged over the three study years. Scores greater than 1 reveal improved yield over previous years and those below 1 show diminished productivity. The figure shows a dramatic trend toward diminished yield even in highly productive *rayon* (Fig. 2).

Table 1. Descriptive Statistics on Outputs and Inputs Used in Efficiency Analysis.

	2007				2008				2009			
	Min	Max	Mean	S.D.	Min	Max	Mean	S.D.	Min	Max	Mean	S.D.
Output of cotton per hectare	1.6	3.0	2.5	0.3	1.6	3.0	2.5	0.3	1.5	3.3	2.6	0.5
Input of water*	8.3	16.6	13.4	2.5	7.3	19.3	12.3	3.7	2.4	16.1	12.5	3.3
Input of capital**	291	2888	990	539	344	4135	1101	711	464	2238	1260	458
Input of labor**	46	1211	401	263	76	862	407	196	20	785	483	214

* In thousand cubic meters per hectare; ** in thousands of Uzbek sum per hectare. *Source*: Author.
Editor's note: At the time of editing, in July 2012, the Uzbek sum was valued at 1,888.9 per U.S. dollar.

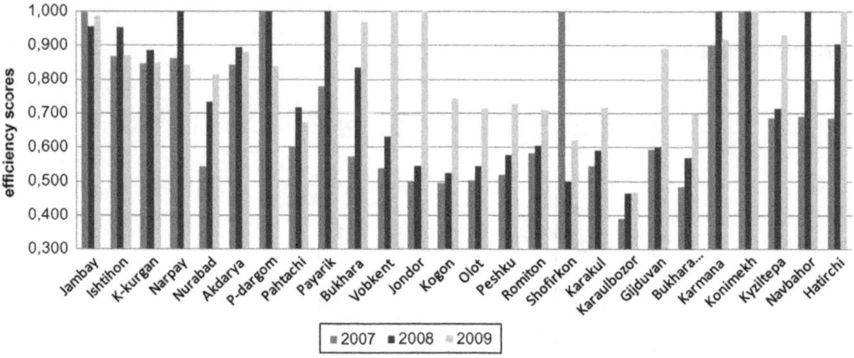

Fig. 1. Technical Efficiency Scores of Cotton Production for 26 *Rayons* in the Zarafshan Valley. *Source*: Author.

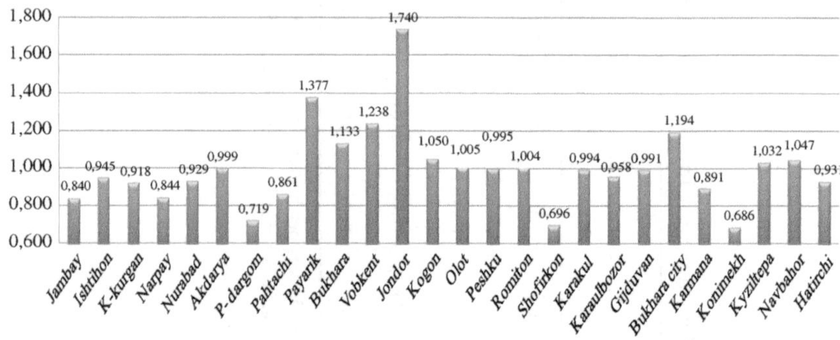

Fig. 2. Mean Cotton Productivity (TFP) Changes Over the Years 2007–2009 for 26 *Rayon* in the Zarafshan Valley. *Source*: Author.

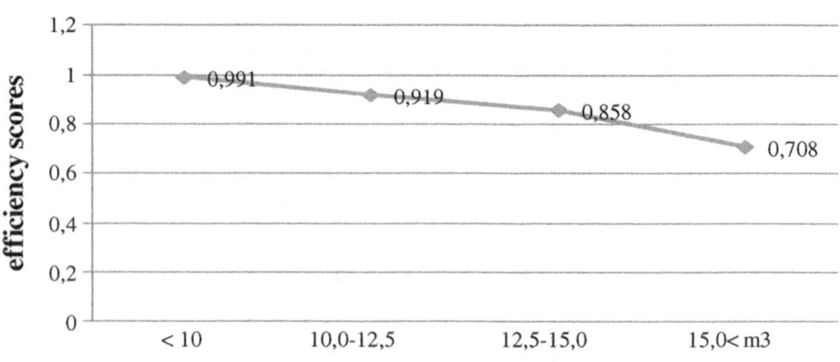

Fig. 3. Efficiency Score Ranges and Water Usage for Irrigation Per Hectare. *Source*: Author.

What variable best accounts for the findings? Fig. 3 demonstrates clearly an inverse relationship between water use per hectare and efficiency scores.

STUDY TWO: METHODOLOGY AND FINDINGS

In order to project future interaction of climate and food production, we employed a computer model. Using classical theory, we assumed that the agricultural portion of gross domestic product is a function of arable land, agricultural employment, technology endowment, temperature, and water use. These factors were mathematically represented, as were productivity shifts over time spurred by technological progress or climate adaption.

One further has to tease out differences between the five different Central Asian nations while recognizing that they are largely homogenous except for factors occurring after independence, such as political change and exact time periods.

The Data Set

A data set for the selected set of Central Asian countries was employed covering four years: 1992, 1997, 2002, and 2007. Because local data is often of low quality, we drew information from reliable international sources, blending data from the World Bank, the Food and Agriculture Organization, and recent data from Central Asian ministerial sources. A balanced panel data set was thus created to represent the five Central Asian countries Kazakhstan, Uzbekistan, Kyrgyzstan, Tajikistan, and Turkmenistan. It covered the major aspects of agricultural production, including land, labor, technology, and water, with land and technology jointly representing capital. These explanatory variables allow for the isolation of impacts due to changes in the water extraction ratios on agricultural production.

There were five key variables used in the study. These were every additional $2,000 in agricultural value added, fertilizer used per unit of land for all crops, technological use as represented by tractors, the mean annual temperature, and overuse of water (where extraction exceeded flow).[2]

A key variable of interest, defined as "overuse," was generated using the total available renewable internal water resources and the extraction by the respective countries, specifically extraction rates connected to the Aral Sea Basin, with the Amu and Syr Darya as the main water providers.

Results

The detailed method and results of the model runs will be published separately. But the results and their implications are of interest for the current publication.

Results clearly depict the adverse effects of a disproportionate use of the region's renewable freshwater resources. An increase in overuse by 1% translates into a significant decline in agricultural contribution to the GDP. Income loss can be inferred from the results. Largely the water scarce downstream countries are expected to suffer. The estimated response function has been designed to be able to draw conclusions about future developments related to climatic change in Central Asia. Moreover, there were tentative indications of negative impact from a 1°C increase in temperature on agricultural production. Although this impact could not be significantly verified and tested given missing data and insufficient observations, the direction of the impact remained stable throughout tests of different model formulations.

CONCLUSIONS AND OUTLOOK

From the first study, our major finding is that one of the main agricultural cotton production areas in Uzbekistan, the Bukhara region, is characterized by inefficiency. In many *rayon* there, increased irrigation levels are being used to push up cotton crop yields that have historically fallen off because of adverse environmental change. In other words, this conventional efficiency analysis conducted within the parameters of economic study shows that current practice is uneconomical. Other papers in this volume demonstrate some of the external costs not captured by the approach employed here that show an even higher cost indeed of current agricultural policy.

Suggested is a need for implementing technological efficiencies such as drip irrigation that will allow for the required water inputs without sacrificing efficiency. To implement this shift, the active involvement of government agencies is required. The current situation where cotton production continues to predominate but with significant inefficiencies reflects the general trends in public policy in Uzbekistan, especially the fall in real public development expenditure intended for agricultural support projects. Given the rapid population growth in the nation, a rethinking of agricultural policy according to principles of sustainability is needed. In order to successfully implement adaptation strategies, the respective agencies, such as agricultural extension services, will have to be strengthened and their outreach will have to be significantly increased. A closer cooperation between government agencies and research institutions will have to be established to guarantee a thorough understanding of climate change and consequently a sound scientific basis for the implementation of either adaptive or mitigative measures.

The second study used a computer model to approximate the effects of climatic change, water use, and thus extreme events on the agricultural productivity in Central Asian agriculture. The model results confirm the agricultural productivity findings from the more geographically limited Zarafshan valley, demonstrating the adverse effects of a disproportionate use of the region's renewable freshwater resources, where an increase in overuse forced decline in agricultural contribution to the GDP. Income loss can thus be anticipated to occur largely in the water-scarce downstream countries. More data is required to show a significant impact of increased temperature on agricultural production. However, if the climate findings are reliable, the additional factor of increased temperature must be integrated into the adaptive frameworks discussed above.

NOTES

1. Specifically, the Malmquist Index of Total Factor Productivity (MITFP) growth was used to detect patterns of agricultural productivity and identify more efficient means of using resources.
2. Sources for this data were, respectively, FAOSTAT (2011), MAWR (2011), and Williams and Konovalov (2008).

REFERENCES

Abduganiev, A. (1983). *Increasing economic efficiency use of irrigated lands in Uzbekistan* (pp. 17–35), Tashkent, (in Russian).

FAOSTAT. (2011). Food and Agriculture Organization of the United Nations, 2011. Available at http://faostat.fao.org/.

Gemma, M. (2003). Water management and agricultural development in Central Asia: Case of Uzbekistan. *Proceedings of 14th International Farm Management Congress, Western Australia.* Retrieved from http://ageconsearch.umn.edu

MAWR – Ministry of Agriculture and Water Resources of the Republic of Uzbekistan (2011). *Development yearbook*. Retrieved from www.agro.uz

SCS Uz. (2009). The State Committee of the Republic of Uzbekistan on Statistics, 2009 and 2011. Tashkent, UZB.

Williams, M. W. & Konovalov, V. G. (2008). *Central Asia temperature and precipitation data, 1879–2003*. Boulder, Colorado: USA National Snow and Ice Data Center. Retrieved from http://nsidc.org/data/docs/noaa/g02174_central_asia_data/index.html

CHAPTER 13

WATER FOOTPRINTS: INTEGRATED WATER RESOURCE MANAGEMENT TO THE RESCUE IN THE ARAL SEA BASIN

Inna Rudenko, Utkur Djanibekov,
Kudrat Nurmetov and John P. A. Lamers

ABSTRACT

Intensive agricultural production in the countries of the Aral Sea Basin has resulted in undesirable ecological and social consequences, including the drying of the Aral Sea. Water has become scarce due to a score of internal and external factors including the growing demand for water resources by the upstream countries, expansion of the irrigated areas to ease food insecurity, and the poor condition of irrigation and drainage networks. To cope with environmental consequences and regional water challenges, it is vital to look for pathways of improved integrated water resource management, higher water use efficiencies, and reducing overall water use.

A combination of value chain and water footprint analyses of the dominant crop, cotton, was applied to assess water use in different sectors

Disaster by Design: The Aral Sea and its Lessons for Sustainability
Research in Social Problems and Public Policy, Volume 20, 197–215

ISSN: 0196-1152/doi:10.1108/S0196-1152(2012)0000020023

of the Uzbekistan economy and to seek water saving and improved water management and efficiency options.

The findings show that reduction in water use could be achieved by diversifying the economy and moving from water intensive agricultural production to less water consuming industrial sectors by introducing water saving irrigation technologies and by raising awareness of the population about the real value of water.

The combined findings of the economic based value chain analysis and ecologically oriented water footprint analysis gave an added value for better informed decision-making to reach land, water, and ecosystem sustainability and to contribute to the Millennium Development Goals of eradicating poverty and hunger and achieving food and water security.

INTRODUCTION

Irrigation is an age-old technique that plays a major role in crop production worldwide. With an estimated 86% share of the global freshwater use, agriculture is the largest freshwater consumer worldwide (Hoekstra & Chapagain, 2007). Concurrently, the share of all freshwater uses in irrigated agriculture has amounted to 70% and used to produce about 40% of the global foods on about 280 million hectares (Mha) of irrigated cropland, or the equivalent of 18% of the world's total cropland (UNESCO-WWAP, 2006). However, irrigated agriculture has simultaneously become a source of concern due to the high water losses, secondary soil salinization, and land degradation. Environmental degradation has reached alarming dimensions in the five Central Asian republics Kazakhstan, Kyrgyzstan, Tajikistan, Turkmenistan, and Uzbekistan where 40–60% of roughly 11Mha of the total irrigated croplands are degraded due to decades long mismanagement and misuse of the freshwater resources (Lal, 2007). In the agrarian republic of Uzbekistan, irrigated agriculture uses 90–95% of all freshwater resources (UNESCO-WWAP, 2006).[1] At the same time, the contribution of agriculture to gross domestic product (GDP) in Uzbekistan in 2006 was only an estimated 25–30% share. This not only represents a much lower contribution to GDP than its share of freshwater use, but in addition, it varies depending on the agro-ecological regions within the country. In the Khorezm region, for instance, agriculture contributes 45% to the regional

GDP, whereas industry contributes up to 9% of GDP while consuming annually less than 3% of total available water, thus representing a more efficient use of freshwater resources in the region.[2]

Recently Martius, Froebrich, and Nuppenau (2009) argued that given the complexity of the "intertwined problems of demographic growth, climate change, water demand and land degradation" in Central Asia, integrated water resources management (IWRM) is perhaps a suitable approach to address the existing challenges of freshwater use in this region. The indicators of water use reflect the overall anthropogenic pressure on freshwater resources. They also give an indication of human vulnerability to water shortages and the need for adjustments in water management policies. This paper uses IWRM as the framework for analyzing the cotton value chain and water use along the chains in greater detail.

The overall objective of IWRM is to satisfy sometimes competing freshwater needs by various users and to ensure sustainable water management. From this point of view, the concept of *water footprint* can enrich IWRM approaches. Water footprint refers to the total volume of water used to produce the product, summed over the various steps of the production chain (Hoekstra & Chapagain, 2007). The water footprint is a geographically explicit indicator, showing not only volumes of water use and pollution, but also the locations. The related concept of *virtual water* helps us to estimate how much freshwater is "embodied" in the product, not in real sense, but in virtual sense.[3] Using these two concepts together can help us perform a more nuanced analysis and gain a critical understanding why Central Asian water is in crisis. To this end, the estimation of virtual water and water footprints for different commodities has gained worldwide interest since their presentation in the late 1990s (Allan, 1998; Hoekstra, 2003). In particular, triggered by the growing concern for global warming and advancing climate change, future freshwater supply and demand are of key interest to water users and administrators alike. This is illustrated in Central Asia where the freshwater supply is under threat of increasing water shortages, decreasing water quality, and increasing demands. The synergy of IWRM and water footprint analyses can be conducive to improved political decision-making.

Uzbekistan, with an annual production of about 3 million tons of raw cotton (*Gossypium barbadense* L.), is the sixth largest cotton producer worldwide. It ranks third among exporters after the United States and the consortium of African cotton producing countries (UNCTAD). With an annual export weight of 1 million tons of fiber, this results in a contribution of about 18% to the national GDP. Among the cotton producing countries in Central Asia, Uzbekistan ranks first. About 2.5–3 million tons of raw cotton are annually grown on its territory with the average input of 610–750mm of irrigation freshwater (excluding leaching water) depending on regional location (given an average yield of 2.6 tons per hectare of raw cotton, or about 0.85 tons per hectare of cotton fiber) (Djanibekov, 2008; Müller, 2006; Rudenko, 2008). Although there are detailed estimates available on the amount of freshwater used for irrigated raw cotton production, less information is available for the amount of freshwater used during the various processing and manufacturing stages of raw cotton. These quantities of freshwater are however significant, as evidenced by the cotton value chain (Rudenko, Grote, & Lamers, 2008). A recent estimate by Chapagain, Hoekstra, Savenije, and Gautam (2006) indicated about 4,460 liters per kilogram of virtual water for raw cotton and 10,410 liters per kilogram of virtual water for cotton fiber, which seem too high when compared to some other cotton producing countries worldwide and when compared to the world average indicators.

The quantity of *virtual water* has a temporal and spatial dimension that must be considered. There is a huge variation in virtual water content of similar products around the globe, due to the way the products are produced and the type of inputs used for producing these products. For example, while the world average virtual water content of beef stands at around 15,500 liters per kilogram, it ranges from 11,000 liters per kilogram in Japan to 37,762 liters per kilogram in Mexico. The value added of the water footprint is therefore not an absolute outcome, but rather, it is a means to present figures in an easy and transparent form to a broad public, gain their attention and interest and raise their awareness about the hidden use of water. This helps the public put this in relation to future water availability and draws their attention to the idea that it is perhaps wiser to cultivate other crops or increase water use efficiencies in certain places. This is true in particular for agricultural export-oriented countries and for highly tradable commodities such as cotton in Uzbekistan.

THE TASK AHEAD

To cope with severe regional water challenges expected in the near future, it is necessary to look now for options to reduce freshwater use on a nationwide level. This could be achieved either by diversifying the economy and moving from water intensive agricultural production to less water consuming industrial sectors (from raw cotton to cotton products, for example) or by improving technical efficiencies (mostly in agriculture) and raising awareness in the population about the likely water deficit and the real value of water.[4] The point is that there is no time to waste in determining the best conversions, because implementing the designated conversions will take time.

This study therefore combined the value chain analysis (VCA) (Rudenko, 2008; Rudenko et al., 2008) with a water footprint analysis (WFA) of the cotton value chain, which provided realistic estimates of virtual water usage of the entire range of cotton products. Next, the economic, social, and environmental gains as a function of replacing agricultural production (blue water reduction) with an intensified production of value-added products (increasing gray water use) were explored, based on the case study region of Khorezm in northwestern Uzbekistan. By investigating the amount of freshwater used at every production stage along the cotton value chain, the study was guided by IWRM approaches to assess both blue and gray water saving options. To the best of our knowledge, this is the first time the combination of the value chain, water footprint, and IWRM approaches is applied in Central Asia as a concept to create water shortage awareness among the population and governments alike, and to form a basis for sound policy recommendations.

MATERIALS AND METHODS

Description of the Case Study Site

The Khorezm region is a 6,400 km^2 sized administrative district located in northwestern Uzbekistan at approximately 250 km upstream of the present shore of the Aral Sea, and between 60.05 and 61.39 N and 41.13 and 42.02 E of the Greenwich meridian. It has become one of the most problematic areas in the world regarding salinity, irrigation water availability, and overall crop performance (Martius, Lamers, Ibrakhimov, & Vlek, 2004). The continental climate, with its typical long and hot summers and short cold winters, is responsible for the high potential evapotranspiration which exceeds the annual precipitation of circa 100 mm by about 15 times. The mean annual air temperature is approximately 13°C and 205 frost-free days are recorded annually (Chub, 2000). Predominantly cotton (about 50% of the area), winter wheat (about 20%), rice, and alfalfa are cultivated under irrigation.

Agriculture in Khorezm is the predominant occupation of the region. It contributes to food security, supports livelihood continuity of the large – up to 70% of the total – rural population, and provides inputs, such as cotton, for the agro-processing industry. This indispensable sector of the local economy is however only made possible by irrigation. Large amounts of water are diverted from the Amu Darya each year (about 4.6 km^3 in 2006) to irrigate the vast agricultural fields in the region.[5] At the same time, the probability of sufficient water supply for irrigation in the region has been decreasing in the last decades as evidenced by the more frequent (every seventh year compared to every tenth year in the last century) occurrence of droughts. Thus, water has become scarce in Khorezm due to many external but also internal factors, such as the growing demand for water resources by the upstream countries, the expansion of the irrigated areas to feed the growing population, and the poor condition of the irrigation and drainage networks that cause high losses. All these factors not only threaten agricultural production and thus the entire setup of the local system, but have already resulted in unprecedented ecological consequences, of which the demise of the Aral Sea is only one example. The increasing scarcity of water may also lead to increased political and social tensions; even trans-boundary conflicts with Turkmenistan over access to water may soon become evident.

Methodological Framework

The methodological framework builds on two approaches: (i) value chain analysis and (ii) water footprint analysis, each emphasizing a specific perception and objective. The value chain analysis of cotton has been completed for the period 2005–2007 in an earlier study.[6] Given that the water footprint concept is a consumption-based indicator of water use that can provide useful information in addition to the traditional production-sector-based indicators of water use, the concept of virtual water has become an important component of sustainable water use and water management assessment in water-scarce regions worldwide, and thus is expected to lead to a more efficient resource use as previously argued (Allan, 1998; Chapagain, Hoekstra, & Savenije, 2005a). In 2008, the value chain assessment was expanded with a water footprint analysis.

The achievements of the value chain analysis included mapping (Kaplinsky & Morris, 2002) of the cotton chain. A closer look at the involved stakeholders and their functions with regard to the value chains helped sketch the flows of the products along the chains as well as estimate the economic turnover and costs incurred (FAO, 2006a, 2006b, 2006c; Rudenko, 2008).

Virtual water was calculated based on the so-called blue (for irrigation) and gray water contents for all products starting from the field level to the ready-made products such as textiles and a simple T-shirt for the cotton value chain.[7]

Calculating Virtual Water and the Water Footprint Value Index

The virtual water content of cotton products was calculated according to the method first proposed by Hoekstra and Hung (2005) and Chapagain and Hoekstra (2004). First, the virtual water content (m^3 per ton) of raw cotton yields was calculated based on its water requirement specific for the study region Khorezm. Next, the virtual water content of all processed cotton products was calculated based on product fractions.[8]

Because of its flexibility and the provided detailed description, the value chain method ("bottom-up calculation") was effective in identifying the flows of products and for the estimation of water use at each stage of the chain. In total, virtual water content, the costs, and revenues of 11 types of cotton products were identified. From the field to semi-finished or finished products, cotton passed through a number of production stages, which are here subdivided into agricultural (raw cotton production at the field level) and industrial stages (processing of cotton into final products).

Agricultural water use (AWU) in realms of irrigated agriculture in the Aral Sea Basin and Uzbekistan includes (i) water used for leaching or washing out salts from the soil; (ii) water used for irrigation during the entire vegetation cycle; (iii) approximate water losses in the main and on-field canals; (iv) and the so called gray water, or water virtually needed to dilute water polluted by pesticide and fertilizer seepage.[9]

$$\text{AWU} = \text{Leaching} + \text{Vegetation} + \text{Losses} + \text{Gray} \quad (1)$$

Total virtual water (TVW) of each product processed in the cotton value chain was calculated as the sum of agricultural water use and water use at each industrial stage (IWU). The gray water was calculated based on methodology suggested by Chapagain et al. (2006).

$$\text{TWA} = \text{AWU} + \text{IWU} \quad (2)$$

Only that fraction of agricultural (or industrial) water was assumed for each subsequent processed product, the corresponding fraction of which was used to produce this subsequent product. For example, total virtual water of cotton fiber was calculated as the sum of 33% of agricultural water use[10] plus industrial water use at ginning and so forth.

$$\text{TVWfiber} = (\text{Leaching} + \text{Vegetation} + \text{Losses} + \text{Gray}) \times 33\% + \text{IWUginning} \quad (3)$$

The water footprint value index was calculated as a ratio between value added of a certain cotton product and its virtual water content. Based on virtual water estimates embedded in various cotton products and on volumes and values of exported cotton, the total amount of exported virtual water was estimated, as well as its value both in absolute terms and for one unit of exported water.

Data Collection

Several data collection methods originating from formal and informal survey methods were used to generate the required information on value chains. Semi-structured interviews with questionnaires were conducted with the main, or direct, actors of the chains. After a preliminary data check for possible blanks, revisits occurred when necessary, and the additional technical parameters were integrated.[11]

RESULTS

Cotton Products' Flows and Fractions

Based on the value chain approach, a visual representation of the entire value chain was constructed including the connections between actors. The cotton value chain was demarcated for tracing the flows of raw cotton grown by the farmers in Khorezm, to either processing by local industrial enterprises (ginneries) or to exporting some products, such as cotton fiber or textiles, abroad; or finally to consumption by households (Fig. 1).

The cotton value chain in Uzbekistan, by example of Khorezm, resembles the chains in other cotton producing countries. In its simplest form, it consists of cotton growing farms, ginneries or cotton refining plants, textile enterprises, and oil extracting plants (as shown in Fig. 1). The flow of cotton products along the chain begins with raw cotton produced by farmers and transferred to the ginneries. Cotton fiber, which amounts to about 33% of raw cotton, then flows from the ginneries to the textile enterprises and to a larger extent is exported. Cottonseed, constituting 54% of raw cotton, partly flows back to farmers as seeding material for the next agricultural

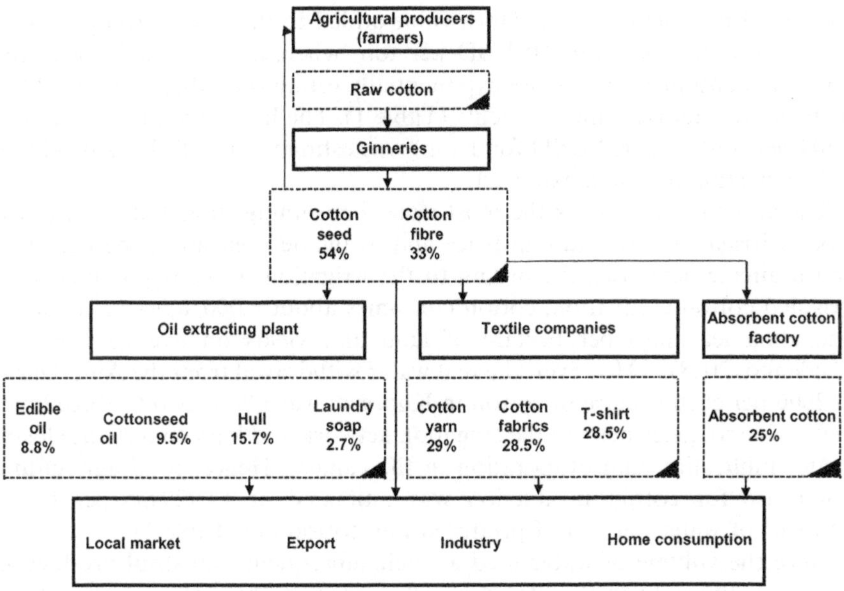

Fig. 1. Cotton Product Flows and Their Output Fractions.

season and to a larger extent is transported to the oil extracting plant (as shown in Fig. 1). Cottonseed oil and cottonseed meal and cake from the oil extracting plant are then purchased by the population or exported to neighboring countries. Finally, textile products such as yarn (29% of raw cotton), fabrics (28.5%), and ready-made garments (28.5%) from textile producers are consumed within the region or exported.

Understanding the flows of production is only part of the story however. If we want to select the most water efficient parts of the cotton value chain and make the best recommendations for Uzbekistan's situation, we need to also understand how value enhancement works for cotton. Experience from developing countries shows that nations specializing in exports of primary commodities are vulnerable to the fluctuations in the world markets. This is no different in Uzbekistan. Building up a diversified export and industrial structure has been necessary in the newly industrialising countries and has been dependent on access to large, diversified markets and the economies of scale and the competitive effects that were thereby possible (Stamm, 2004). The logic is that the more processing stages raw materials pass through, the higher prices can be obtained and the higher value is added with every additional production cycle. Thus, value added to the raw cotton produced in Uzbekistan was about 50 USD per ton, whereas it doubled for cotton fiber and continued to increase exponentially for all the subsequent products such as yarn, fabrics, and garments (Table 1). The highest value added (0.66 USD per unit or 2,000 USD for 1 ton) was estimated for T-shirts produced in Uzbekistan and then exported.

The agricultural stage is the most water consuming stage both due to the type of irrigated agriculture in place and to the deficiencies of the irrigation and drainage networks. According to the irrigation water requirements for irrigated cotton cultivation, cotton consumes about 6,000–8,000 m^3 of water (including leaching) per hectare of land and yields on average 2.6 tons (Rudenko, 2008).[12] However, the real water withdrawal from the Amu Darya in 2006 per each hectare of cotton in Khorezm was about two to three times higher due to deficiencies in the irrigation network and high conveyance losses (water infiltration and evaporation in the canals). Hence, total agricultural water use for cotton production was estimated at 17,729 m^3 per ha or 6,819 m^3 of water per ton of produced raw cotton (see Table 1).

Once the volume of water used at each subsequent industrial production stage, i.e. the industrial water use, was added to the virtual water of the preceding cotton product, the findings showed that the processing of raw cotton into cotton fiber consumed about 1 liter of water per kilogram, rendering the virtual water content of fiber equal to ca. 6,820 m^3 of water per ton (see Table 1). The most water intensive processing stage along the cotton chain (considering only textile production phases) was weaving, or producing cotton fabrics. Industrial water use in the weaving process amounted to 789 liters of water per 1 kilogram of fabric, since the bleaching, washing and dyeing, consumed much water and produced much wastewater.

Table 1. Value Added and Virtual Water of Cotton Products in Khorezm, Uzbekistan.

	Value added (USD per ton)	AWU (m^3/ton)	IWU (cumulative) (m^3/ton)	TVW (m^3/ton)	Water Footprint Value Index (USD per m^3)
Raw cotton	50	6,819	0	6,819	0.007
Fiber	112	6,819	1	6,820	0.016
Yarn	284	7,759	(1) + 0.7	7,761	0.037
Fabrics	313	7,895	(1 + 0.7) + 789	8,686	0.036
T-shirt	2,000	2,074	(1 + 0.7 + 789) + 0	2,865	0.698

AWU, agricultural water use; IWU, industrial water use; TVW, total virtual water.
Source: Based on study results by the authors in 2005–2008.

DISCUSSION

The two approaches applied in the study yielded somewhat contrary recommendations. On one hand, from a purely economic point of view and assuming Uzbekistan will continue producing cotton (which is most likely), the findings of the value chain analysis favor adoption of a value added strategy. It would, therefore, be highly recommended for Uzbekistan to develop industries branching off its agricultural production. Industries producing textiles would enable a shift from the export of cotton fiber to the export of higher priced and higher value added textile products.

On the other hand, water footprint analysis allowed assessing the real water use during the production of raw cotton and for all the subsequent processing cycles. Based on the water footprint findings, one would need to caution water-constrained Uzbekistan against engaging in all textile manufacturing steps, as some steps are too water intensive and wastewater producing. When the two approaches are combined, a hybrid strategy emerges that is both economically desirable and environmentally sustainable. This hybrid strategy would either entail excluding the last, water intensive, steps in the cotton value chain or making them contingent upon state or private investments for wastewater treatment facilities and improved handling of the gray water component.

The study confirmed that the highest water use along the cotton value chain in Uzbekistan happened at the field level through leaching and irrigation related losses.[13] Furthermore, the real water withdrawal from the primary water source, the Amu Darya, was estimated to be two to three times higher than the recommended irrigation norms due to deficiencies of the irrigation network and high conveyance losses along the entire irrigation network. The agricultural sector of the cotton value chain thus holds the greatest potential for increasing water use efficiency and in turn water saving in the water-scarce Aral Sea Basin. Effective implementation of water saving options and irrigation technologies would lay a foundation for land, water, and ecosystem sustainability. For a country such as Uzbekistan, where most of the entire territory is desert and only about 10% is arable, decreasing overall water losses and increasing water use efficiencies particularly in the cotton value chain is necessary in order to safeguard the natural resources base. Present agricultural practices cannot provide a sustainable basis for maintaining livelihoods and the environment. This further necessitates that field level and management practices be improved. This means not only addressing modern technical solutions such as drip or sprinkler irrigation, but also economics oriented (e.g. water pricing) and management oriented solutions (e.g. irrigation scheduling) (Bekchanov, Lamers, & Martius, 2010).

What makes matters more challenging is that agriculture in Uzbekistan is considered a matter of state. The government determines policy for the production of state target crops like cotton and winter wheat, which annually cover about 60–70% of all irrigated cropland. Hence, in the short and medium term, it is unlikely that an option such as increased crop diversification, featuring less water intensive crops, is an option acceptable to the state. However, since irrigated agriculture is by far the main sector driving water consumption in the country, investing in the irrigation and drainage networks can yield immediate improvements in water use efficiency.[14] The existing irrigation and drainage networks in Uzbekistan were developed and constructed to satisfy the needs of the large-scale agricultural production units established during the Soviet Union era. At present, less than 30% of these are lined and only 12% are considered "waterproof" resulting in high seepage losses and rising groundwater tables.[15] Furthermore, the average irrigation system efficiency is also considered low.[16] Recurrent estimations also indicated that less than 30% of the river water extraction even reaches farmers' fields. Rehabilitating and renovating the irrigation and drainage systems, for example, by lining the earthen channels with concrete, could reduce irrigation water losses in the whole Aral Sea Basin by 10–22 km^3 annually, but would require an estimated portfolio of 16 billion USD to cover human resources and materials.[17]

Additional or alternative gains are plausible by reducing the area of irrigated agriculture, improving watershed management, or introducing water-wise technologies, in particular, to decrease agricultural water use (e.g., Bekchanov et al., 2010). Although the potential benefits of water wise innovations are estimated to be huge, the impact strongly varies with the technology and in fact an inverse relationship exists between the water use improvement potential and financial feasibility: the more water-efficient technologies usually are more capital intensive as well (Bekchanov et al., 2010). When for instance existing innovative irrigation technologies would be introduced in the Khorezm region, it could reduce water demands by 1.5–3.0 km^3 of water annually. But this is less appealing to the farmers under the present conditions due to the need for additional investments and added labor costs. The present underdeveloped links between farmers and markets are reinforcing the negative feedback loop and keeping farmers from gaining the necessary capital to innovate.

With an estimated 70% water saving potential, drip irrigation has the highest water saving potential, but there are some accompanying challenges to conversion in Uzbekistan. Due to the high initial investments, drip irrigation pays off for cash crops such as vegetables, but not for major state crops such as cotton and wheat.[18] In addition, changes in water use by farmers at the field level will require agricultural extension services to be broadly expanded. These include effective farm level consultations that both disseminate instruction on modern irrigation techniques and provide financial advisory services to encourage broader adoption. The currently existing support services for farmers do not meet these needs (Bekchanov, Kan, & Lamers, 2009).

Thus, it is worth discussing less expensive and less knowledge intensive irrigation methods. Considerable efficiencies for irrigation water (between 0.4% and 0.9%, annually) could be achieved with a widespread application of improved irrigation methods such as double-sided furrow irrigation, alternate dry furrows, and shorter furrows. These methods are simple to implement and require relatively low investments. A reduction of between 25% and 30% in water use can be expected when employing a laser-guided land leveler (Egamberdiev, Tischbein, Lamers, & Martius, 2008).[19] This tool illustrates one example of how extension services are intended to function. Farmers' cooperatives or extension service providers can make it available to their members, since this equipment is too expensive for individual farmers. Finally, although paddy rice has been a lucrative alternative crop for farmers in recent years, introducing less water intensive crops, including maize or aerobic rice, can lead to substantial savings in irrigation water, albeit with lower yields and thus profits than compared to paddy rice (Djanibekov, 2008).

Keeping in mind that the priority task is to increase overall water use efficiency, this can be achieved by other means, including diversifying the economy by stimulating the domestic processing of agricultural output within the manufacturing or industrial sectors. Previous analysis of the cotton value chain in the Khorezm region underlined for instance the potential for reducing total water use by encouraging the manufacturing of yarn or ready-made garments.[20] Such options have good potential since focusing on maintaining export revenues from cotton, while shifting from the present strategy of exporting cotton fiber to one of exporting higher valued manufactured products, would simultaneously reduce the land needed for cotton production by up to almost 70%. Additionally, such a shift bears the potential to free about 0.5 km^3 of irrigation water annually and reduce the present state subsidies by about 14 million USD. Moreover, despite the low water use (IWU) in the total cotton chain, which is below 10% of total water use, water pollution is an important issue in light of the environmental problems.

The global networks and interdependence that we are familiar with introduce other quirks into a complete analysis. Worldwide, countries try to save domestic water resources by importing products with a high water footprint and high virtual water content, and in turn exporting commodities with a low water footprint and low virtual water content (Chapagain, Hoekstra, Savenije, & Gautam, 2005b). The export of unprocessed or slightly processed agricultural products (like cotton fiber) assumes export of the domestic water resources which have been used (Chapagain et al., 2005b) for the production of these exported commodities. With the export of cotton fiber, Uzbekistan thus exports significant volumes of virtual water, up to a quarter of its share from the Amu Darya in the Aral Sea Basin, or around 7 Gm^3 of virtual water. However, since much of irrigation or blue water is lost or wasted in the agricultural sector, the overall water footprint of cotton production in Uzbekistan was estimated to be as much as 26 Gm^3 of water, almost double of the same indicator as reported by Aldaya, Muñoz, and Hoekstra (2010).

When decreasing water use in water-scarce regions such as Central Asia is held as the goal to be achieved, reducing water intensive crop production and developing processing industries are a wise approach to keeping economic performance constant. However, since these countries have little capacity for treating and reusing wastewater, it is prudent to avoid further ecological degradation from industrial pollutants. New industrial processes have to be chosen with care.

For example, exploring the value chain up to cotton yarn would be feasible and reasonable for Uzbekistan according to the economically oriented value chain analysis and the ecologically oriented water footprint analysis. This does not require much industrial water usage, nor does it produce wastewater, while at the same time bringing in higher export earnings compared to cotton fiber, and higher export earning per unit of water used.

In order to achieve feasible and long-term changes along the cotton chain both for the economy of the individual Aral Sea Basin countries like Uzbekistan, and a healthier overall environment, concerted efforts are required from all the involved stakeholders, each in their field of jurisdiction and expertise. Moreover, some of these stakeholders will have to upgrade and develop additional capacities and skills, like certification, storage, and marketing of the yarn, as opposed to the fiber.

The value of conceptualizing and realizing the integrated water resource management (IWRM) approach in the Aral Sea Basin is a timely and necessary modernization effort to replace the previously used isolated measures on various water use levels. Furthermore, it would offer policy options simultaneously targeting environmental problems stemming from agriculture, industry, and trade structures. Understanding and estimating the real water use by different sectors will help policy makers to develop balanced water resources management options (Abdullaev, De Fraiture, Giordano, Yakubov, & Rasulov, 2009). All riparian states of the Aral Sea Basin will, however, need to exhibit strong political will and commitment to adhere to the developed IWRM options, which can inter alia lead to a reduction in irrigated areas in some countries, introduction of water saving technologies, switching to less water-demanding, but possibly less profitable crops, and developing environmentally friendly industrial production.

CONCLUSIONS

The combined findings of the economic based value chain analysis and ecological based water footprint analysis provide the necessary details and accuracy for better-informed decision-making in order to reach land, water and ecosystem sustainability in the Aral Sea Basin as compared to the recommendations based on the two approaches separately. The added value of the water footprint analysis to the value chain analysis pointed out two options for decreasing water use and coping with water scarcity in the case study region Khorezm, which is representative here for large parts of the irrigated croplands in the Aral Sea Basin. One is to reduce agricultural water use through upgrading irrigation and drainage networks, and introducing innovations that have a high potential for increasing water use efficiencies. The other is to shift water use from the highly water intensive agricultural sector to a more efficiently water consuming industrial sector, but with some caution. Although the development of processing industries in particular holds considerable hope for decreasing overall water use in the Aral Sea Basin, decision-makers need to be cautioned, since some industries create an undesired output – large quantities of wastewater. Under the present conditions, it is therefore suggested to develop the local textile industry in general and the cotton value chain in particular up to cotton yarn, because of its negligible industrial water use and considerable economic returns as a priority, unless decision-makers would promote investments in water treatment infrastructure.

Three other strategies can make significant improvements in Khorezm's water usage reports. First, increasing water use efficiency at the agricultural level through irrigation technology improvements; second, moving toward the production of processed cotton derivatives with higher value added; and third, raising water users' awareness of real water shortage in the region and the real value of water. As an added benefit, once implemented, these measures should help the Central Asian countries to contribute to the Millennium Development Goals of eradicating poverty and hunger and achieving food and water security. Last but not the least, it would also prepare the Central Asian nations and populations to cope with the predicted water shortage problems and learn to adapt to the changed and deteriorated environment that is likely already irreversible.

NOTES

1. Up to 95% of freshwater from the main feeding rivers in Uzbekistan is used for irrigating vast agricultural fields of water consuming crops (cotton, wheat, rice), whereas 2% is consumed by municipalities and the remaining 3% by industry.

2. According to the Regional Division of the Ministry for Macroeconomics and Statistics in Khorezm.

3. The "virtual-water content of a product" is the same as "the water footprint of a product," but the former refers to the water volume embodied in the product alone, while the latter term refers to that volume, but also to which sort of water is being used and to when and where that water is being used. The water footprint of a product is thus a multidimensional indicator, whereas virtual-water content refers to a volume alone. http://www.waterfootprint.org/?page = files/Glossary

4. See earlier work by Rudenko (2008) for a discussion of value-added cotton products as alternatives to raw cotton production.

5. According to calculations by BUIS, the Lower Amu Darya Basin Department of Irrigation Systems.

6. See, for example, Rudenko (2008).

7. The water footprint methodology is detailed in Chapagain and Hoekstra (2004).

8. For example, 1 ton of raw cotton gives 330 kg of cotton fiber and around 540 kg of cotton seed, so the fraction would be 33% for fiber and 54% for cotton seed (see Figure 1).

9. Water, part of which after irrigation seeps into the ground water table, may get polluted by pesticides, herbicides, or fertilizers. Details on the estimation of gray water are reported by Chapagain et al. (2006).

10. Ginning ratio in Uzbekistan stands at 33%, thus the same fraction of agricultural water is assumed to be transferred to cotton fiber.

11. Additional data sources for technical parameters for industrial water use and estimating virtual water content included:

   1. for agricultural water use (crop water requirements and irrigation water withdrawal from Amu Darya) – Regional Division of the Ministry for Macroeconomics and Statistics in Khorezm, and Basin Management of Irrigation Systems (BUIS);
   2. technical coefficients, parameters, and water use requirements at various processing stages – processing organizations along the cotton and wheat

value chains in Khorezm as detailed previously for the study region by Rudenko (2008).

12. According to calculations by BUIS, the Lower Amu Darya Basin Department of Irrigation Systems.
13. Aldaya et al. (2010).
14. For a longer discussion, see Micklin (1988).
15. See, for example, Ibragimov et al. (2007).
16. Irrigation efficiency in Uzbekistan has just a 30–40% efficiency rate (Kijne, 2005).
17. As calculated, for example, by Micklin in Micklin (1988) and Micklin (2002).
18. Vegetable and fruit farming in Uzbekistan are not controlled by a state production system and bring farmers a good source of income, whereas state procurement prices farmers receive for cotton and wheat are often insufficient for investing in infrastructure improvement.
19. Laser-guided land leveling is a process of smoothing the land surface (± 2 cm) from its average elevation using laser equipped drag buckets. Smoothly leveled land surface allows an even and fast distribution of irrigation water, and also more efficient fertilizer application and even seed germination.
20. Rudenko et al. (2008).

REFERENCES

Abdullaev, I., De Fraiture, C., Giordano, M., Yakubov, M., & Rasulov, A. (2009). Agricultural water use and trade in Uzbekistan: Situation and potential impacts of market liberalization. *Water Resources Development*, *25*(1), 47–63. doi: 10.1080/079006 20802517533.

Aldaya, M. M., Muñoz, G., & Hoekstra, A. Y. (2010). *Water footprint of cotton, wheat and rice production in Central Asia* (p. 31). Value of Water Research Report Series, No. 41. UNESCO-IHE. Delft, The Netherlands. Retrieved from http://www.waterfootprint.org/Reports/Report41-WaterFootprintCentralAsia.pdf

Allan, J. A. (1998). Virtual water: A strategic resource, global solutions to regional deficits. *Groundwater*, *36*(4), 545–546. doi: 10.1111/j.1745-6582.4.1998.tb02825.x.

Bekchanov, M., Lamers, J. P. A., & Martius, C. (2010). Pros and cons of adopting water-wise approaches in the lower reaches of the Amu Darya: A socio-economic view. *Water*, *2*(2), 200–216. doi: 10.3390/w2020200.

Bekchanov M. B., Kan, E., & Lamers, J. P. A. (2009). Options of agricultural extension provision for rural development in Central Asian transition economies: The case of Uzbekistan. In *5th Annual Conference Proceedings Research for Sustainable Development*. Westminster International University, Tashkent, May 14, 2009; pp. 72–83. Retrieved from http://uni-hohenheim.academia.edu/ElenaKan/Papers/422591/Options_of_agricultural_extension_provision_for_rural_development_in_Central_Asian_transition_countries_the_case_of_Uzbekistan

Chapagain, A. K., & Hoekstra, A. Y. (2004). *Water footprints of nations. Volume 1: Main report*. Value of Water Research Report Series No. 16. UNESCO-IHE, Delft, The

Netherlands, p. 75. Retrieved from http://www.waterfootprint.org/Reports/Report16Vol1.pdf

Chapagain, A. K., Hoekstra, A. Y., & Savenije, H. H. G. (2005a). *Saving water through global trade* (p. 36). Value of Water Research Report Series No. 17. UNESCO-IHE, Delft, The Netherlands. Retrieved from http://www.waterfootprint.org/Reports/Report17.pdf

Chapagain, A. K., Hoekstra, A. Y., Savenije, H. H. G., & Gautam, R. (2005b). *The water footprint of cotton consumption* (p. 39). Value of Water Research Report Series No. 18. UNESCO-IHE, The Netherlands. Retrieved from http://www.waterfootprint.org/Reports/Report18.pdf

Chapagain, A. K., Hoekstra, A. Y., Savenije, H. H. G., & Gautam, R. (2006). The water footprint of cotton consumption: An assessment of the impact of worldwide consumption of cotton products on the water resources in the cotton producing countries. *Ecological Economics*, *60*(1), 186–203. doi: 10.1016/j.ecolecon.2005.11.027.

Chub, E. V. (2000). *Climate change and its impact on natural resources potential of the Republic of Uzbekistan.* Main Administration on Hydrometeorology under the cabinet of Ministers of the Republic of Uzbekistan. Central Asian Hydrometeorological Research Institute, Tashkent, Uzbekistan.

Djanibekov, N. (2008). A model-based analysis of land and water use reforms in Khorezm: Effects on different types of agricultural producers. In P. Wehrheim, A. Schoeller-Schletter & C. Martius (Eds.), *Continuity and change: Land and water use reforms in rural Uzbekistan. Socio-economic and legal analyses for the region Khorezm* (pp. 41–60). Halle/Saale, Germany: IAMORetrieved from http://www.iamo.de/dok/sr_vol43.pdf

Egamberdiev, O., Tischbein, B., Lamers, J. P. A., & Martius, C. (2008). *Laser land leveling: More about water than about soil. ZUR No. 1.* Center for Development Research (ZEF), University of Bonn. Retrieved from http://www.zef.de/fileadmin/webfiles/downloads/projects/khorezm/downloads/Publications/ZUR/ZUR_No1.pdf

FAO. (2006a). *Commodity chain analysis. Constructing the commodity chain, functional analysis and flow charts. EASYPol.* Online resource materials for policy making. Analytical tools. Module 043, 19. Retrieved from http://www.fao.org/docs/up/easypol/330/cca_043EN.pdf

FAO. (2006b). *Commodity chain analysis. Financial analysis. EASYPol.* Online resource materials for policy making. Analytical tools. Module 044, 18. Retrieved from http://www.fao.org/docs/up/easypol/331/cca_044en.pdf

FAO. (2006c). *Commodity chain analysis. Impact analysis using market prices. EASYPol.* Online resource materials for policy making. Analytical tools. Module 045, 1-20. Retrieved from http://www.fao.org/docs/up/easypol/332/CCA_045EN.pdf

Hoekstra, A. Y. (2003). Virtual water: An instrument in relieving the pressure on the world's water resources? *Stockholm Water Front*, 2(June).

Hoekstra, A. Y., & Chapagain, A. K. (2007). Water footprints of nations: Water use by people as a function of their consumption pattern. *Water Resource Management*, *21*(1), 35–48. doi: 10.1007/s11269-006-9039-x.

Hoekstra, A. Y., & Hung, P. Q. (2005). Globalization of water resources: International virtual water flows in relation to crop trade. *Global Environmental Change*, *15*(1), 45–56.

Ibragimov, N., Steven, R., Evett, Y., Esanbekov, B., Kamilov, S., Mirzaev, L., & Lamers, J. P. A. (2007). Water use efficiency of irrigated cotton in Uzbekistan under drip and furrow irrigation. *Agricultural Water Management*, *90*(1–2), 112–120. doi: 10.1016/j.agwat.2007.01.016.

Kaplinsky, R. & Morris, M. (2002). *A handbook for value chain research.* Prepared for the IDRC, 119. Retrieved from http://www.fao.org/fileadmin/user_upload/fisheries/docs/Value_Chain_Handbool.pdf

Kijne, J. W. (2005). *Aral Sea Basin initiative: Towards a strategy for sustainable irrigated agriculture with feasible investment in drainage* (Vol. 72). Synthesis report. International Programme for Technology and Research in Irrigation and Drainage (IPTRID) at FAO. Retrieved from ftp://ftp.fao.org/docrep/fao/009/a0010e/a0010e00.pdf

Lal, R. (2007). Soil and environment degradation in Central Asia. In R. Lal, M. Suleimenov, B. A. Stewart, D. O. Hansen & P. Doraiswamy (Eds.), *Climate change and terrestrial carbon sequestration in Central Asia* (pp. 127–137). London: Taylor & Francis. Retrieved from http://www.ars.usda.gov/research/publications/publications.htm?seq_no_115=210261.

Martius, C., Froebrich, J., & Nuppenau, E. A. (2009). Water resource management for improving environmental security and rural livelihoods in the irrigated Amu Darya Lowlands. In H.G. Brauch, Ú. O. Spring, J. Grin, C. Mesjasz, P. Kameri-Mbote, N. Chadha Behera, B. Chourou, & H. Krummenacher (Eds.), *Facing global environmental change: Environmental, human, energy, food, health and water security concepts* (pp. 749–762). Hexagon Series on Human and Environmental Security and Peace, 4. Berlin: Springer-Verlag. Retrieved from http://dx.doi.org/10.1007/978-3-540-68488-6_57

Martius, C., Lamers, J.P.A., Ibrakhimov, M., & Vlek, P. (2004). Towards a sustainable use of natural resources in the Aral Sea Basin. In H. Bogena, J.-F. Hake, & H. Vereecken (Eds.), *Water and sustainable development* (Vol. 48, pp. 117–134). Schriften des Forschungszentrums Jülich. Reihe Umwelt/Environment. Retrieved from http://juwel.fz-juelich.de:8080/dspace/bitstream/2128/351/1/Umwelt_48.pdf

Micklin, P. (1988). Dessication of the Aral Sea: A water management disaster in the Soviet Union. *Science*, *241*(4870), 1170–1176. doi: 10.1126/science.241.4870.1170.

Micklin, P. (2002). Water in the Aral Sea Basin of Central Asia: Cause of conflict or cooperation? *Euroasian Geography and Economics*, *43*(7), 505–528. doi: 10.2747/1538-7216.43.7.505.

Müller, M. (2006). *A general equilibrium approach to modelling water and land use reforms in Uzbekistan.* PhD Dissertation, ZEF, Bonn University, Germany (p. 170). Retrieved from http://hss.ulb.uni-bonn.de/2006/0801/0801.htm

Rudenko, I. (2008). Value chains for rural and regional development: The case of cotton, wheat, fruit and vegetable value Cchains in the lower reaches of the Amu Darya River, Uzbekistan. PhD Dissertation, Hannover University, Germany (p. 222). Retrieved from http://edok01.tib.uni-hannover.de/edoks/e01dh08/571076203.pdf

Rudenko, I., Grote, U., & Lamers, J. P. A. (2008). Using a value chain approach for economic and environmental impact assessment of cotton production in Uzbekistan. In J. Qi, & K. T. Evered (Eds.), *Environmental problems of Central Asia and their economic, social, and security impacts* (pp. 361–380). NATO Science for Peace and Security Series – C: Environmental Stability. Dordrecht, The Netherlands: Springer.

Stamm, A. (2004). *Value chains for development policy. Challenges for trade policy and the promotion of economic development. GTZ concept study* (p. 38). Eschborn. Retrieved from http://www2.gtz.de/dokumente/bib/04-0270.pdf

UNESCO-WWAP. (2006). Water a shared responsibility. *The United Nations world water development report* (Vol. 2, 584 pp.). Retrieved from http://unesdoc.unesco.org/images/0014/001444/144409e.pdf

SECTION FOUR ADAPTING TO CATASTROPHE: CASCADING SOCIAL IMPACTS OF THE ARAL SEA DISASTER

From the lead editor's nearly forty years of work on environmental disaster, there is a basic rule of thumb that has never disappointed (Edelstein, 2000, 2004). No matter how severe the direct impacts of a disaster are, at least half the stress comes from the secondary psychosocial impacts involved in dealing with the aftermath. In the case of the Aral Sea, most of the stress is back loaded. The population of Karkalpakstan, particularly those living by and working on the sea, was literally left high and dry, suffering substantial psychosocial and health impacts.[1]

Take the case of Muynak. Why where the people we saw on the streets still there? Were they perhaps stuck with family or other obligations by fear of leaving homes that they are used to living in when they might have no rights elsewhere, or by the lack of an idea about where else to go and what they will find there? In my work in Russia, I have encountered entire organizations that stopped paying salaries yet people continued to show up and work, for lack of a better alternative. At least if they held the firm together, they had a social role and colleagues and some place to go every day. Muynak as a city felt just this way, except here the fish canning factory was fenced up and we were asked not even to snap a photo of the derelict building. When we ate lunch in the only restaurant, a small room attached to a convenience store, chickens ran about the place and omelet was the only menu item. Those remaining in Muynak were holding it together, but barely.

Absent from the streets were the large numbers of residents who have packed up and gone, part of the growing number of environmental refugees worldwide. In this case, the disaster is chronic; there was no flood or tsunami. The water left instead of coming. For the departed, perhaps it wasn't the falling sea, but the abandoned boats, or the closing of the factory that signaled the point of no return. But those departing clearly went through a process of deciding to leave much as those remaining had not. Each faced their challenges ahead. In Karakalpakstan, the neighboring area of Khorezm and a broader stretch of the Aral disaster zone, and most certainly in neighboring Kazakhstan as well, we find a destabilized social situation to which people were forced to make hard adaptations. This section is our window into this place of hard choices.

In Chapter 14, "The Tragedy of the Aral: Counting on Cotton, A Region Loses Its People," cultural geographer and coeditor Astrid Cerny wades into the complex social impacts of the disaster. Cerny, a field geographer with long-term experience in China and Central Asia, was part of our Aral expedition. On several occasions our bus had to pull over so that she could disappear into a yurt that appeared like a mirage on the horizon or down some twisted path into a village center. Here she addresses the social impacts with a strong sensitivity to the cultural diversity of the region, an understanding that there were differential effects for urban and rural populations and careful attention to the substantial out migration of Karakalpakstan's population.

One of the interesting questions about the psychosocial impacts of disaster is how far does it reach, and then, does it reach all equally. We were well aware of the distance of the Aral Sea from Tashkent, Uzbekistan's capital city in the east, and from Samarkand, its historic and cultural capital that borders the eastern and central regions of the country. It is a long and still difficult trip to Nukus, capital of Karakalpakstan. People in Tashkent are far away from the Aral Sea disaster, physically and psychologically. But how do those distances apply to Nukus, more than 200 km to the south of the former Aral Sea shore? Chapter 15, "Reflections on Growing Up in the Karakalpakstan Region," is a short biographical piece by a young Karakalpak, Aziz Murtazaev. No longer living in Uzbekistan at all, he gives us a sense of what it was like to grow up at the edge of disaster, not too close to be engulfed but not too far to be unaffected.

Of course, the quality of life overall is heavily influenced by health. The health impacts of this disaster are one of its major negative legacies. As the Aral Sea ebbed, so did the wealth and institutional resources of the region. Diet was affected. Exposure to windblown contaminated dusts increased. Potable water became scarce and consumption of saline water necessary. Poor sanitation magnified the problems so that disease became a constant challenge. And health care, one of the staples of Soviet society but never highly developed here, largely vanished in the wake of the disaster and independence. Thanks to the work of Doctors without Borders, baseline services have been offered and the attention of the international community engaged. But the health legacy of the disaster remains one of its key effects for the region. Two chapters offer perspectives on this legacy.

In Chapter 16, "An Unhealthy Place to Live: Prioritizing Public Health and Addressing Environmental Contamination in Karakalpakstan," public health expert Ramona Lall surveys the literature and relates it to what she observed on the ground as a member of our exchange team. Lall summarizes the major health problems and challenges and their environmental causes and reflects on some of the steps required to address this humanitarian disaster, a secondary and continuing impact of the Aral disaster, itself. Her familiarity with disasters and her global perspective allow important insights to be made.

By luck, in Uzbekistan, we had the fortune to meet, receive treatment from and include in our discussions an Uzbek physician renowned for his attention to the plight of Karakalpak people. In Chapter 17, "A Physician's Observations of Karakalpak Health," Rakhmon U. Arzikulov, working with Laylo L. Almatova and Saodat Safarova, discusses long term observations about key indicators of health for which there are no reliable statistics. Here we see the Aral story projected onto pregnancy outcomes and the health challenges faced by the young and teenagers. These stark findings complement Lall's broader scope and allow for an appreciation of this important part of the Aral Sea disaster legacy, one that represents a continuing challenge that demands attention.

And then there are the non-human victims of the Aral Sea disaster. The removal of one of the world's most significant bodies of water could not come without tremendous ecological disruption. The showcases at the cultural center in Nukus are full of stuffed or preserved animals and fish that once populated the waters, wetlands and skies of the region, species driven to extinction among them. The ecological carnage was unbelievable. Walking on the Aralkum, the spreading desert that has replaced the sea, one realizes just how extraordinary and traumatic this transformation of ecosystems has been. Of course some species were able to survive by extending their range and finding new habitat. How have they fared? In Chapter 18, "Ecological Change in the Aral Region: Adaptations by the Spoonbill and Black-Crowned Night Heron," we examine this process of adaptation. Bukhara-based Ecologist Mukhtor Muradovich Turaev, whom we had the pleasure to meet there, shares his work on avian species that have populated new emergent lakes created spontaneously in areas to the south and east of the Aral desert by the displacement of water that has leaked into the ground rather than reaching the sea. His examination of the endangered Spoonbill and the Black Crowned Heron provides insight into the adaptive success of Aral Sea survivors.

Chapters 19 and 20 address the cultural heritage of Uzbekistan. Following the work of geographer David Lowenthal (1998), we can understand how heritage has been consciously used as a basis for reestablishing place identity across the world. And Uzbekistan has emerged as a perfect place in point. The architectural legacy of Uzbekistan is remarkably preserved and, in many cases, restored. One truly understands by traveling the countryside that in an earlier epoch, Central Asia was the center of trade in goods, culture and people. The bounty of the Silk Road was translated substantially into architectural monuments and utilitarian forms that have lost neither their fascination nor beauty. Many such monuments were lost as waves of conquering armies swept through, sometimes destroying what they found. Early ecological disasters presaged the current time as salt intrusion from irrigation destroyed many a community's food sources. Yet, overwhelmingly, one is struck less by what has been lost than by how much remains. As we look at the current conditions of ecological disaster, it is important to remember that the value of the threat is in no small part the value of what is threatened.

In Chapter 19, we turn to the World Heritage cities of the west that are front and center to the Aral disaster. Uzbekistan's sweeping effort to capture a global Silk Road identity is based on the integrity of its architectural heritage. In "Environmental Change as a Threat to the Khorezm Heritage," Gavhar Salaevna Durdieva addresses the threats to this heritage stemming from the salinization of the region. A gracious host to the exchange team in Khiva, Gavkhar Durdieva not only participated in a wonderful two-day exchange seminar there at Mamun Academy but also offered an extensive tour of Ichan Kala, the old city, documenting the points she makes in this chapter. She is a technical building specialist with particular expertise in architectural preservation. She grew up in the living museum of the walled old inner city of Khiva, the Ichan Kala, where she continues her work. She approaches her topic not only as a building scientist, but also as a person of place, proudly continuing the lineage of ancient Khorezm, once, not that long ago, a great and prominent independent state of the known world. But, as she describes in detail in this chapter, the vestiges of the broader Aral disaster threaten the monuments of that era.

In Chapter 20, "Whose Disaster is it Anyway? Romancing the World Heritage Status in Uzbekistan," Flavia Alaya provides a critique of the inherent perceptual conflict between Uzbekistan's effort to brand itself as the authentic Silk Road destination, on one hand, and the Aral Sea disaster, on the other. Alaya, part of our exchange team, charts the systematic use of cultural heritage as a consciously selected value for defining and uniting the young state of Uzbekistan. In contrast to the active promotion and search for historical legitimacy for the Silk Road identity, the identity of the now devastated but once heralded Aral region far to the West is in no small part neglected and forgotten. Alaya discusses at length the effort led by Mickey Glantz to seek World Heritage recognition for the Aral Sea as a valiant attempt to address this disconnect and keep attention on Karakalpakstan.

Michael R. Edelstein
Editor

NOTE

1. Some excellent work has come out of Karakalpakstan in conjunction with the overall health efforts of Medecins Sans Frontieres (Doctors Without Borders). See for example, Crighton, Elliott, Upshur, van der Meer, & Small (2003).

REFERENCES

Crighton, E. J., Elliott, S. J., Upshur, R., van der Meer, J., & Small, I. (2003). The Aral Sea disaster and self-rated health. *Health & Place*, *9*, 73–82.

Edelstein, M. R. (2000). Outsiders just don't understand: Personalization of risk and the boundary between modernity and postmodernity. In M. Cohen (Ed.), *Risk in the modern age: Social theory, science and environmental decision-making* (pp. 123–142). New York: St. Martin's Press.

Edelstein, M. R. (2004). *Contaminated communities: Coping with residential toxic exposure* (2nd ed.). Boulder, CO: Westview Press.

Lowenthal, David. (1998). *The heritage crusade and the spoils of history*. New York, NY: Cambridge University Press.

CHAPTER 14

THE TRAGEDY OF THE ARAL: COUNTING ON COTTON, A REGION LOSES ITS PEOPLE

Astrid Cerny

ABSTRACT

This chapter briefly discusses an often overlooked impact of the loss of the Aral Sea, the social costs to the population in the surrounding areas. While much focus has been on the shrinkage of the Aral in the past half century, the story is much older and includes a diversity of people, ecosystems, and productive ways of living in a harsh climate. This chapter is an overview of how ecological parameters have influenced life for the local population in Karakalpakstan. The research is based on a short site visit and tests ideas for a longer project that will look at who is migrating for employment, based on what criteria, and the dynamics of that choice. Seminal writings about human activity around the Aral when it was full, along with recent reports on livelihoods, health, and migration patterns are reviewed. The chapter supports the idea that migration can be an adaptation to climate. Migration is overall a negative social sustainability indicator for Karakalpakstan.

Disaster by Design: The Aral Sea and its Lessons for Sustainability
Research in Social Problems and Public Policy, Volume 20, 223–250

ISSN: 0196-1152/doi:10.1108/S0196-1152(2012)0000020025

INTRODUCTION

We have seen that the virtual disappearance of the Aral Sea from the landscape of Central Asia is an environmental change of epic proportions. The drainage of the lake happened largely with good intentions: to water a crop of significant economic value and which provided an important raw material for making textiles that were used throughout the Soviet Union. The growing of cotton provided work opportunities for many families and communities for several decades, and Karakalpakstan was integrated into the productivity of the utilitarian worker states that the bureaucrats in Moscow envisioned for the Soviet Union. Unfortunately, therein lays the rub. The 5 year plans were the standard strategy for administratively organizing agricultural production on paper and then orchestrating it in reality. But, as we now know, they operated in rigid lockstep coupling with the promotion structure for bureaucrats and were, as a result, also fraught with deception and falsified reporting. Worse still, they were a myopic planning tool (see Edelstein chapters, Chapters 9 and 28).[1] The goal, well into the 1980s, was increasing output, expanding agriculture in the Aral Sea region, even as the sea was visibly shrinking.[2] While some outside observers initially believed that the loss of the Aral Sea was an unintended consequence, something akin to "they didn't know better," we now understand that the planners, mainly the Russians in Moscow, as well as local bureaucrats in Tashkent knew very well, as early as the late 1950s that the irrigation demands for cotton would deplete the Aral Sea.[3] They were willing to make this sacrifice in the interests of the greater success of the Soviet Union and to obtain more of that highly desirable commodity, cotton. From his perspective as an engineer and as a native son of the Aral region, Kamalov writes passionately about how the planning for output was coupled with a fondness for grand scale projects.[4]

In this chapter, we will probe the consequences of the planners' decisions in the Aral Sea region for the people who live in Karakalpakstan, the region surrounding the lake on the Uzbek side, and Khorezm, the region south of Karakalpakstan that shares use of the Amu Darya with Karakalpakstan. For the Aral Sea's loss is not simply the loss of a body of water. The effects include social impacts to families, their livelihoods, and the fabric of rural society, which applies to 70% of the country's population. The fate of Karakalpakstan seems linked inexorably to the condition of the Aral Sea, but is also bound by its ethnic composition. This research is a relatively new and ongoing body of work for the author, so it reflects initial observations from a tour of the country blended with existing research on nomadic, especially Central Asian cultures and the stresses of adapting to global environmental change. In particular, my remarks will address the modern realities of Karakalpak or Kazak people, reviewing how migration patterns have changed and what that has to do with adapting to changing ecologies.[5]

The very fabric of society has been forced to change since Uzbekistan's independence as a result of two major forces, one precipitous – the collapse in Soviet style social services, and one creeping – the poisoning of soil, air, and water as the Aral Sea disaster unfolds. By some estimates, up to 5 million people are currently affected by blowing sand and salt residues alone.[6] There are reports of dry villages with no water, and villages where everyone has left, migrating out in search of employment.[7] In some areas, water is a constant topic of conversation, as people discuss what the weather and the irrigation channels are likely to deliver to their farms.[8] Unemployment is rampant, with underemployment in seasonal, underpaid, or casual labor as well as incomplete government statistics masking how pervasive labor idleness is.[9]

Poor healthcare and education systems destabilize the well-being of people today as well as the future. This dramatic upheaval of communities had already begun in the 1980s, when the first fisheries dried up.[10] It continued in the 1990s with the massive outmigration to Kazakhstan, addressed below, and the departure of ethnic Russians. With education and healthcare systems in disrepair, grave health consequences of several dimensions continue to unfold and challenge those who are working to stem the suffering.[11] The pervasive and difficult task is to stop the problem of blowing salts. These threaten human and livestock health directly, call into question the future viability of agriculture in the area by blanketing fields with residues toxic to plants and further exacerbate the problem of livability.

This chapter thus approaches the Karakalpakstan region from three angles: the historical record, recent livelihood and social indicators, and evidence of migration. The evidence will suggest the ways in which social and economic sustainability of the entire Karakalpakstan region is in the balance. One-sided economic decision-making contributes a cause; the fate of the Aral Sea contributes the effects. There is instability for health, education, and the total population at large when those who are educated, young, and able have left by the tens of thousands. Fishing and hunting as professions have virtually disappeared with the species that once supported them abundantly. Livestock breeding and herding, for which the region was once ideally suited, are activities almost impossible to sustain. With this chapter, by considering the richness and vitality of life before and after, I hope to contribute to a discussion of what we, all of us, learn from the Aral Sea disaster.

ADAPTATION HAS BEEN CONSTANT SINCE ANCIENT TIMES

> These circumstances incontestably prove that Lake Aral has formerly been a much larger and deeper body of water, and a very few figures would show that about double the supply of water which it now receives would soon again raise its level to overflowing, while, on the other hand, if the present supply were cut off, about ninety years only would suffice to dry up the lake entirely.
>
> (Wood, 1876, p. 114)

Living in the twenty-first century as we do, inundated with media attention to human-environmental issues, the immediacy of floods, famines, earthquakes, and other crises are easily etched in our minds through television or the Internet. We can be transported to faraway places in our mind's eye, and we think we see the story as it is. But what is real today has its own history. Despite how we view it and think we understand, it is in reality a summation of all that has come before. And so it is with the Aral Sea Basin. It is drying out now, and we are concerned it is forever. This excerpt above, from Wood's tales of an overland journey, surprises us with his perception of a lake that is susceptible to great changes in volume, and, as much of his book examines, the concerns of the local people about getting flooded out. The Aral Sea then, or now in the form of its own desiccated lakebed, looms large over the region and the populations who try to make a living around it.

To understand the region at a deeper level, it behooves us to understand the composition of the ethnically rich population today and the significance of population movement in the historical context of the region. We should also better understand that the inhabitants of the region are not witnessing a change in their water tables for the first time, nor are the scorching temperatures and dried up channels today something they are adapting to for the first time.[12] Moreover, the irrigation systems have been built, rebuilt, destroyed, and abandoned for millennia.[13] What is new are the overlay of the Soviet agricultural production template on one hand, and the political control exercised on the information available to people throughout the twentieth century on the other. The withholding or hiding of information pertaining to Aral Sea hydrography has continued after independence, solidifying the pre-existing do-nothing approach over the resilient adaptation approach that existed previously.[14]

These factors have brought the Aral Sea region to the edge of collapse, as Jared Diamond aptly named situations (across time and continents) where people have used up their available natural resources and thus collapsed their own basis for existence (Diamond, 2005). McLeman's analysis of ancient and modern abandoned settlements tells a very similar story. His analysis includes Karakalpakstan as one of over 200 examples of settlement abandonment. Classifying the "settlement abandonment process" in three stages, Karakalpakstan meets the criteria for stage three, abandonment, where "autonomous outmigration grows, temporary migration becomes indefinite, population decline accelerates and institutional decisions (or the lack thereof) become critical in determining outcome."[15]

To review the literature on the Aral Sea disaster, and to examine the details of the many ways in which life in Karakalpakstan is faced with collapse, is to be confronted with the question, has the point of no return been reached for the Aral? There is a historic scale to what we are witnessing in our lifetimes, which is tragic. The fourth largest body of fresh water has shriveled to salty, peanut-shaped fractions of its former self, and the local population has had the rug of sustenance pulled out from under them. The tragedy is anthropogenic; it did not need to happen this way. The increments of human manipulation in recent decades generated responses from nature – over and over again. Humans built faulty irrigation systems, and the soils became waterlogged, and then salinized. Humans dusted the land with defoliants, and the contaminants linger silently, year upon year. This is a reversal of fortune in the Aral region, where previously the ecology of the wetlands, the lakes, the swamps, and the fields guided people in making decisions and adapting the abundance (or shortage) into their way of life. Ironically, in attempting to manipulate nature for greater abundance, humans engineered unintended consequences of deprivation and loss. Much of Central Asia is considered harsh in climate and difficult to live in, yet the ancestors of today's population built civilizations in the desert, inventing agricultural methods that were so successful, the ideas were exported across Eurasia.[16]

The tipping point will not be crossed if the existing framework of growing more (more cotton for export and more wheat or rice to feed a growing population) is challenged and dismantled. Before the 1950s, the scales had not tipped. It remains to be seen whether important lessons are learned from the Aral Sea disaster, and the do nothing response of delay, more studies, and inaction is replaced by sustainable solutions in planning and technology. To better understand what actions make sense, a holistic approach to understanding the Aral is a good place to start.

ECOSYSTEMS BASED ON WATER, HUMAN LIFE BASED ON MOVEMENT

If we expand our viewpoint to include a larger historical scale, the geological record of marine fossils and other evidence substantiate sea level fluctuations in the tens of meters over at least 10,000 years.[17] Modern scientific research converges with the written accounts of geographer travelers to confirm that to some extent the ecological vulnerability of the people today has been there since ancient times. So has human adaptability.[18] The Ust Yurt Plateau between the Aral and the Caspian Seas has been ankle deep or more in water, and Arundo reeds proliferated in what were not desert but wetland conditions.[19] Even well into the twentieth century, the Amu River delta could flood over an area of 7000 km^2, creating new temporary lakes in the process (Schlueter et al., 2005).

During times of stable waters, the local people (including Karakalpaks, Kazaks, Turkmen, and Uzbeks) expanded both agriculture and free range livestock herding with the help of irrigation, but not without new risks. For centuries, particularly during the middle ages, the oases and fortified settlements along the Amu Darya and many other waterways were vulnerable to raiding and utter destruction of their carefully cultivated fields and their increasingly elaborate irrigation canals and hydrologic ingenuity.[20] Here as elsewhere in the steppes, the Mongols and other nomadic tribes came to raid; they knew that to destroy the irrigation systems was to destroy the settlement.[21] Yet the people stayed. What supported them was the water – the availability of clean, fresh water for themselves, their livestock, their growing plants. The desert steppe allowed for grazing, the wetter areas allowed for irrigated agriculture. Migration was a choice based on an understanding of the ecosystems, and, I would assert, a respect for them. Today we have gotten to a place where those who speak on behalf of the water, or nature, are silenced or ignored in the name of economics.[22]

In sum, the history of large water in the region, both naturally flowing and harnessed by humankind, has been intimately tied to the ability of civilizations to thrive, settlements to prosper, and individual families to provide for themselves.

ETHNIC COMPOSITION OF THE REGION: HISTORICAL AND CONTEMPORARY

Central Asia has the distinction of having been, until fairly recently in human history, a vast geographical area inhabited largely by migratory peoples, mingling or not mingling over the centuries, who then found their territory taken over and carved up by outsiders, primarily Russians, both under the late Russian empire and the Soviet Union. United in name as republics under the Soviet Union, where borders were domestic, today's international borders of the independent Central Asian states divide land and people more rigidly than before. This has led to the phenomenon of many ethnic groups being intermingled and represented in each of today's independent countries, even though the country's name would suggest otherwise. For the -stan suffix means "land," such that Tajikistan is the land of the Tajiks, and Uzbekistan is the land of the Uzbeks, and so on. Uzbekistan is not uniformly made up of Uzbeks, however.

Uzbekistan, like its surrounding neighbors, is an amalgamation, a nation state coalesced out of ethnic groups in recent modern history. Today, the Uzbeks are indeed in the majority, but they share their nation with Kazaks, Karakalpaks, Tajiks, Khorezmians, and even Koreans – each population having arrived in this territory, migrated in, at some point in the past. From early eras of human inhabitation to the mid nineteenth centuries, the territory of what is Uzbekistan today was part of a loose steppe and desert terrain inhabited here and there by large tribes and their sub-branches.[23] Some were at war, some at peace. Some traded, some did not.

Kazaks, for example, differentiate three large tribes and quite a few sub-branches. Many Kazaks can name seven generations of ancestors, and begin a new conversation with a stranger or a guest by asking what sub-branch the person belongs to. The large Kazak tribes (upper, middle, and lesser) extend from Mongolia through China and through all the current Central Asian republics. The Karakalpaks are similarly structured, with two divisions, and tribes, sub-groups, and clans branching off within these.[24] The way to determine ethnic identity varies; indeed many Central Asian peoples only formed their present appellation in the last hundred years.[25]

Our best aid in distinguishing tribes and identities until recently has been languages spoken. Very recently, genetics is helping us to understand the rich patterns of ethnic variation across the entire region of Central Asia.[26] As politics and nationalistic ideas are taking hold in Uzbekistan, Kazakhstan, and Turkmenistan, shaped and molded by the ideology of their autocratic leaders, the science of genetics may quietly undermine at least part of their agendas, and could eventually, prove helpful to the ethnically diverse regions for asserting their rights.[27]

From Tribes to Nationalities

The power to make decisions in Central Asia became concentrated in the hands of a few during the twentieth century. Tribal belonging was no match for the expansive Soviet regime. If we understand Uzbekistan as the land of the Uzbeks and Karakalpakstan as the land of the Karakalpaks, and remember that Uzbekistan was an outlying republic in the Soviet Union, we can begin to appreciate the vulnerability of the ethnic minority within a minority. The hinterland of the Soviet Union included the Uzbek SSR, and the hinterland of Uzbekistan today is still cotton producing Karakalpakstan. Even a passing familiarity with the many examples in world history where "minorities" do the bidding for their colonial masters – or the land they live on is seen as a treasure chest of natural resources to be exploited – are enough to reveal the "how could they" of why the Soviets planned to let the Aral Sea run dry.[28] This is the underlying importance to understanding the nuances of ethnic composition in the region, and the predicament that Karakalpakstan is in. Positive social and environmental change in the Aral Sea region is as tied to an appreciation of Karakalpakstan's ethnic composition as it is to the political willpower to manage water differently.

KARAKALPAKSTAN'S MISFORTUNES

The Aral Sea region was chosen in the name of economic development by faraway decision-makers, and it is a poor, underdeveloped region today. It has been forsaken, even as cotton, wheat and rice are still eked out of the impoverished soils. Karakalpakstan today is ranked fourteenth out of fourteen regions in economic development.[29] There is hopelessness in the people and salt in the drinking water. People, with a few ethnic distinctions, are leaving in droves and not coming back. Those who remain have few opportunities; poverty in the region was 37% in 2003, the most recent year for which reliable figures could be confirmed.[30] The political will to save the Aral Sea has not been there, though there was no shortage of money spent and studies written on "what if" scenarios to try it.[31] Indecision and delay mark the passage of time, as even the studies in our volume show the many competing ideas and technological solutions, each proposing ideas for solving a different aspect of the crisis that has befallen the Aral Sea region.

Much more was lost besides the Aral Sea in the past several decades. Human civilization was compromised as an exodus of the best and brightest occurred, and much biodiversity was lost. Flora and fauna have been decimated as ecosystems changed, and knowledge disappeared with the people who held it. While the planners calculated the export value of cotton, they forgot to calculate for these impending losses, and many other subsidiary costs. The following sections describe some of the earlier and ongoing human impacts unfolding from the one-sided economic calculation that was made in the 1950s.

LIVING IN THE ARAL SEA REGION

Less than 100 years ago, the region surrounding the Aral Sea to the east, south, and west still supported the descendants of a population that had lived with the risk of extreme ecological changes over time. Karakalpakstan supported ethnic groups that identified more with farming (Uzbeks) and nomadic herding (Karakalpaks, Kazaks, and Turkmen) without running out of water. Fishing and hunting were also practiced, not only in the Amu Darya Delta, but on the many small lakes that dotted the landscape. The Ust Yurt plateau and the Kara Kum desert regions were too dry or too boggy at times to support farm fields, but they were ideal for extensive grazing of cattle and sheep. Thus, here as elsewhere in remote parts of Central Asia, farmers and herders created their own livelihoods and support structure.[32]

According to Wood, there is evidence from the accounts of travelers since at least Herodotus and Pliny that such populations as existed in this stark region did so on terrain whose water balance was constantly shifting. Particularly in Khorezm, the region to the south of Karakalpakstan, the local people went to great lengths to control the river. The Amu Darya was diverted by villagers for vineyards, crops and orchards, and into channels that both supported the settlements, and inundated them. The Turkmen nomads herded a region from the Caspian shores to Khorezm in the south and the Kizilkum, or red desert, to the east. Wood makes repeated reference to the nomads supporting large, robust herds of camels, cattle and flocks of sheep.[33] This industriousness and adaptability of the local people, who knew the ecology of the area intimately, began to change when cotton production expanded.[34]

That picture looks totally different today. Because fodder and water are scarce, livestock are difficult to keep, much less maintain in large herds. With the expansion of cotton production, outsider decision-making and collectivized agriculture, individual knowledge, and initiative were necessarily compromised. Knowledge about water storage and efficiency was lost, and people lack even a linguistic concept for "saving" water.[35] The cost of water was reduced to nothing in Soviet days, and still costs far too little today, providing no incentive for people to value it.[36]

Even as the water sloshes through the irrigation channels and soaks the fields, initiatives to improve water management have not taken hold, with people retaining their hierarchical understanding of their rights to water (and the idea that it is someone else's job to do something).[37] A Water User Association model has proven problematic if not ineffective.[38] Worse still, the pervasive use of persistent organic pollutants (POPs) has not been safely managed, nor the public in contact with them adequately informed of the dangers of handling these materials.[39] The life-giving quality of agricultural products is severely compromised today, and the life-supporting function of healthcare, education, and institutional integrity are a shambles.

From a short-term perspective, it appears that Karakalpakstan is in a downward spiral – though it has been cyclical in the past. If we rely on eyewitness accounts as historical sources (such as Major Wood reviewing over one thousand years of travel accounts), we understand that this region is actually wealthy and adaptable. Fishing has died out, but it is not the first time. Much agricultural land has been salinized, but as recently as the nineteenth century, parts of Karakalpakstan have been unproductive or invaded by blowing sand. The Amu Darya, its tributaries, and the irrigation channels have been overused before, have dried out before, have become replenished before. People have been ingenious about diverting the Amu Darya to suit their purposes of providing towns with drinking water and irrigating fields.

Whether Karakalpakstan is in a downward spiral (for example towards water wars), or merely in the downward slope of a cycle that will eventually reach its nadir and slope upward to abundance again, remains to be seen. The region is autonomous in name, yet firmly embedded within Uzbekistan, which in turn is politically wrapped within Central Asia. Central Asia must grapple with water management sooner or later, as being an upstream or downstream country is already becoming a source of power conflicts.[40] The fate of the Aral Sea has been contentious for decades but seems sealed at this point in time. The fate of Karakalpakstan and the people who live there is still in the balance. There are signs that a certain trend beginning in 1960 remains negative. That trend has the people and the sea ineluctably intertwined.

THE IMPORTANCE OF THE ARAL SEA TO REGIONAL LIVELIHOODS[41]

The Aral Sea once had two major ports, Aralsk in the north, now in Kazakhstan, and Muynak in the south, now in Uzbekistan. Both had thriving fishing industries, with trawlers, high capacity fishing boats of various sizes, and employing up to 60,000 people in the heyday of the 1950s.[42] The cannery in Muynak was producing 20 million cans of fish a year in those days.[43] Very quickly though – after those decisions to increase the hectares and the irrigation capacity devoted to growing cotton in Karakalpakstan were made, the tides began to turn on the sea itself, and with it, on the fishing industry. The fishing industry hung on into the early eighties, even when, incredibly, the Soviets imported fish for the local factory to can after the stocks in the Aral Sea were depleted. Eventually, the collapse of the fishery resulted in over 100,000 people losing their jobs.[44] Fishing villages were abandoned as people left by the tens of thousands.[45] But it was not only Muynak that was affected. The shrinkage of the sea has affected the water salinity, depth, and the hydrological cycle itself, thus impacting the productivity of agriculture in the regions south and east of the former lakebed. As the sea retreated north, the problems fanned out to the south.

Muynak, when we visited, was an iconic town of crumbling prefab apartment blocks and white houses with sky blue window trim, lining a dusty street with no traffic. The government building was the busiest in town, a trickle of people walked in and out with papers in hand, to get signed, to turn in. The glare from the sun was pervasive, and the shade trees few and far between. The town has seen better days. One can only imagine the trucks rumbling through, laden with a cargo of canned fish, prized in faraway markets, as one stares through the crooked, thin metal gates at the dilapidated and abandoned buildings of the town's erstwhile employer number one. No pictures, please, we were told, it is not allowed.

The fish cannery was a major source of employment for the women of the region, as we learned from browsing the bulky photo albums in the town museum. The women were employed on the assembly lines while men worked the fishing boats. Huge albums filled with black and white photos, hundreds of photos, showing piles of fish, busy workers and well-fed townspeople. The walls of the museum are hung with large-size framed photos showing men on their boats in all seasons, knee deep in their catch of the day. Surrounding the town, we learn from the rest of the museum's displays, was rangeland. Cattle were kept for ample access to dairy products, and wildlife was hunted as a traditional basis to the diet. Most of the people up here have been Karakalpaks and some Kazaks.

The most comprehensive study of the Karakalpak people, recently appeared in book form, has been an ongoing project of personal interest for David and Sue Richardson.[46] In their estimation, the occupations were traditionally divided among the different branches of the Karakalpak people. One branch was engaged in irrigated agriculture while another tended to cattle, while yet another more northern branch was associated with fishing.[47] In a desolate region, this division of labor ensured cooperation and access to a more diversified diet and materials for their homes.

Today, the employment portfolio has shifted. Fishing exists but is a marginal existence based on small lakes, some man made. Lucky are those with a government job. Although underpaid, theirs are considered stable jobs in a region where large families are common and many people find no work at all. The largest sector continues to be agriculture, but attrition comes from three directions: the days of universal employment are long gone as the collectivized structure has crumbled, the declining yields of cotton means less workers are needed to pick it, and the switch to rice or wheat is less labor intensive, again employing less people.[48] Whether fishermen or agricultural workers, they were the backbone of the local economy, and with the drying up of the Aral Sea, the stability for their livelihoods deteriorated sharply, and for all the men and women employed in fisheries, disappeared completely.

KARAKALPAKSTAN TODAY

International development and humanitarian agencies, working in Uzbekistan, have focused their activities since the 1990s on health and employment generation. Their approach in developing countries is generally through promotion of small business, based on ideals for a thriving market economy. In Karakalpakstan, the development work has had a different focus. The health of the entire population, and most disturbingly, new mothers and children, has been so severely compromised due to the effects of poverty as well as the poor state of water and dangerous dust blowing off the Aral Sea bed that Doctors without Borders have been active there for over 10 years. Their 2003 report, *Karakalpakstan: A Population in Danger*, spells out the grim reality of living in an area where the water cannot be trusted and the soil is contaminated. The people, they reported, were in dire need of mental health and other support services.

While the presence of this level of medical support is reassuring on one level, it is disquieting on another. The organization's ongoing presence speaks to an unabated crisis and a compromised society. An unhealthy adult population with low capacity, and youth with an incomplete education are difficult to train and put to work in higher skilled jobs. Recognizing that human capacity is low, UNDP continues its presence in the region by funding directed projects to alleviate poverty and provide meaningful sustainable work opportunities – the underlying contributing factors to the health crisis. These improvements are happening with full support and in cooperation with the government of Uzbekistan (Embassy of Uzbekistan to the United States, 2012).

UNDP has several recent projects in the area, recognizing the need both to build capacity and to foster occupations that make sense. A project completed in 2005–2006 sought to improve living standards and empower communities, with direct funding support from the Uzbekistan government and the European Union.[49] Ironically, one new project is called Sustainable Livestock Production and Pasture Management. While teaching modern techniques of livestock breeding, this project also teaches skills that were known to the Kazaks over a hundred years ago (MAWR, GEF, & UNDP, 2009).

When considering the total picture for aid and development in Karakalpakstan as reported for over a decade now, one gets the sense that the blowing toxic sands and the salinized soils are spreading faster than solutions. The land has no way to support the people. As the ecological problems continue to worsen, the same health issues reappear and require treatment. Without institutional support and investment from Tashkent, the agricultural sector is limping along and the people of Karakalpakstan suffer silently.[50] Political criticism is not much tolerated domestically, and attempts to agitate for more freedom or independence in Karakalpakstan have not been well received.[51] Lately, there is some discussion of and preparation for oil and gas drilling in the Aral Sea bed. This would be one explanation for the lack of political interest in seeing the Aral fill with water again, but it does not engender hope that the local people will be the beneficiaries of a boom in employment.

There is no shortage of people in Karakalpakstan, but there is a shortage of educated, trained and healthy workers. Drilling is highly technical and physically demanding work and based on examples from China and elsewhere, skilled workers will migrate in to do that work. What we currently see happening in Karakalpakstan instead, is an ongoing and persistent trend in out-migration, to be discussed in greater detail below.

For now, the situation in Karakalpakstan remains an area for further study and needs deeper attention by the research community. For example, population figures for Karakalpakstan vary, and are generally retrievable only in round numbers. In part this is due to the fact that no census has been done since 1989, when the last Soviet census was taken. We will know exactly what is going on when we benefit from transparency in health and population data, and when the need for more on-site research has been satisfied.

TWO REPUBLICS, UNEQUAL OPPORTUNITIES

We have looked at the history of the Aral Sea Basin and the modern conditions for the people who inhabit the former delta. To really understand how differently this region functions within Uzbekistan, a comparison of demographic data illustrates the basic problems. Table 1 compares a set of indicators for Karakalpakstan with Uzbekistan as a whole. Using the CIA World Factbook's statistics as a guide, we see that Karakalpakstan accounts for less than 6% of the total population of Uzbekistan, with a population of 1.6 million almost equally split between Karakalpaks (518,415), Uzbeks (529,720), and Kazaks (436,050) (CIA 2012).

What is more troubling, and where the inequalities of being born in Karakalpakstan become apparent, is the mid-section of the table. We saw that employment is chronically unstable in Karakalpakstan in recent decades. The destabilization has become systemic in addition to being economic. The loss of livelihoods, such as fishing, that supported the population with high quality food and a balanced diet is reappearing in the heath outcomes of the population. Life expectancy is almost ten years shorter for both males and females here than for Uzbekistan as a whole. The birth rate is higher but both infant and maternal mortality are high, too. The difference is striking numerically; the Aral Sea region has the highest child mortality rate in the country (75 children per 1000 newborns). Compared to global rankings of nations for child mortality, Karakalpakstan is similar to Algeria, ranked 63rd out of 172 countries ranked in the world.[52] The

Table 1. Karakalpak and National Demographics Compared.

	National	Karakalpakstan
Population	28,128,600	1,615,000
Ethnic population	Uzbek (80%), Russian (5.5%), Tajik (5%), Kazakh (3%), Karakalpak (2.5%), Tatar (1.5%), other (2.5%)*	Karakalpaks (32.1%), Uzbeks (32.8%), Kazakhs (27%), other (8.1%)**
Languages	Uzbek, Russian, Tajik, other	Karakalpak, Uzbek, Russian
Life expectancy (male)	69.4*	60.6**
Life expectancy (female)	75.7*	67.7**
Birth rate	17.43/1,000*	30/1,000**
Maternal mortality rate	30 deaths/100,000 live births*	120 deaths/100,000***
Infant mortality rate	21.92 deaths/1,000 live births*	75 deaths/1,000****
Net migration rate	−2.65 migrants/1,000 people*	−2.18 migrants/1,000

Prepared by author.
Data Sources: **CIA world factbook 2011*, ***JDA International* online; ***UNECE; ****Columbia University.edu.

maternal death rate is even more disturbing, about 120 women per 100,000 – comparable in ranking to Ethiopia and Equatorial Guinea which are ranked 14th and 15th out of 222 countries measured.[53] Uzbekistan as a whole ranks much better on maternal health and safety, 93rd out of 222. These comparative indicators support the assertions in the literature that the collapse of the healthcare system, in combination with the loss of livelihoods for so many people, has left Karakalpakstan in desperate straits. Rehabilitation of the Aral Sea is undoubtedly of secondary concern to a population that wants immediate relief from its struggles for survival.

The final indicator in the table gives us a segue into the last section of this chapter's discussion, out-migration of the people of Karakalpakstan as a strategy for finding work. The higher rate for Uzbekistan as a whole needs further research, but may be explained by factors such as skilled labor leaving urban areas, students going abroad to study, and established routes for temporary labor. What is known about the rate is that a negative migration rate has been in effect since 1989, from the very beginning of the collapse of the Soviet system. In general, it measures the following reality: that, in Karakalpakstan, outmigration of individuals and families exceeds the natural population growth rate.

UNFAVORABLE CONDITIONS PROMPT MIGRATION

Like several Central Asian republics, Uzbekistan as a whole has seen economic hardship following the collapse of the Soviet Union. Labor data for most of the newly formed states reflect that independence has been good for Kazakhstan, rich in oil and gas, but more turbulent for the others. Kazakhstan and Russia are both destinations for labor migration from the poorer neighboring Kyrgyzstan, Tajikistan, and Uzbekistan.[54] In Karakalpakstan, it also reflects – besides the regionally lower wages – the more pointed realities of the collapse of the Aral Sea fisheries, the decline of agriculture, healthcare and education systems. These have contributed to robust trends in labor and ethnic migration, which will be explained below. Some estimates put the migration rate at an average of 4000 people per year, but upwards of 6000 during the drought years 2000–2001.[55] Figures for the total number of people that have left the region since the 1990s vary, one estimate counting 50,000 (UNHCR) while the Uzbek government counted 63,000 for that time period.[56]

The decision to grow more cotton and the ensuing Aral Sea disaster has had consequences for the social stability of the region and the livelihoods of the people. What began with the loss of the fishing industry from the 1960s onward was compounded by unemployed agricultural workers and a growing, but underemployed, population. The loss of incomes and the impoverishment of the diets made people more susceptible to poverty-related illnesses. The poor health of women of childbearing age is a notable, ongoing concern.

The region is remote and has always been sparsely populated. The local people were adapted to the continental climate and had arranged their social organization along tribal lines with subgroups inhabiting certain terrains and involved with particular professions, primarily herding, farming fishing and hunting. In the past, these professions were strongly bound by the limitations of the landscape. In the twentieth century, planning and a concern for participating in global economics sought to override the limits of the landscape. The Aral Sea was the biggest victim of this flawed thinking. Throughout human history, Karakalpakstan has seen migration in, out, and around.

MIGRATION OUT OF KARAKALPAKSTAN

The last section addresses how migration remains a characteristic strategy of the region, how the attributes of migration have changed, and what that tells us.

Leaving in Droves: Three Migrations

There are three kinds of human migration we need to understand about Karakalpakstan. All are based on practical considerations of the local people for how best to meet their daily needs and, at all times, to ensure the survival of their families. The first, landscape adapted migration is the classic seasonal migration with livestock for food and water that we associate with nomadism. The second, ecologically adapted migration results from the loss of the Aral Sea. And, the third, economically adapted migration, arises from a still unfolding set of factors. All three modes of migration share some attributes. Each should be understood as distinct, for they are all three practiced by different populations.

A Geography of Containment

Because migration is always relative to the space perception of the people exercising their vision for mobility, before it manifests in behavior or a cultural norm, it is worthwhile to remember the location and limitations of Karakalpakstan on the map.

The vastly larger size of the autonomous republic, compared to the other *viloyat,* as the other regions of Uzbekistan are called, gives an indicator that something is different about the terrain here. It is desolate and sparsely populated. The distances are huge. Karakalpakstan encompasses the far northwest of the country, its angular edges forming a border to Kazakhstan on three sides. That border, one can surmise, was once drawn on the map in an office far removed, with a straight edge and a pencil, by a person with limited time and even more limited appreciation for following either contours of landscape or habits of culture. The eastern border slashes straight down through the Ust Yurt Plateau, the former swamp, now dusty livestock rangeland between the Caspian and the Aral Seas. This places a single plateau half in Uzbekistan and half in Kazakhstan. The northern border's line segments cross the former Aral Sea, then angles back south through the Kara Kum, the black desert, for an internal border with Novoiy *viloyat*. The southern reach of what are the borders both of Uzbekistan and Karakalpakstan are more fluid, mostly following the meandering contour of the Amu River or the delta, and suggesting a more thoughtful pencil stroke, for a shared border with Turkmenistan.

Landscape-Adapted Migration

These domestic and international borders establish limitations on the local people today that seem to be stricter all around but also more porous to the north.[57] This fact of enclosure within nation-states is a struggle wherever nomadic people still exist today. In the old days, that is, in the days of the Russian Empire and before, virtually no borders existed, and only the first category, landscape adapted migration, existed.[58] The Karakalpaks, the Kazaks, the Turkmen and the Kirghiz as we call them today, are descendants of nomadic tribes that stretched their presence across the steppes and deserts of greater Central Asia for centuries.[59] As recently as the late nineteenth century, Wood describes the efforts to count population by families according to the number of yurts the Russian surveyors saw dotting the landscape. Much of it was too dry or too wet for farming, he commented, but ideally suited for extensive grazing of cattle. The isolated surveyors were a sign of more control to come, as the Russian Empire yielded to the Soviet Union and then to an independent Republic of Uzbekistan. The nomads were encouraged to settle into villages, then coerced into farming followed by collectivized labor, and then finally abandoned to their fate today.

Partially essentialized as colorful ethnic "nationalities," partially vilified as backward people, the yurt dwelling tribes had survived each other's raids for centuries, the sacking by the Mongols in 1221, forced resettlements and political disruptions. It is the current state of affairs that may at last do them in. The historical record shows that local people were perpetually vulnerable to outsiders from far away, wielding power at regular intervals.[60] It was only the manipulation of the landscape in vast Karakalpakstan for industrialized agriculture and the contamination of some stretches of the grazing and farming lands, even, in the late twentieth century, which have left much of the rural population with no place to work.

Ecologically Adapted Migration

The second migratory population consists of nomadic Kazaks adapting to the ecological disaster in the Aral Sea Basin. When these Kazak out-migrants from Uzbekistan become "returnees" to Kazakhstan, they are referred to as the "oralman." Their emigration is spurred by the inability to graze their livestock properly or to otherwise make a living in the severely compromised landscape of Karakalpakstan. As documented in my past interviews with Kazaks in China, the Karakalpak Kazaks are drawn "home" by the opportunities for free education for their children, high quality grazing land, and the freedom to speak their own language.[61]

Kazakhstan dangles appealing prospects before them. Uzbek Kazaks benefit from a nationalistic and politically astute program which actively seeks to welcome Kazaks from the diaspora as new immigrants (and workers) into the oil and gas wealthy, but under-populated, independent Kazakhstan, where grazing land is also plentiful. Kazaks living anywhere in the world may apply, though they mainly come from the surrounding former Soviet Republics and China. Uzbekistan has consistently had the most number of "oralman" since the program began in the early nineties. According to sources inside Kazakhstan, 285,409 ethnic Kazaks left Uzbekistan, mostly from Karakalpakstan and Navoiy between 1991 and 2005.[62]

The UNDP detailed in-migration records are valuable in two ways. They reflect the uniquely poor environmental conditions in the Aral region districts, and the compounded vulnerability of the population when weather becomes more extreme. For example, in-migration figures to Kazakhstan went from just under 25,000 in 2001 to over 87,000 in 2002, an extremely difficult year of intense drought ridden summers and a brutal winter.

UNDP has been able to catalog a majority of the arrivals from Uzbekistan as coming from the disaster-affected areas.[63] Once arrived, the Uzbek Kazaks tended to settle in southern Kazakhstan, in the *oblasts* or districts just to the north of the border. In 2005, 73% of the new *oralmans* were reportedly from Uzbekistan.[64]

There have been some hiccups with administering the resettlement program over the years. For example, when demand far exceeded the supply of quota slots in some years (affecting several countries, not only the Uzbek Kazaks), some people were left stateless. They gave up citizenship when they left their home country, but by not being granted status under the quota, they were left without papers of any kind.

Nevertheless, the program has achieved what it was intended to, and lessons have been learned. The Kazakhstan government is expanding the *oralman* employment assistance through a new program called "Employment 2020."[65] The program advertises that there are plenty of jobs to be had. The Uzbek Kazaks are thus also economic migrants, relocating out of a desire to improve their life circumstances through the perceived economic advantages of Kazakhstan. Kazakhstan continues to benefit when new immigrants help to mitigate the demographic nosedive the country suffered after ethnic Russians left in large numbers with independence.

Economically Adapted Migration

The third migration, economically adapted migration, is the most recent flow of people to affect the population stability of Karakalpakstan. They are the most desperate. Driven to emigrate in search of work, these are mostly young, mostly male and both documented and undocumented workers (Abdulaev, 2008). They tend to be less educated and find work in construction or agriculture. They may be vulnerable to crime and unfair working conditions in their destination country.[66] Their primary destinations are Russia (60%) and Kazakhstan (29%), with South Korea placing third (5%).[67] Most of the out-migrants from Karakalpakstan move into neighboring Kazakhstan, likely because the transportation costs and language barrier are lower. This whole migration category is driven by poverty and lack of opportunity back home, not ethnicity. These migrants are primarily labor migrants, compelled to leave Karakalpakstan and other parts of Uzbekistan for reasons of high unemployment, no job prospects and the lure of higher wages for the same work. According to a study done for the International Organization of Migration by Chupik, estimates of the volume of immigrants vary greatly, showing a minimum of 200,000 to as many as 1 million Uzbek nationals working in Kazakhstan.[68]

In summary, migration patterns for people in the Karakalpakstan region have changed dramatically over the last 100 years. Before the Aral disaster, migration was rooted in the nomadic way of life, driven by the demands of weather and seasons and carried out in a voluntary, culturally normative manner. The majority population were once yurt and mudbrick house dwelling Karakalpaks and Kazaks. Today, most of the Kazaks have relocated permanently. They are compelled to migrate by the limitations brought on by anthropogenic changes to the waters and the soils. Previous generations were previously ecologically adaptive to the needs of livestock and capable of adapting to the landscape. Now they have to adapt by migrating out, or risk illness, or even death to themselves or their livestock, from contact with pesticide contaminated blowing sands and salty earth. It seems there is nowhere to go, no pastures that are safer in the Aral Sea region. For livestock herding Kazaks, the damaged ecology and the lack of progress in remediating the soils is one reason that compels the migration out of Uzbekistan to Kazakhstan, where the government proudly markets itself with nomadic imagery and offers land to qualified *oralman* settlers. The third group is responding to any number of factors, including the isolation of the Aral Sea region, the disaster itself, or the insufficient political will to rehabilitate the Aral Sea or even to restore the delta that once supported a diverse array of livelihoods. In their remote region, livelihoods based upon nature are lost, necessitating a willingness to migrate in that most modern form: labor migration. The uncertainty of the seasonal rains has been replaced with the uncertainty of fair pay and job security.

CONCLUSIONS

As we have seen in this chapter, Karakalpakstan has been on the losing end by all counts. Its losses reflect the true cost of draining the Aral Sea to grow cotton. It is measured in the brain drain of doctors and teachers, the collapse of the health system, the deaths of babies, the loss of young and able workers, the loss of established livelihoods, and finally, the challenge to get educated people from elsewhere in the country to move to Karakalpakstan to redevelop the education and healthcare infrastructure.

Uzbekistan could make more of its independence. It could do more to prepare for a stable, productive environment and healthy future for the people of the Karakalpakstan region. The plan for growing cotton that began in the late 1950s has long ago ceased being uniquely positive for economic development. The true costs of raising this cash crop in an arid region with hazardous inputs and inefficient irrigation include not only the cost of water, but also the repercussions to society.

The population of Karakalpakstan, once largely migratory by cultural preference, is now migratory by desperate economic necessity or to avoid the negative social and health consequences of remaining in the region. This has resulted in negative population growth, which further hampers the chances for recovery of social stability.

Political willpower could have been put to greater service by now than it has been. Instead, the decades long contemplation of options has not been followed up by much action. As Philip Micklin, who has written for many years about the Aral Sea disaster, put it in his 2007 report, "The fundamental aims, ... have remained remarkably consistent since that time: better medical and health services, greater access to safe drinking water supplies, improved food supplies, and diversification of the economy for the people living near the sea; mitigation of negative ecological trends in the delta of the Amu Darya; and rebuilding irrigation systems to raise their efficiency to deliver more water to the Aral Sea."[69] The persistence of denial and inaction pervades both domestic and international decision-makers.

At a minimum, investments made in the late 1980s, or at any time since then, could have employed thousands of people across the region to re-dig, renew, insulate, line, and otherwise modernize the irrigation system from its highly inefficient and wasteful state to a more water efficient transport network. This would have been costly, but not as costly as starting today. This would have provided people with the income they need to make better food choices and support their local economy, thus stemming the severity of some of the serious repercussions of the poverty that came after the Aral Sea disaster, and the loss of the people's original livelihoods.

NOTES

1. These systemic flaws in the incentive structure have been much written about in the parallel case of Chinese Communism. See, for example, Williams (2002). *Beyond great walls: Environment, identity, and development on the Chinese grasslands of inner Mongolia.*

2. Micklin (1988) is one of the earlier studies of the unfolding Aral Sea disaster we have, which is remarkable to read today with the benefit of hindsight.

3. Understanding taken from discussions among Uzbek and American project members, May 2011. Furthermore, Philip Micklin's Aral status report, in *Science* (1988) reviewed Soviet materials from the 1960s to the 1980s, illustrating the depth to which the planners had researched the possible negative consequences, and calculated the lake attrition, salinity increases, salt residues, and the effect on the biota they anticipated as outcomes.

4. See Kamalov, this volume.

5. There is another herding population, the Turkmen.

6. Achieving Ecosystem Stability on Degraded Land in Karakalpakstan and the Kyzylkum Desert, (UNDP, 2012b).

7. There are reports of empty fishing villages, farming villages and dry (waterless) villages.

8. Oberkircher and Hornridge (2011) have written a thought-provoking discussion, based on ethnographic fieldwork, of how water exists as a constant presence in the minds and experiences of local residents.

9. Richardson, *Demographics*, Karakalpak.com.

10. Micklin, 1988.

11. See, for example, Lall and Arzikulov, this volume, as well as Medicins Sans Frontieres (2003) and Government of Uzbekistan, UNDP, the Council of Ministers Republic of Karakalpakstan, Uzbekistan (2005).

12. See for example Boroffka et al. (2005) for a discussion of archaeological evidence of changing water levels and human settlement.

13. For an extensive discussion of irrigation systems in Central Asia since ancient times, see Andrianov, (1995).

14. See, for example, Asia Briefing nr. 67 by International Crisis Group (2007) which discusses the economic and political cycles that seek resolution but instead stagnate.

15. McLeman (2011). *Settlement abandonment in the context of global environmental change*. The first two stages are rising vulnerability based on social or environmental risk factors, followed by emerging population attrition. With regard to the relationship of outmigration and institutions, McLeman found that in the case of Karakalpakstan, outmigration peaked when trust in institutions was lowest. In contrast, in another more acute environmental scenario, a volcanic eruption on Montserrat, he credits the consistent and robust institutional support with the recovery of the community.

16. Perhaps the most famous among these are *karez* (alt. *qanat*), the gravity fed underground irrigation channels sourced from mountain springs.

17. Micklin (1988), expanding on Soviet studies, reports "fluctuations of at least 20 m and possibly more than 40 m," p. 1170.

18. See Micklin (1988) and Wood (1876). Wood, in his classic study *On the Shores of Lake Aral*, summarizes the landscape and habitation observations of Pliny, Herodotus, and Ibn Battuta, among others.

19. Wood, throughout his travelogue referred to the Ust Yurt Plateau as well as other Aral Sea proximate land as marsh, wetland, bog, and vast tracts covered in Arundo reeds. These more or less firm lands being home to nomads who sometimes were flooded out; he realized the practicality of yurts, or some kind of movable abode.

20. See Andrianov (1995) for a perspective on the ebb and flow of civilization tied to irrigation in the Amu Darya region.

21. Oberhaensli, Boroffka, Sorrel, & Krigonogov (2007) integrate natural and human indicators in their thought-provoking discussion of climate variability.

22. Yusup Kamalov, who lives in Nukus, Karakalpakstan has been a vocal advocate and prolific writer on the subject of Aral as a common good. Mickey Glantz (2007) has argued for the Aral Sea to be considered for World Heritage status. Philip Micklin has written about the Aral consistently for over two decades, often framing his discussion as what to do *for* the Aral Sea, rather than *about* the Aral Sea (disaster).

23. It should be mentioned that the Khorezmians are a notable exception. They came from the south, they have Iranian roots; their language, gene pool, and cultural

habits are distinct and unrelated to those of the Turkic peoples. The Khorezmians are not nomadic, but sedentary. They built the fortress cities like Khiva.

24. Richardson, *Tribal Structure*, Karakalpak.com.

25. This has led to the unfortunate situation for readers, that historical (eastern and western) sources applied labels liberally, using Kazak, Kirghiz, or Kazak-Kirghiz without careful study or accurate distinction of territory and tribe. Some sources even dispute the Karakalpaks' claim to their own tribal identity, considering them a variant on the Kazaks. This latter assertion has been disproven, although our understanding of appellations and self-identifications remains complicated.

26. Martínez-Cruz et al. (2011) found that genetic diversity in Central Asia is shaped by languages spoken, and has multiple clusters. More intriguing is that genetic breakdown of individuals in their study (my interpretation) shows patterns of nomadism and sedentarism, *in contemporary urban populations*. The Indo-Iranian speakers (Tajik, Bukhara Uzbeks) who cluster in the data come from a sedentary agriculturalist background, while the Turkic speakers (Karakalpak, Kazak, Kirghiz, Uzbek, Turkmen) who cluster come from nomadic backgrounds. The Uzbek and the Turkmen are the pivot populations in the study ranking, which is intriguingly consistent with Uzbeks having originally been nomadic but becoming sedentary and the intermingling of the Turkmen with other peoples from the south. The Turkic speakers showed Asian markers (such as Mongolian) in their clustering, while the Indo-Iranians show European markers.

27. Efforts by ethnic minorities, such as the Karakalpak in Uzbekistan or the Uighurs in China, to assert human rights or some degree of freedom have shown how disenfranchised they are from the national agenda of the leadership.

28. The chapter on Vozrozhdenye in this book details the more sinister side of Soviet thinking with regard to its hinterland of non-Russian peoples.

29. Richardson, *Modern Life*, Karakalpak.com.

30. This figure is based on World Bank data cited in Richardson, Karakalpak.com. UNDP Uzbekistan is using a figure of 27% for the same year on their current webpage "economic governance." This information is sensitive and the figures are likely to have fluctuated in recent years.

31. When I told a senior colleague last year I was about to go to the Aral Sea region for research, she brushed it off saying, "The Aral Sea? There have been several board feet of studies produced, nothing is going to help save the Aral."

32. This is not to romanticize the past, for there was certainly poverty; travelers often commented on the people's lack of material goods and even tattered clothing. Life was not easy in the steppes. The point to be made is that the land, this same territory surrounding the Aral Sea, supported farmers, fishermen, nomads, and merchants alike and they found ways to sustain their families from generation to generation.

33. For more on the industrious activities of farmers and nomads, see, for example, chapter XXIII in Wood.

34. See, for example, Dunn and Dunn (1967) for an extensive discussion of the cultural implications of the shift to agricultural expansion, or Stadelbauer (1974, in German) for the impacts of railroads and industry.

35. See village level ethnographic work by Oberkircher and Hornridge (2011).

36. See, for example, Kamalov and other discussions, this volume.

37. Oberkircher and Hornridge (2011).

38. Wegerich (2001).

39. Abdulaeva (2006) makes a case for safer handling; the danger results from negligent handling and disposal.

40. Kyrgyzstan and Tajikistan and considered water rich, due to mountainous rain fed environments predominating, while Turkmenistan and Uzbekistan are dry and getting drier.

41. I recognize that the region includes Nukus and Urgench as regional cities, and that these have universities and other features of urban infrastructure. However, given that 70% of the population is rural and the agriculture is the dominant sector still today, my remarks in this chapter are limited in scope to these.

42. Micklin (1988).

43. Lally (1999).

44. *Karakalpak Demographics*, Karakalpak.com.

45. It is hard to find reliable numbers for population of villages and towns in the region. For a sense of the emptiness that exists where once there were active communities, Micklin reports an exodus in the eighties from the Kyzl Orda district on the Kazakkhstan side of the Aral Sea of more than 40,000 people.

46. The Richardsons' comprehensive research on all aspects of Karakalpak culture, history, and more has appeared on their Web sites, Karakalpak.com, or qaraqalpaq.com. Their just published book, *Qaraqalpaqs of the Aral Delta*, is available through Prestel. There are two variant spellings, Karakalpak and Qaraqalpaq.

47. Richardson, *Karakalpak tribal structure*, Karakalpak.com.

48. Richardson, *Karakalpak demographics*, Karakalpak.com.

49. This project was funded at a relatively modest 1.1 million Euros, and was carried under the name "Enhancement of Living Standards in Karakalpakstan."

50. Karakalpakstan, like the other viloyat has been included in the government's long-term poverty alleviation and economic growth planning Government of the Republic of Uzbekistan (2005).

51. Radio Free Europe: Shadowy group agitates for "Free Karakalpakstan."

52. This entry gives the number of deaths of infants (under 1-year-old) in a given year per 1,000 live births in the same year.

53. CIA World Factbook Child Mortality and Maternal Health rankings worldwide are retrievable from any country's main page and the demographic section.

54. See for example: Quinn-Judge (1989), Chupik (2005) and Tahir (2011).

55. Chupik (2005) and Medicins sans Frontieres (2003).

56. *Karakalpak Demographics*, Karakalpak.com.

57. This is to say nothing of the police controls and manned barriers across roads at unspecified intervals across Uzbekistan that form temporary borders.

58. See for example Tsvetsinskaya, Vainberg, & Glushko (2002) for a discussion of ancient nomadic patterns.

59. The Uzbeks have been oases dwellers and settled farmers – not nomadic – in this area and in the Amu Darya delta also for many centuries.

60. For the sedentary Khorezmians and residents of Khiva, for example, the city walls were their defense, tested many times over.

61. Cerny (2010) is a study of the China Kazaks who leave Xinjiang and emigrate to Kazakhstan. Here too, a significant push factor was exhausted grazing land.

62. UNDP Kazakhstan (2006), *Status of Oralmans*, is one of the only reports of its kind on this massive movement of people, Kazaks from the diaspora, as permanent migrants to Kazakhstan.

63. UNDP Kazakhstan (2006) although dated is one of the few useful reports in English about the *oralman*.

64. There is much debate about the numbers of *oralman* immigrants. In part this is due to the split nature of immigration into Kazakhstan. The *oralman* program is based on a quota per year and per country, and this number is typically greatly exceeded. Kazakhstan lets people in beyond the quota, but the *oralmen* are entitled to benefits like housing, land, and stipends, whereas the beyond quota arrivals are entitled to no benefits. Accounting discrepancies are also a sensitive matter where the countries of exodus are reluctant to admit or release figures for how many people have left the country. UNDP and IOM (International Organization of Migration) can therefore be seen as more reliable and impartial numbers.

65. Kamziyeva (2011).

66. See Ilkhamov (2006) for a discussion of how economics and politics influence migration choices.

67. Chupik (2005) has a comprehensive discussion of Uzbekistan to Kazakhstan migrant outflows.

68. The complexity of counting the labor migrants arises because of factors including seasonal fluctuations of temporary workers, unofficial border crossings, and the fact that migrants being a vulnerable population are apt to "disappear" or go into some form of hiding once in Kazakhstan. The receiving countries (Russia and Kazakhstan both) have been unprepared for accommodating and integrating these influxes, although by and large everyone finds some kind of work. The last 10 years have seen migration flows grow to a massive scale of people in and out, and the attempt by international agencies to count, integrate, and protect them from crime (especially human trafficking) is a rising concern.

69. Micklin (2007, p. 57).

REFERENCES

Abdulaev, E. (2008). *Labor migration in Uzbekistan: Social, legal and gender aspects*. Tashkent: UNDP Uzbekistan.

Abdulaeva, N. (2006). *Informing residents of the Republic of Karakalpakstan (RK) living near disposal sites of persistent organic pollutants (POPs)*. International POPs Elimination Project. Retrieved from www.ipen.org

Andrianov, B. V. (1995). The history of economic development in the Aral region and its influence on the environment. *GeoJournal*, *35*(1), 11–16. doi: 10.1007/BF00812617.

Boroffka, N., Obernhänsli, H., Achatov, G., Aladin, N., Baipakov, K., Erzhanova, A., & et al. (2005). Human settlements on the northern shores of Lake Aral and water level changes. *Mitigation and Adaptation Strategies for Global Change*, *10*, 71–85. doi: 10.1007/s11027-005-7831-1.

Cerny, A. (2010). Going where the grass is greener: China Kazaks and the Oralman immigration policy in Kazakhstan. *Pastoralism: Research, Policy and Practice*, *1*(2).

Chupik, V. (2005). Labor migrants from Uzbekistan in the southern regions of Kazakhstan (K. Alimov, Trans.). In IOM Almaty. International Organization of Migration. Retrieved from http://public.julias.promessage.com/IUED/CentralAsia/QuestionsRef/Uzbek-migrants.pdf

CIA. (2012). *The World factbook*. Retrieved from https://www.cia.gov/library/publications/the-world-factbook/geos/uz.html

Diamond, J. (2005). *Collapse: How societies choose to fail or succeed*. New York, NY: Penguin.

Dunn, S. P., & Dunn, E. (1967). Soviet regime and native culture in central Asia and Kazakhstan: The major peoples. *Current Anthropology*, *8*(1).

Embassy of Uzbekistan to the United States. (2012, July 5). Aral Sea area returns to life: Former fishermen of Karakalpakstan learn new skills. Retrieved from http://www.usbekistan.org/news/archive/1365/

Glantz, M. (2007). Aral Sea basin: A sea dies, a sea also rises. *Ambio*, *36*(4), 323–327. doi: 10.1579/0044-7447(2007)36[323:ASBASD]2.0.CO;2.

Government of the Republic of Uzbekistan. (2005). *Welfare improvement strategy paper of the Republic of Uzbekistan for 2005–2010, 105*. Government of the Republic of Uzbekistan.

Government of Uzbekistan, UNDP, the Council of Ministers Republic of Karakalpakstan, Uzbekistan. (2005). Enhancement of living standards in Karakalpakstan.

Ilkhamov, A. (2006). *Geographic mobility of Uzbeks: The emergence of cross-national communities vs. nation-state control*. National Bureau of Asian Research. Retrieved from http://www.nbr.org/Downloads/pdfs/PSA/Uzk_Conf06_Ilkhamov.pdf

International Crisis Group. (2007, August 22). Uzbekistan: Stagnation and uncertainty. *Asia Briefing*, *67*. Retrieved from http://www.crisisgroup.org/en/regions/asia/central-asia/uzbekistan/B062-uzbekistan-stagnation-and-uncertainty.aspx

Kamziyeva, G. (2011). *Astana develops new Oralman repatriation programme*. Retrieved from http://centralasiaonline.com/en_GB/articles/caii/features/main/2011/11/02/feature-01

Lally, K. (1999). As the Aral Sea retreats, dust and disease flourish. *The Baltimore Sun*. Retrieved from http://articles.baltimoresun.com/1999-02-14/news/9902140143_1_aral-sea-uzbekistan-tuberculosis

Martínez-Cruz, B., Vitalis, R., Ségurel, L., Austerlitz, F, Georges, M., Théry, S., … Heyer, E. (2011). In the heartland of Eurasia: The multilocus genetic landscape of Central Asian populations. *European Journal of Human Genetics*, *2011*(19), 210–223. doi:10.1038/ejhg.2010.153.

MAWR, GEF, & UNDP. (2009). Achieving ecosystem stability on degraded land in Karakalpakstan and the Kyzylkum Desert. *Information Bulletin Nr. 1*, 1–16.

McLeman, R. (2011). Settlement abandonment on the context of global environmental change. *Global Environmental Change*, *21S*(2011), S108–S120. doi:10.1016/j.gloenvcha.2011.08.004

Medicins sans Frontieres. (2003). Karakalpakstan: A population in danger. Retrieved from www.msf.org/aralsea

Micklin, P. (1988). Desiccation of the Aral Sea: A water management disaster in the Soviet Union. *Science*, *241*(4870), 1170–1176. doi: 10.1126/science.241.4870.1170.

Micklin, P. (2007). The Aral Sea disaster. *Annual Review of Earth and Planetary Sciences*, *2007(35)*, 47–72. doi:10.1146/annurev.earth.35.031306.140120

Oberhaensli, H., Boroffka, N., Sorrel, P., & Krigonogov, S. (2007). Climate variability during the past 2,000 years and past economic and irrigation activities in the Aral Sea basin. *Irrigation and Drainage Systems*, *21*(3-4), 167–183. doi: 10.1007/s10795-007-9031-5.

Oberkircher, L., & Hornridge, A.-K. (2011). "Water is life" – farmer rationales and water saving in Khorezm, Uzbekistan: A lifeworld analysis. *Rural Sociology*, *76*(3), 394–421. doi: 10.1111/j.1549-0831.2011.00054.x.

Quinn-Judge, P. (1989). Environmental damage seeps into Central Soviet Union. *Christian Science Monitor*. Retrieved from http://www.csmonitor.com/1989/0706/onuk.html

Richardson, D., & Richardson, S. *Modern life*. Retrieved from http://www.karakalpak.com/demographics.html

Richardson, D., & Richardson, S. Tribal structure. Retrieved from http://www.karakalpak.com/tribalstructure.html

Richardson, D., & Richardson, S. Demographics. Retrieved from http://www.karakalpak.com/demographics.html

Richardson, D., & Richardson, S. Karakalpak genetics. Retrieved from http://www.karakalpak.com/genetics.html

Saidazimova, G. (2008). Uzbekistan: Shadowy group agitates for 'free Karakalpakstan'. *Radio Free Europe*. Retrieved from http://www.rferl.org/articleprintview/1079744.html

Schlueter, M., Savitsky, A., McKinney, D., & Lieth, H. (2005). Optimizing long-term water allocation in the Amudarya River delta: A water management model for ecological impact assessment. Available at http://www.cawater-info.net/amudarya/papers/schluter.pdf

Stadelbauer, J. (1974). Die wirtschaftliche Regionalentwicklung zwischen dem Amu Darja Delta und Westkazachstan under dem Einfluss des Eisenbahnbaus (The influence of railway construction upon economic development in the region between the Amu Darya Delta and the Caspian Sea). *Erdkunde*, *28*(4), 282–295Retrieved from http://www.jstor.org/stable/25641551.

Tahir, M. (2011). As Central Asia dries up, states spar over shrinking resources. *Radio Free Europe*. Retrieved from http://www.rferl.org/content/central_asia_states_spar_over_shrinking_water_resources/24275157.html

Tsvetsinskaya, E., Vainberg, B., & Glushko, E. (2002). An Integrated assessment of landscape evolution, long-term climate variability, and land use in the Amudarya Prisarykamysh delta. *Journal of Arid Environments*, *51*(3), 363–381. doi: 10.1006/jare.2001.0951.

UNDP. (2012a). *Achieving ecosystem stability on degraded land in Karakalpakstan and the Kyzylkum Desert*. Retrieved from http://www.undp.uz/en/projects/project.php?id = 123. Accessed on July 6, 2012.

UNDP. (2012b). *Sustaining livelihoods affected by the Aral Sea disaster*. Retrieved from http://www.undp.uz/en/projects//project.php?id = 179

UNDP Kazakhstan. (2006). *Status of oralmans in Kazakhstan*. Retrieved from http://www.undp.kz/library_of_publications/files/6838-29587.pdf

UNECE. (2012). *Reproductive health*. Retrieved from http://www.unece.org/stats/gender/genpols/keyinds/health/concerns.html. Accessed on June 15, 2012.

Wegerich, K. (2001). *Not a simple path: A sustainable future for Central Asia*. Occasional Paper No. 28, Water Issues Study Group, School of Oriental and African Studies (SOAS).

Williams, D. M. (2002). *Beyond great walls: Environment, identity, and development on the Chinese grasslands of inner Mongolia*. Palo Alto, California: Stanford University Press.

Wood, H. (1876). *The shores of Lake Aral*. London: Smith, Elder and Co. Retrieved from http://digicoll.library.wisc.edu/cgi-bin/EastEurope/EastEurope-idx?type = header&id = EastEurope.ShoreAral&isize = M&pview = hide

CHAPTER 15

REFLECTIONS ON GROWING UP IN THE KARAKALPAKSTAN REGION☆

Aziz Murtazaev

ABSTRACT

In this chapter, I reflect on growing up in Karakalpakstan during the 1980s and 1990s. I review my childhood memories, and also my learning about the Aral Sea disaster since that time. Mostly, even growing up in Nukus, we were not very aware of the total extent of the disaster or the shrinking sea. It was still far away for us. Today we read about the shrinking Aral Sea and we wonder, how could it be true? Knowing what we know now, I hope my story can provide insight into understanding the region and why it is important to plan for environmental safety.

☆In presenting one individual's story, we make no representations that his experience is necessarily indicative of others in Karakalpakstan. Note that the author lived in Nukus and not on the shores of the Aral Sea.

Disaster by Design: The Aral Sea and its Lessons for Sustainability
Research in Social Problems and Public Policy, Volume 20, 251–259

ISSN: 0196-1152/doi:10.1108/S0196-1152(2012)0000020026

INTRODUCTION

My name is Aziz and I am a native of Karakalpakstan. There has been much interest from foreigners especially about my region since the Aral Sea began to shrink a few decades ago. Many people wonder what it was like to grow up in this area to the south of the Aral. It was once the fourth largest lake in the world, and they wonder what it was like to be there while it was slowly disappearing. Well, the short answer is, we hardly noticed it. We just went about our lives. However, when I think about it, the Aral Sea has touched our lives in many ways, and I would like to reflect here on some of those.

I was born in 1985 in the capital of Karakalpakstan, Nukus, and lived there until I graduated from high school and got enrolled at the university in 2003. As it is the case in many countries, the universities in the capital city provided more opportunities than the local ones, so I decided to study in Tashkent at National University of Uzbekistan. Because the distance between Nukus and Tashkent is significant (around 1,000 km), it was not easy to go back to Nukus frequently. I used to go to Nukus to visit my parents twice a year, usually for the new year holiday and for summer vacations. In the days when I was an undergraduate I was seeking various options for studying abroad and that always motivated me to learn English. After studying four years and working two more years in the computer software field in Tashkent, I got a chance to study computer engineering at a South Korean University for a master's degree. Now I am located at a city near Seoul, South Korea and have been living here for the past three years. After graduating with my master's in two years, I am lucky enough to work in one of the biggest consumer electronics, companies, Samsung Electronics as a software engineer. Even though i live in Korea, I am still connected with my hometown and my country. Now I visit Nukus around once a year.

Nukus is the capital of the northwest region of Uzbekistan – the Republic of Karakalpakstan. It has a population of about 260,000 people. Nukus is surrounded by the Amu Darya to the west and the beginning of the Kizilkum desert to the east. Nukus is located around 220 km from the original port city on the Uzbek side of the Aral Sea, Muynak.

First of all, I would say that I did not have the feeling that I was living in an ecological disaster zone. We used to hear about the Aral Sea problem from our teachers in middle school and high school and we would write essays about this problem. We used to write about how precious every drop of water is and were urged to use water wisely. But still, it was as if the ecological problems were happening apart from us, somewhere too far away. We did not connect those ecological transformations to our daily lives. Up to now I have not had any chance to see the Aral Sea with my own eyes, even though I lived relatively close to it during my childhood. After the shore of the sea receded away from the former port city of Muynak, people did not consider going there, as there was no reason to do so. My father told me that people used to go to the Aral as a resort place in the 1960s and 1970s. But other than that, I was too young to ask deeper questions myself. People just did not talk much about this issue. Today I have a different perspective, of course, but back then, it just happened without us thinking about it much.

WATER CONTAMINATION

In my childhood, we used to play outside with the neighborhood kids. Sometimes we would get thirsty and go to the water tap at the other kids' houses (for some houses the water taps are located outside the home to help water small gardens outside) and drink from them. However, at that time in the mid-1990s (and continuing until now) elders would always warn us that it is not good to drink from the water tap directly, and that it would cause stomachache. People usually poured the water from the tap into some container and let whatever heavy particles (I do not know exactly what it was, but it looked like sand at the bottom of the container) settle to the bottom by storing it for several hours. Then people would use only the water above those heavy particles. But in many cases, kids would ignore those kinds of warnings and drink directly from the water tap and just continue playing their games. Whatever those particles were, it seems to not have damaged our health as kids. I would say that I did not hear of any case where a person got sick after drinking from the water tap directly. I guess the human organism gets adapted to the surrounding environment, including the water and food available there. Now, as an educated adult, I have a broader perspective of course. I do not know the long-term effect of drinking such low quality water as comes out of those same taps today. I think it could somehow affect the immune system; the contaminants or particles could make the immune system weaker.

Looking back, I realize now that I was very lucky overall because my family has been drinking filtered water since 1994. At that time, some French people from Greenpeace visited Nukus – I guess they visited Nukus on Aral Sea related issues. At any rate, they installed huge water filtering equipment at two hospitals in Nukus: First, the Republican Children's

Hospital, which is the biggest one for children's healthcare in Karakalpakstan, where my mother has been working as a pediatrician for many years. The second one was installed at the infectious diseases hospital. Since that time when the equipment was installed in the first hospital, first my older brother, and several years later I, used to bike to that hospital with empty plastic canisters and bring this filtered water home by transporting the canisters on our bicycles. We used to bike around 2.5 km from our home to my mother's workplace where that filter was installed and fill the several (between three and six) water containers of 1.5 L capacity each and take them back home. Even though we had access to this filtered water, the interesting point was that I had the feeling that many people did not realize that drinking filtered water versus drinking from the tap, even after storing it in a container, would make a big difference. My family, however, used to always consume the filtered water we had brought home from my mother's workplace and, when we were guests at another family's home, we could clearly taste the difference with the ordinary water. The water from the tap did not taste good. I remember it as being a little bit bitter and salty. Our family could definitely taste the difference, and we knew that the quality of the filtered water was better than from the tap. Otherwise, why would we transport filtered water across the town for so many years from my mother's workplace to our home? I cannot say for sure how other people in my neighborhood understood the issue of the water. Some people might have known about the filtering equipment at the hospital. But probably they did not find much difference between it and the tap water, or they simply might not have had access to it.

For example, schools did not have filtration systems. There was no money to make filtration systems available in every building in the old days, so it was only those few systems that the French installed as a development project that filtered our water. These hospital machines worked for many years, about 15, before they finally could not be fixed or used any more. I am not sure how the filters got replaced in the meantime, but I think there were some locally made replacements and the hospitals paid for the maintenance with their own resources.

In recent years, getting water has changed like in so many places; people buy it in the supermarkets. Now it is not only my parents, but the people who used to drink the tap water before. The last time I visited Nukus, I noticed large advertisements in the streets offering home water filter systems. Some people might have already started using water filters right at home, I am not sure, since I do not go back regularly. But of course not everyone can afford to buy water or filtration systems. I have also heard that now the water quality coming from the tap is even worse than it was several years ago.

WATER SUPPLY FOR AGRICULTURAL USE

Even though we have had significant problems with the quality of water for many years, the situation seems to be different with the quantity. From my experience growing up in Nukus, I would not say people lack water for everyday use or for watering their small household gardens. As I understand it, farmers are also able to raise cotton, rice, cereal grains, and other crops every year. My uncle is a farmer. I have not visited him often in his village, but he grows cotton, rice, watermelons, and some other kinds of vegetables. Since I work in the city, I am not totally sure, but I think farmers flexibly choose what to cultivate based on the water supply, such as growing fewer water-intensive crops such as rice during seasons with decreased water supply, or vice versa.

I can remember only one year when things were really different. In 2001, there was an unprecedented case with an extremely short water supply for agricultural purposes. This was a year with bad droughts in many places. Most farmers in Karakalpakstan had very tough days and some had to sell their assets to compensate for their crop losses. But these days, I think, the situation with water supply is somewhat stable.

CLIMATE CHANGE

While the water situation was quite difficult for me to understand when I was younger, I distinctly remember the changes in the weather patterns. Climate change is one of the obvious consequences of the Aral Sea drying up that I could feel. Compared to my childhood during the 1990s, from the beginning of this first decade of the new millennium, I could feel the change of climate because the winter and the summer weather are becoming more extreme. Winter used to be really, really cold sometimes, the temperature going down to minus 25°C. And when this temperature was combined with a strong wind, it was better to stay home. I remember once I was walking outside in such temperature and I was wearing gloves and even though I put my gloved hands into my pockets to make them warmer, they got stiff and almost frozen.

Summer used to be very hot, reaching +45°C or even higher in some cases. When I visited Tashkent, the capital of Uzbekistan, and compared the feeling of sunlight on the skin in the summer, it felt like there was a filter in the sky over Tashkent which reduced the effect of the sunlight. The feeling of sunlight in Nukus was always stronger, as if it was unfiltered. This intense sunlight continues, but the temperatures are even higher. Parents usually do not recommend for their kids to go out during midday unless we have an unavoidable reason to do so, from around noon to 4pm. Everyone understands that it is easy for weaker people, including children, to suffer sunstroke on such days.

Our region is considered to have acutely continental climate and most of Karakalpakstan is a desert, so that kind of weather should not be surprising. However, it seems that the Aral Sea Basin was some kind of softening factor to the weather, and continuous drying out of the lake caused the end of that softening effect.

TREES

One thing that is very vivid in my experience is the fate of the trees. I noticed that currently we have very few trees in our city of Nukus. I remember in my childhood, we used to go to the parks with dense tree coverage. And also along the streets we used to have more trees than we do now. Elders used to tell us that once upon a time, some parts of Nukus were a forest. I know that the overall ecological situation in Karakalpakstan contributed to this change. The soil is less productive because it contains more salts than before. At the same time, the municipal government is not actively planting more trees. I believe people would suffer less in the very hot days in summer, if we had more trees across the city. Worsening soil productivity cannot be a proper excuse for having very few trees, as there must be some sort of trees which can grow in less favorable conditions such as ours. That is actually important for quality of life, as during the very hot days, which we are getting more of recently, people need the shade to rest and stay cooler when walking down the street.

PEOPLE'S AWARENESS OF THE PROBLEM

Since the time I have been outside of Karakalpakstan I have a feeling that the outside world is more aware about the real tragedy and the real possible consequences of the drying up of the Aral Sea than the local people (Note, that by local people I mean the people living in Nukus. Those living in Muynak, for example, might have different perspectives yet again.) When I was younger, we did not really discuss the Aral Sea problem, for example, in conversation with my relatives, my friends, or other acquaintances. Nor was the media interested in raising awareness of this problem to the public. This continues today. And also, I have a feeling that people are much more concerned about the socioeconomic issues of their lives than the issue of the Aral Sea dying. Some typical concerns local people might have include organizing the wedding of their child or finding a good job or preparing their child for universities. When I was living in Karakalpakstan, I did not feel that we were in big ecological trouble. I guess that people might have known that this was a big issue, however, they could do nothing about it. They took it for granted that the Aral Sea had a smaller and smaller volume in the present than it used to have several decades before.

Two years ago the United Nations General Secretary, Ban Ki-moon, visited Karakalpakstan and the Aral Sea Basin area. He described the drying up of the Aral Sea as one of the planet's most shocking disasters. I cannot agree more. Furthermore, I think, it is one of the major ecological disasters in the world which is caused entirely by human decisions.

Recently, I looked at pictures of the Aral Sea online using satellite images on the website for the NASA Earth observatory. I also looked at a time lapse, which shows how much the sea has shrunk, just in the last 11 years. It used to be one massive lake back in the 1960s, but today, it has only three small parts: a northern part in Kazakhstan, an elongated western part, and a very shallow part in the middle on the Uzbekistan side. The small section in the middle almost disappeared in 2009, but the next year it partly recovered.

Overall this is tragic because a lot of effort and money went into trying to reverse the damage. I have heard about a lot of cases when some kinds of foreign organizations came to our region and undertook projects to alleviate the consequences of the steady shrinking of the Aral Sea resulting from the Amu Darya and Syr Darya rivers no longer reaching the Aral. I know of a program called DOTS which provided medications and treatment for tuberculosis. Also I heard that some organization a few years ago had a vegetation project to plant a windbreak near the Aral Sea. This was designed and intended to reduce the spread of salt and other chemical residues to inhabited areas. They used special plants, commonly called *saksaul*, which thrive in the desert.

Despite the foreign organizations' interest and awareness of the problem, I cannot see considerable steps taken by our political leaders to change the current status. I think that unless the local people start to understand the real situation and take some steps to change the situation they themselves live in, or at least proactively cooperate with foreign organizations, the foreigners alone cannot make a big difference. That is, I think, the actions need to be started from inside Karakalpakstan. However, the economic difficulties of the region make this problem even harder, because in the presence of economic problems, other problems tend to get ignored.

MEDIA COVERAGE AND REALITY

After I left the region, I sometimes read foreign articles and news about the consequences of the Aral Sea disaster to the health of local people. The articles described the situation as very serious – as if this area is not suitable for human habitation. They would write that many people have serious health issues, such as anemia, tuberculosis, and heart-related problems. Furthermore, they reported that babies are born with many nutrition deficiencies. I would say that these reports are not entirely true, at least not in the case of Nukus as of four or five years ago, and they are somewhat exaggerated overall. Among the people whom I know in Nukus, I cannot say I knew anyone with serious health problems caused exclusively by the ecological changes in the Aral Sea region. However, I can only speak confidently of Nukus. As I used to live in the city and very few times visited rural areas, I do not know the current situation in such areas and it could be quite different. There should be studies conducted to investigate the impact of Aral Sea drying to the health of the local people in the very rural areas. There are some people who are moving to cities from villages for a better life. Also there are some ethnic Kazaks who are emigrating to Kazakhstan.

Several years ago it was found that a large reservoir of gas and some oil is located in the former basin of the Aral Sea. Nowadays the government shows interest in actively performing geological exploration work in the area, such that local and foreign oil and gas companies are working in the basin to find oil and gas fields. Finding these energy sources could improve the economic situation in the region and we hope that some parts of the profits coming from oil and oil extraction will be allocated to solving the ongoing Aral Sea problems. These include improving drinking water quality, reducing dust and salt spread across the region, planting more trees, and improving the irrigation systems.

In conclusion, I think that the problem of the Aral Sea should be viewed as a complex problem requiring coordination with all the surrounding countries. The question of why two rivers, the Amu Darya and Syr Darya, are not supplying enough water to the Aral has to be addressed from the mountains where these rivers have their source (in Afghanistan), to their dried out deltas in the Aral region. All along the flow of the two rivers until they end in the Aral basin, the irrigation systems need to be reconsidered and possibly redesigned. The first consideration is to reconsider the building material, such as using concrete channels to reduce water loss during transportation to crop areas. Then, to combine that with using advanced watering mechanisms, such as drip irrigation, a common method in hot, arid climates elsewhere, which will require much less water to accomplish the same crop watering. These are the technological solutions. But that still leaves us with the problem of people: on the one hand the quality of life for the local people and on the other hand, working together to solve all the problems.

Of course, one possible problem with achieving the willpower to implement these changes could be that the neighboring countries, because they are upstream, are not as affected by the loss of the Aral Sea. They might not show interest or political will to modernize the ways irrigation worked during the Soviet Union days. Furthermore, these days Tajikistan is trying to build a hydropower plant, called Rogun, which will take water away from the Amu Darya far upstream. That would again be very destructive to the Aral, because the Tajiks need to collect a large amount of water into the reservoir over several years, meaning that less water will be available for other uses downstream, and the inflow of Amu Darya into Aral, which is already meager and too little to replenish the Aral, might cease altogether. These days, our government is trying to get support from developed countries to persuade Tajikistan to alter their plans.

The real tragedy here is that there are many options for solving the problems associated with the shrinking Aral Sea. The water quality issues can be fixed. The irrigation systems can be improved for efficiency and conservation. The two rivers can be dammed and channeled with better, modern methods. Most of the ways are already known to the general public. The only thing left to do is to initiate political will, adopt sensible plans, and take action.

CHAPTER 16

AN UNHEALTHY PLACE TO LIVE: PRIORITIZING PUBLIC HEALTH AND ADDRESSING ENVIRONMENTAL CONTAMINATION IN KARAKALPAKSTAN

Ramona Lall

ABSTRACT

This chapter provides a brief overview of our understanding of major public health challenges and environmental concerns in Karakalpakstan today, and highlights questions that still remain unanswered. As seen in the case of Muynak, the fishing town on the southern edge of the former Aral Sea, ecological disasters do not happen alone – they spur socioeconomic disasters that only heighten the health disasters. The loss of the sea, the loss of local livelihoods, and mass out-migration of the population, along with economic depression following the collapse of the Soviet Union, have adversely affected the community living in Muynak. They face major public health challenges, such as tuberculosis, multidrug resistant tuberculosis, and anemia as a result of their impoverishment. The desiccation of the Aral Sea is but one of the many disasters linked to

Disaster by Design: The Aral Sea and its Lessons for Sustainability
Research in Social Problems and Public Policy, Volume 20, 261–274

ISSN: 0196-1152/doi:10.1108/S0196-1152(2012)0000020027

intensive cotton cultivation in Uzbekistan. Pesticide contamination and the salinization of drinking water in Karakalpakstan are yet other environmental disasters that further threaten the health of the population and of future generations. Currently, there is an urgent need for greater international involvement and collaboration with Uzbeks to reverse the poor public health trends and to study the extent of environmental contamination in communities across Karakalpakstan, in order to reduce the health threats presented by these.

INTRODUCTION

From the Uzbek capital of Tashkent to Muynak, the little town that still sits along the southern edge of the former Aral Sea, it is a distance of about 1,300 km. On the road to Muynak, we passed through the ancient cities of Samarkand and Bukhara, before making a twelve hour long journey through the hot and desolate Kyzyl Kum desert to arrive at Khiva. Over the next two days, we pushed into the depths of the Karakalpakstan Autonomous Republic, passing through Nukus and the Kara Kum desert and onto the remote town of Muynak and the newly formed desert, the Aral Kum.

The loss of the sea has forced many to leave, but many have chosen to stay. Far removed from Tashkent, separated from the major urban centers of Uzbekistan by vast expanses of desert, with mass out-migration of its population and lost access to other port communities by sea – Muynak exists in isolation today.

The poignancy of this town's drastic decline is best captured by the local museum – filled with a rich collection of artifacts, stuffed birds, and other animals many now extinct, and heavy photo albums with black-and-white photographs of a once thriving fishing industry. Altogether, these provided a glimpse of a rich ecosystem and of a functioning society that once existed here.

At lunch we learned of the "Surgil Nàtural Gas Chemicals" plant, a joint venture between Uzbekistan and South Korea that plans to extract gas from the Aral Sea bed and produce polyethylene and polypropylene. These new economic prospects bring hope to the people of Muynak, who have long suffered from the loss of their sea. Hope is a rarity here.

This chapter focuses on the public health consequences and major environmental concerns in Karakalpakstan today. I do not have environmental or health data to add to the literature, but I hope a discussion framed by the experience of our visit to this remote area of Uzbekistan might provide a perspective that sometimes is hard to gain through the reading of scientific and medical material on the Aral Sea and the health of the population in the adjoining region.

As the world continues to grapple with this catastrophic disaster, we are becoming more acutely aware of a series of other disasters that too are linked to intensive cotton cultivation in Uzbekistan; in particular, pesticide contamination and the salinization of drinking water. These other disasters are not as well studied as the Aral Sea disaster nor are their boundaries as distinct as the satellite images outlining the desiccation of the Aral Sea; rather, these disasters stretch well beyond the shoreline of the former sea and across Karakalpakstan posing a threat to the health of the 1.5 million people living in the region today.

This chapter is divided into two sections. The first section provides a brief history of the people of Muynak, describing the chain of events that led to their current fate and the major health issues faced by the present-day community. Of all the towns and cities we visited, it was evident that Muynak had suffered most directly from the physical loss of the sea.

The second section details our current understanding of pesticide contamination and increasing salinization of drinking water faced by the population of Karakalpakstan, and highlights some areas that require further research. These are not the only environmental concerns in the region, but are possibly the two most critical issues with respect to the health and well-being of the people and the future generations of Uzbekistan.

MUYNAK: PHYSICAL LOSS OF THE ARAL SEA, ITS IMPACT ON COMMUNITY AND LIVELIHOOD

Muynak presents a special case in this discussion on health in Karakalpakstan, as it illustrates in parallel the unfolding Aral Sea ecological disaster and the sudden and rapid breakdown of a once functioning community. Some historical details are provided in this section for a greater appreciation of the community that once existed in Muynak.

The first impact of the decision to divert water from the Amu and Syr Darya rivers for irrigation purposes was a decline in the water levels of the Aral Sea. This decline coupled with the increasing salinity (up to 12–14 gL^{-1}) of the Aral Sea Basin Region affected spawning and development of roe (Ermakhanov, Plotnikov, Aladin, & Micklin, 2012). In a short span of 5 years, the annual fish catch plummeted from 40,000 tons in 1962 to 20,000 tons in 1967 (Glantz, Rubinstein, & Zonn, 1993). By 1981, salinity exceeded 18 gL^{-1} and in 1983 only 50 tons of fish were caught from the sea, resulting in the complete demise of commercial fisheries in Muynak (Karimov, Lieth, Kurambaeva, & Matsapaeva, 2005). Local subsistence fishing was likewise severely affected.

Historically, the people of Muynak were fishermen or employed in the Muynak fish-canning factory. Karimov provides estimates indicative of the scale of the fishing industry: "1200 fishermen were involved in 12 fishery collective farms. They had 113 fishing ships and caught about 75% of total fish in the country" (Tleuov, 1981). Other forms of employment, ship maintenance, and recreational health spas, also depended on the sea.

The dramatic and sudden loss of livelihood resulted in mass out-migration from Muynak. Local government statistics indicate that the population of Muynak had already declined from 30,100 to 21,600 between 1960 and 1970 (Crighton, Elliott, Van der Meer, Small, & Upshur, 2003), approximately a 30% decrease within the first decade of the disaster. Today the total population of Muynak is approximately 14,000.

The Aral Sea disaster has not been the only factor in continued out-migration. Following the collapse of the Soviet Union and the independence of Uzbekistan, there was a severe economic depression. This caused further out-migration, not just from Muynak, but from the larger Karakalpakstan region and included the emigration of members of the medical and health professions. Overall out-migration from Karakalpakstan, including the abandonment of fishing settlements and out-migration of ethnic Russians (due to Soviet collapse) is estimated at 300,000 (McLeman, 2011).

Deteriorating Health-Care Infrastructure

Post-independence, revenue and subsidies received from the former Soviet Union ceased along with benefits of free access to health care and education. Uzbekistan's health system has replaced the multitiered Soviet model of primary care with a two-tiered system consisting of rural physicians and outpatient clinics (Rechel et al., 2012). The official health financing is collected and spent predominantly at the regional and district levels, which creates disparities in regional health financing and places localities within the Aral Sea region at a great disadvantage requiring greater out-of-pocket payments by individuals to cover health needs (Ahmedov, Azimov,

Alimova, & Rechel, 2007a). A lack of funds also makes it difficult to attract primary care workers to rural areas (Ahmedov, Rechel, Alimova, & Azimov, 2007b).

Médecins Sans Frontières (MSF) has played a vital role in providing emergency assistance in the affected region and has also provided much of our understanding of the health situation in Karakalpakstan since Uzbekistan's independence. A 2003 MSF report described the state of the health infrastructure as: "Services are in rapid decline in Karakalpakstan, as they are throughout much of Uzbekistan. Hospitals suffer from years of under-investment, and lack basic amenities, medical supplies, essential medicines, and diagnostic equipment. Health-care workers that remain are poorly paid and without basic tools for treating diseases and improving services. Government spending on health care in Karakalpakstan was only $US 6.5 per capita in 2002" (Médecins Sans Frontières, 2003).

The current desperate state of the overall health-care system in Uzbekistan is further evident by a recent government decree that enforces stricter requirements for travel by medical personnel (*Moscow Times*, March 30, 2010). Uzbekistan is the only former Soviet Union nation that still requires its citizens to obtain an exit visa before they can travel abroad. The new requirements that have been put in place are aimed at preventing trained doctors from seeking employment outside the country. These measures also discourage medical professionals from traveling to international conferences.

The impoverishment of Muynak's society as well as the people of the surrounding region in Karakalpakstan over the past decades has weakened any ability to provide for even the most basic health care. This has been one of the more disastrous consequences for health as a result of the Aral Sea disaster. Under present conditions in Muynak, maintaining or reinstating a health-care infrastructure, consisting of health professionals from within the same community, would be extremely difficult. And, studying diseases caused by environmental contamination and creating a momentum for conducting such studies is all the more challenging.

Public Health Challenges

Tuberculosis (TB), multidrug resistant tuberculosis (MDR-TB), and anemia – diseases that are associated with poverty – were some of the most significant public health concerns for Muynak in the early 1990s. This section presents some epidemiological data for these diseases.

Tuberculosis (TB) and Multidrug Resistant Tuberculosis (MDR-TB)

With the collapse of the Soviet Union, a dramatic rise in rates of TB and MDR-TB were observed in several countries due to widespread breakdown of TB control systems and declining socioeconomic status (Cox et al., 2005). MSF and Uzbekistan's Ministry of Health implemented the WHO's Directly Observed Treatment Short (DOTS) treatment protocol in 1998 to combat TB in the Aral Sea region. At the time, TB incidence in Muynak was the highest in the region, with more than 250 cases per 100,000 persons per year (Kittle, 2000). In comparison, Karakalpakstan was estimated to be 89 per 100,000 persons (estimates from 2002). The WHO overall estimates for TB in Uzbekistan were 41 per 100,000.

A survey conducted by Médecins Sans Frontières in 2003 found 13% of new patients presenting to TB clinics in Karakalpakstan to have MDR-TB – one of the highest rates in the world (MSF, 2012).

Drug-resistant forms of TB continue to spread in Karakalpakstan, with only 1 in 10 patients receiving treatment (MSF, 2012). Treatment for TB lasts for 2 years, but one in five patients are not cured. MDR-TB is no longer developing only in patients who have failed an earlier treatment, rather 40% of newly diagnosed patients are infected directly via person-to-person transmission.

The wider spread of MDR-TB due to out-migration from Karakalpakstan is considered a major public health threat (Kolb, 2003).

Iron-Deficiency Anemia

At the time of Uzbek independence, the Uzbek Ministry of Health Protection identified anemia among young children as a major public health concern and invited Crosslink International to conduct a population-based survey of anemia in Muynak (Giebel, Suleymanova, & Evans, 1998). This study, conducted in May–June 1993, reported 72.5% of the children of ages 1–4 as having anemia; 26.3, 38.8, and 7.4% were characterized as having mild, moderate, and severe anemia, respectively (cited by Crighton et al., 2003; Morse, 1994). These prevalence rates were noted as being among the highest in the world. Iron-deficiency anemia was the primary form of anemia among the children.

Anemia among women was also found to be extremely high, with a reported 60% of women of reproductive age having anemia. A 2002 UNICEF report also found 98% of pregnant women to be anemic (UNICEF, 2002).

Iron-deficiency anemia is known to cause maternal mortality, low birth weight, poor intellectual development in children, poor brain development in toddlers, poor resistance to infections, and weak concentration and memory resulting in poor learning achievements (UNICEF, 2004). In an effort to address the problem, the Government of Uzbekistan initiated a program in 2003 for iron fortification of flour.

A study published in 2004 among school-aged children in the Aral Sea Region in Kazakhstan looked at anemia related to the sufficiency of dietary iron (Hashizume et al., 2004). The authors estimated iron bioavailability based on the child's consumption of absorption enhancers (e.g., meat, fish and poultry, and vitamin C) and inhibitors (e.g., tea) in the same meal, and their findings suggest that absorption inhibitors in the diet could partly play a role. It was suggested that nutritional education be provided and that iron fortification should not be the lone strategy.

A better understanding of strategies adopted to reduce anemia in the region and the effectiveness of these programs is necessary. Although, environmental contamination has also been hypothesized as a cause for increased anemia, review of current literature does not point to any study that has evaluated this.

Statistics on health in Karakalpakstan, prior to Uzbekistan's independence are hard to come by, and therefore, it is not possible to assess historically the health conditions in the Aral Sea region to comment on what conditions in the region were previous to the 1990s or how precipitous the decline in health conditions have been. One can only assume that the rapidly worsening health conditions in the region were affected, at least in part, by the impact of the ecological disaster, the large waves of out-migration, and the weakened health infrastructure. The situation in Karakalpakstan following the Aral Sea ecological tragedy, and in particular, the case of Muynak that depicts the impoverishment of a community and the resulting deterioration of health, provide important lessons as to the consequences of unsustainable development.

ENVIRONMENTAL HEALTH CONCERNS IN KARAKALPAKSTAN

Some of the greatest concerns to health include the poor quality of drinking water, the increasing salinization of drinking water sources, frequent dust storms, and the presence of agricultural chemical pollutants in the environment and food chain (Small, van der Meer, & Upshur, 2001). Despite these serious environmental impacts, research examining community exposures to contaminants and links to observed health conditions has been sparse, with researchers noting a declining trend in the number of publications covering these areas in more recent years (Crighton, Barwin, Small, & Upshur, 2011). There continues to be an urgent need for greater environmental health research in the region. The following sections focus on two areas of major environmental health concern: pesticide contamination due to agricultural practices and the salinization of drinking water sources.

Pesticide Exposures via Water, Food, and Air-borne Dust

Cotton production in Uzbekistan has relied on the intensive use of pesticides, namely dichlorodiphenyltrichloroethane (DDT), aldrin, dieldrin, and lindane (HCH) (United Nations Economic Commission for Europe, 2001). According to the World Bank, persistent organochloride pesticides (such as DDT) made up about 50% of 100,000–200,000 tons of pesticides applied annually (World Bank, 2009). Organochloride pesticides are known to be persistent in the environment; for example, DDT is reported to have a half-life greater than 15 years (US Environmental Protection Agency, 1989). The half-life of DDT varies depending on the media it is present in: in soil the pesticide can persist for 15 years, in surface waters 350 days, and in groundwater 31 years (Howard, 1991).

Although, DDT is mostly immobile in soil, contaminated soil can enter surface water through runoff as well as seep into groundwater. Such transport of DDT and its breakdown products would thus account for the presence of pesticides in the former Aral Sea bed.

With the Aral Sea bed now exposed, it is considered to be a large open deposit of pesticides, including DDT. Strong winds are thought to transport the pesticide contaminants comingled with the estimated 43 million tons of salt and dust that are dispersed each year from the lakebed (Micklin, 1988). Large plumes of desert dust being transported eastwards are frequently captured by NASA satellite images (NASA, 2006).

While much attention has been placed on the historical use of pesticides, less is known about current pesticide use in the country. Uzbekistan is not party to the Stockholm Convention to reduce or eliminate the use of persistent organic pollutants. The World Bank states that efforts to secure 13 burial sites of obsolete pesticides to prevent public access have been made by national authorities (World Bank, 2009).

The Environment Justice Foundation (EJF) suggests that "highly toxic cotton pesticides that characterized Soviet cotton production" have continued to be used in recent times (2007). Findings of a more recent study of surface water along the Amu Darya also seem to suggest this (Törnqvist, Jarsjö, & Karimov, 2011). The study observed the presence of

DDT and its breakdown products in ratios that suggest deposition of these as being more recent and on-going rather than due to past applications of the pesticides. The authors point out that it is unclear whether this is due to the illegal use of older stockpiles, the recent leakages from storage sites, or via atmospheric deposition.

Such widespread dispersal of pesticides increases the risk of human exposure. Special concern with regard to pesticide exposures was first raised in 2000 when a study noted detectable levels of organochlorides (HCH and DDT) in maternal and cord blood (newborns), and in human breast milk (Atanyiazova et al., 2001). Since then, another study evaluating foods commonly consumed in Karakalpakstan has also found persistent organic pollutants, including DDT and their breakdown products in all foods of animal origin and in some vegetables such as onions and carrots (Muntean et al., 2003). This same study also reported high levels of the highly toxic and known carcinogen dioxin, potentially residue from herbicide use, in these food items. Earlier studies examining the accumulation of pesticides in fish in the southern Aral Sea region found levels to be high but within a range found for other hydro-ecosystems in the world (Karimov et al., 2005).

The intensive use of pesticides, particularly the possibility of the continued use of banned pesticides, is of major concern. Reproductive and developmental toxicity are major health concerns for fetuses and small children. Young schoolchildren who are sent mandatorily as part of their school program to work in the fields picking cotton in the summer months are also at increased risk of pesticide exposures (EJF, 2007).

More studies are required to provide a thorough assessment of the pesticides in use, for determining the geographical spread of pesticide contamination (including over nonirrigated areas) and for the identification of communities at increased risk of exposures to the most toxic pesticides. Levels of pesticides in fish caught in remaining pockets of water in the Aral Sea region and consumed by the population today also need to be further investigated. Safety guidelines for the application of pesticides by trained professionals are essential. Such steps could help minimize the risk of exposures and allow for strategies to reduce exposures (e.g., limiting consumption of certain types of foods when possible).

Salinization of Drinking Water

A recent study provides a better understanding of the geographical variability in the levels of salinity in groundwater in Karakalpakstan (Johansson, Aimbetov, & Jarsjö, 2009). Researchers have characterized the groundwater levels and salinity in an area encompassing Muynak and Nukus. A large part of this study area was formerly irrigated, with only the southern sections that surround Nukus currently under irrigation. Shallow groundwater downstream in the region not currently irrigated was found to be far more saline (average 23 gL^{-1}) than the currently irrigated region (average 3 gL^{-1}). Some of the wells in Muynak have been reported as having levels as high as 19–61 gL^{-1}. Overall, the study found an average salinity of 2.7 gL^{-1}. Water is increasingly unpalatable at salinity levels above 1 gL^{-1}.

According to a WHO (2001) report, only 32% of the population in Karakalpakstan use piped water as the primary source of drinking water (WHO, 2001). About 57% of households use hand-pumps and open wells. Some also rely on their irrigation canals for water and some 2% purchase water from vendors. Access to piped water is higher in urban areas (approximately 60–80%) compared to rural (approximately 20–30%) (Small, Falzon, van der Meer, & Ford, 2003). As a result, rural areas of Karakalpakstan have been reported to have the lowest per capita consumption of water in the world (7.5 L/day/person); rates that are extremely low even when compared to urban Karakalpakstan (257 L/day/person) and the average across urban areas in Uzbekistan (461 L/day/person).

In response to the saline water, desalinization plants have been built in the region, including five small desalinization plants that were installed in Muynak to process about 15 cubic meters of water per day at each plant, a fraction of the water needs in the area (World Bank, 2004). There is a need for an assessment of the effectiveness of measures taken to improve water provided by piped water systems and the achieved capabilities of these desalinization plants. A better understanding is also required of the rate of increasing salinity in groundwater in the region over time and the level of salinity of groundwater among areas that rely primarily on wells. An assessment of whether adequate measures are in place to desalinize water being drawn from wells in these areas is also essential.

Increased salinity in water affects palatability of drinking water, though in general salinity of drinking water is not considered as having health effects. However, it is uncertain how increased salinity of drinking water, at levels observed in Karakalpakstan, might be affecting health in the region.

Studies in the Aral Sea region have found hypercalciuria – high levels of urinary calcium (Ca) and sodium (Na) excretion – in almost 40% of children compared to 13% in the study's reference group (Kaneko et al., 2002). Excessive salt intake and renal tubular dysfunction (Kaneko et al., 2003) due to toxic chemical exposures are possible causes for this. Further studies are required for a better understanding of the cause for these kidney conditions in children. Increased salt intake could also present a problem for certain susceptible groups including those with severe hypertension, diabetes, and patients requiring renal dialysis. Therefore, more research on the health impacts of consumption of highly salinized water is necessary.

THE FUTURE OF MUYNAK AND KARAKALPAKSTAN

The satellite images outlining the desiccation of the Aral Sea have come to symbolize one of the greatest ecological tragedies of our times. The cause for the desiccation of the Aral Sea is straightforward and simple to understand. The lack of health infrastructure, deteriorating public health conditions, and increasing environmental contamination, on the other hand, present a complex situation for the health of the people and the future generations in Karakalpakstan. There is an urgent need for the international community to get involved and collaborate with Uzbeks on reversing the poor public health trends.

The geographical extent of the environmental contamination disasters are not as sharply delineated as the outlines of the former Aral Sea, and identifying the populations at risk is not as straightforward. More exposure assessment research studies are needed to examine environmental exposures in communities of Karakalpakstan.

The recent Uzbek–Korean "Surgil Natural Gas Chemicals" venture will produce polyethylene and polypropylene using gas from the Aral Sea bed (Uz-Kor Gas Chemical LLC, 2012). These bring new hope for better economic prospects to the people of Muynak and with greater economic activity the possibilities for improving health infrastructure. However, it is essential that proper environmental controls be adopted for safeguarding the health of the people of Muynak and the surrounding areas, communities that already face a multitude of environmental exposures and diseases, and who cannot afford further threats to their environment and health.

ACKNOWLEDGMENT

I am thankful to Jonathan Reisman for his assistance in compiling medical/ health literature.

REFERENCES

Ahmedov, M., Azimov, R., Alimova, V., & Rechel, B. (2007a). Uzbekistan: Health system review. *Health Systems in Transition, 9*, 1–210.

Ahmedov, M., Rechel, B., Alimova, V., & Azimov, R. (2007b). Primary health care reform in Uzbekistan. *International Journal of Health Planning and Management, 22*(4), 301–318. doi: 10.1002/hpm.897.

Atanyiazova, O. A., Baumann, R. A., Liem, A. K. D., Mukhopadhyay, U. A., Vogelaar, E. F., & Boersma, E. R. (2001). Levels of certain metals, organochlorine pesticides and dioxins in cord blood, maternal blood, human milk and some commonly used nutrients in the surroundings of the Aral Sea (Karakalpakstan, Republic of Uzbekistan). *Acta Paediatrica, 90*, 801–808. doi: 10.1111/j.1651-2227.2001.tb02808.x.

Cox, H. S., Kubica, T., Doshetov, D., Kebede, Y., Rüsch-Gerdess, S., & Niemann, S. (2005). The Beijing genotype and drug resistant tuberculosis in the Aral Sea region of Central Asia. *Respiratory Research, 6*, 134. doi: 10.1186/1465-9921-6-134.

Crighton, E. J., Barwin, L., Small, I., & Upshur, R. (2011). What have we learned? A review of the literature on children's health and the environment in the Aral Sea area. *International Journal of Public Health, 56*, 125–138. doi: 10.1007/s00038-010-0201-0.

Crighton, E. J., Elliott, S. J., Van der Meer, J., Small, I., & Upshur, R. (2003). Impacts of an environmental disaster on psychosocial health and well-being in Karakalpakstan. *Social Science & Medicine, 56*(3), 551–567. doi: 10.1016/S0277-9536(02)00054-0.

Environment Justice Foundation. (2007). *The deadly chemicals in cotton*. Environmental Justice Foundation in collaboration with Pesticide Action Network UK, London, UK. ISBN No. 1-904523-10-2.

Ermakhanov, Z. K., Plotnikov, I. S., Aladin, N. V., & Micklin, P. (2012). Changes in the Aral Sea ichthyofauna and fishery during the period of ecological crisis. *Lakes & Reservoirs: Research and Management, 17*, 3–9. doi: 10.1111/j.1440-1770.2012.00492.x.

Giebel, H. N., Suleymanova, D., & Evans, G. W. (1998). Anemia in young children of the Muynak district of Karakalpakstan, Uzbekistan: Prevalence, type, and correlates. *American Journal of Public Health, 88*(5), 805–807. doi: 10.2105/AJPH.88.5.805.

Glantz, M. H., Rubinstein, A. Z., & Zonn, I. (1993). Tragedy in the Aral Sea basin – Looking back to plan ahead. *Global Environmental Change: Human and Policy Dimensions, 3*(2), 174–198. http://dx.doi.org/10.1016/0959-3780(93)90005-6.

Hashizume, M., Shimoda, T., Sasaki, S., Kunii, O., Caypil, W., Dauletbaev, D., & Chiba, M. (2004). Anaemia in relation to low bioavailability of dietary iron among school-aged children in the Aral Sea region, Kazakhstan. *International Journal of Food Sciences and Nutrition, 55*(1), 37–43. doi: 10.1080/09637480310001642466.

Howard, P. H. (1991). *Handbook of environmental fate and exposure data for organic chemicals* (Vol. III). Pesticides. Chelsea, MI: Lewis Publishers.

Johansson, O., Aimbetov, I., & Jarsjö, J. (2009). Variation of groundwater salinity in the partially irrigated Amu Darya River delta, Uzbekistan. *Journal of Marine Systems, 76*, 287–295. doi: 10.1016/j.jmarsys.2008.03.017.

Kaneko, K., Chiba, M., Hashizume, M., Kunii, O., Sasaki, S., Shimoda, T., … Mazhitova, Z. (2002). Extremely high prevalence of hypercalciuria in children living in the Aral Sea region. *Acta Paediatrica, 91*(10), 1116–1120. doi: 10.1111/j.1651-2227.2002.tb00108.x.

Kaneko, K., Chiba, M., Hashizume, M., Kunii, O., Sasaki, S., Shimoda, T., … Dauletbaev, D. (2003). Renal tubular dysfunction in children living in the Aral Sea region. *Archives of Disease in childhood, 88*(11), 966–968. doi: 10.1136/adc.88.11.966.

Karimov, B., Lieth, H., Kurambaeva, M., & Matsapaeva, I. (2005). The problems of fishermen in the southern Aral Sea region. *Mitigation and Adaptation Strategies for Global Change, 10*, 87–103. doi: 10.1007/s11027-005-7832-0.

Kittle, D. (2000). Dispatches from abroad: Aground in a sea of TB. *Canadian Medical Association Journal, 163*(12), 1594–1595.

Kolb, A. (2003, Spring). *An ominous flip side: Population dynamics in an environmental disaster zone*. A population, environmental change, and security newsletter. Published bi-annually by The Woodrow Wilson Center's Environmental Change and Security Project, Issue No. 8.

McLeman, R. A. (2011). Settlement abandonment in the context of global environmental change. *Global Environmental Change, 21*(Suppl. 1), S108–S120. doi: 10.1016/j.gloenvcha.2011.08.004.

Médecins Sans Frontières (MSF). (2003). *Karakalpakstan: A population in danger. The impact of the Aral Sea disaster and a worsening economic climate on the health and wellbeing of the people of Karakalpakstan*. Retrieved from http://www.aerzte-ohne-grenzen.de/_media/pdf/karakalpakstan/2003/2003-04-karakalpakstan-report-population-in-danger.pdf. Accessed on July 14, 2012.

Micklin, P. (1988). Dessication of the Aral Sea: A water management disaster in the Soviet Union. *Science, 241*, 1170–1176.

MSF (Online Video). (April 2012). *Uzbekistan: Tuberculosis shows resistance*. Retrieved from http://www.doctorswithoutborders.org. Accessed on July 14, 2012.

Morse, C. (1994). *The prevalence and causes of anemia in Muynak District, Karakalpakstan, the Republic of Uzbekistan. A chart book (results of a study)*. Tashkent: Crosslink International.

Moscow Times. (2010, March 30). *Uzbekistan limits foreign travel for its doctors*. Retrieved from http://www.freerepublic.com/focus/f-news/2483058/posts. Accessed on July 14, 2012.

Muntean, N., Jermini, M., Small, I., Falzon, D., Furst, P., Migliorati, G., … Kolb, A. (2003). Assessment of dietary exposure to some persistent organic pollutants in the Republic of Karakalpakstan of Uzbekistan. *Environmental Health Perspectives, 111*(10), 1306–1311. doi: 10.1289/ehp.5907.

NASA. (2006). *Dust storm over the South Aral Sea*. Retrieved from http://www.nasaimages.org. Accessed on July 14, 2012.

Rechel, B., Ahmedov, M., Akkazieva, B., Katsaga, A., Khodjamurodov, G., & McKee, M. (2012). Lessons from two decades of health reform in Central Asia. *Health Policy and Planning, 27*, 281–287. doi: 10.1093/heapol/czr040.

Small, I., Falzon, D., van der Meer, J., & Ford, N. (2003). Safe water for the Aral Sea area: Could it get any worse? *European Journal of Public Health, 13*, 87–89. doi: 10.1093/eurpub/13.1.87.

Small, I., van der Meer, J., & Upshur, R. E. G. (2001). Acting on an environmental health disaster: The case of the Aral Sea. *Environmental Health Perspectives, 109*(6), 547–549. doi: 10.1289/ehp.01109547.

Tleuov, R. T. (1981). *The new regime of the Aral Sea and its influence on the ichthyofauna*. Tashkent: Fan Press. (In Russian).

Törnqvist, R., Jarsjö, J., & Karimov, B. (2011). Health risks from large-scale water pollution: Trends in Central Asia. *Environment International, 37*, 435–442. doi: 10.1016/j.envint.2010.11.006.

UNICEF. (2002). *Update of the situation of children and women: Uzbekistan*. Tashkent: UNICEF.

UNICEF/Micronutrient Initiative. (2004). *Microelements – Macro consequence*. Retrieved from http://www.unicef.org/ceecis/Uzbek.pdf. Accessed on July 14, 2012.

United Nations Economic Commission for Europe (UNECE). (2001). Environmental Performance Review: Uzbekistan. New York, NY: United Nations.

US Environmental Protection Agency. (1989). Environmental Fate and Effects Division, Pesticide Environmental Fate One Line Summary: DDT (p, p′). Washington, DC.

Uz-Kor Gas Chemical LLC (prepared for the Asian Development Bank). (2012). *ADB Social Audit Report: Surgil environmental and social impact assessment. International Business Centre, Tashkent, Uzbekistan*. Retrieved from www2.adb.org/Documents/Supplementary.../44944-01-uzb-sa.pdf. Accessed on July 14, 2012.

World Bank. (2004). *Seawater and brackish water desalination in the Middle East, North Africa and Central Asia*. Final Report Annex 4 Uzbekistan. Retrieved from http://www-wds.worldbank.org/servlet/WDSContentServer/IW3P/IB/2005/09/08/000160016_20050908165143/Rendered/PDF/335150v50Seawater0Annex040uzbekistan.pdf. Accessed on July 14, 2012.

World Bank. (2009). *Uzbekistan: In preparation of Stockholm Convention ratification*. Retrieved from http://go.worldbank.org/K2FVM5ZK20. Accessed on July, 14 2012.

World Health Organization. (2001, July). *Health aspects of the drought in Uzbekistan 2000–2001*. Technical Field Report Series. Retrieved from http://reliefweb.int/report/uzbekistan/health-aspects-drought-uzbekistan-2000-2001. Accessed on July 14, 2012.

CHAPTER 17

A PHYSICIAN'S OBSERVATIONS OF KARAKALPAK HEALTH

Rakhmon U. Arzikulov, Laylo L. Almatova and Saodat Safarova

ABSTRACT

The discussion in this chapter is intended to provide a snapshot of the health picture for western Uzbekistan. The collection of reliable and valid health data is limited and data available is often incomplete. Here we are forced to rely on our observations as physicians. Life expectancy is low in Karakalpakstan, being about 10 years shorter than the national average for both men and women. Changes in the water quality, soil quality, air quality, and food quality, have had negative impact on health and well-being. Life has become more difficult for everyone, especially for much of the population in rural areas, with very limited access to services. Maternal mortality rates are extremely high in these regions. Morbidity rates in the population, in particular blood and blood-related diseases and respiratory disease, are particularly high. Slight improvements are evident in the past few years for these different health indicators, however, further substantial improvements are essential for the health and well-being of the population.

Disaster by Design: The Aral Sea and its Lessons for Sustainability
Research in Social Problems and Public Policy, Volume 20, 275–281

ISSN: 0196-1152/doi:10.1108/S0196-1152(2012)0000020028

INTRODUCTION

In Central Asia, including the Republic of Uzbekistan, the ecological situation is changing for the worse and should be a matter of global concern. The President of Uzbekistan, Islam Karimov, described it this way in his book *Uzbekistan is on the Threshold of the 21st century* (Karimov, 1997), "Complexity of the situation isn't only in the fact that it has been ignored for years, but also in the ecological dangers of daily life for the local population."

In September 1995, under the leadership of the United Nations, an international conference devoted to the developmental and rehabilitation problems of Aral Sea Basin was held in Nukus. The resulting Nukus Declaration was an important milestone for understanding the greater context of the Aral Sea disaster. The declaration pronounced that the solution for the existing multiple ecological problems and their consequences is not in the power of one republic or country and, significantly, that the involvement of international help is very important.

The Nukus Declaration set forth the following principles for improving the current situation and supporting sustainable development in the ecological disaster regions. First, there is recognition of the importance of water, soil, and all biological resources as the basic foundation for regional economic and social development. Second, there is a need to change current agricultural and forest management practices into more balanced and scientifically based systems. Third, it is urgent to change the efficiency of irrigation systems through wise water usage and support conservation of nature in the region. Fourth, the declaration supports the long-term use of the land and water resources by local farmers. Fifth, it supports initiatives for the improvement of living conditions and health in the region and preserving the cultural heritage of the population (Alibekov, 2003).

HEALTH PROBLEMS ESCALATE

Though the Aral Sea itself has shrunken to a fraction of its former size, the drainage basin of the Aral Sea is directly related to ecological and health problems at a large scale. The territory surrounding the basin is now plagued by terrible environmental problems which directly influence the lives and health of the local population.

These health issues are a consequence of economic planning decisions several decades earlier. During the era of the Soviet Union's planned economy, the Republic of Uzbekistan was directed to grow cotton as a monoculture crop on a massive scale across its western region. This region is arid, but has traditionally been fed through irrigation canals diverting water from the Amu Darya, and also from the Aral Sea. The extent of cotton development under the Soviets demanded much greater water extraction from the Aral than before, and neither land nor water was spared. The result was a steadily decreasing water level in the Aral Sea. Within one generation the Aral Sea was allowed to dry out. At present, there are remnant lakes far to the north of the original extent of the Aral, fed by waters trickling in from the Amu Darya and Syr Darya.

Unfortunately, these economic decisions have completely upset the distribution of water. People who live in this region have lost access to their traditional water sources and are only beginning to get clean water again through filter systems. It has been at least a decade where the water has been more or less problematic, according to location. Many people are drinking unclean water, it is difficult to say how many. We are still trying to understand what else, besides the water, is causing the uneven health outcomes in Karakalpakstan.

CONFIDENCE IN HEALTH STATISTICS

This almost complete desiccation of a massive water body is an ecological tragedy, but the social repercussions are extensive and ongoing. The dried up sea has meant direct negative influences on the local population's health – especially that of children – and brought with it a wide range of diseases and debilitating conditions. The discussion in this chapter is intended to provide a snapshot of the health picture for western Uzbekistan. The authors wish to give a sense of the challenges for public health that have arisen in the last few decades as unintended consequences of agriculture and water management. The intention is to discuss some of the main diseases and health impacts.

For us, the challenge of working with health data is a matter of completeness. As practicing doctors, we have access to a broad range of knowledge and observations within our field. Unfortunately, the collection of statistical data is still limited and can be inconsistent. Some of it is published, some is inaccessible even for these purposes. We are a developing country and the healthcare system has gone through many structural changes in the last 20 years; this is reflected in the data. We were fortunate to gain access to health statistics from the official statistical database, Uzstat, and will refer to findings as appropriate.

HEALTH AND AGE

Life expectancy is a function of many variables including diet, political stability, and occupational hazards. In western Uzbekistan, it is very much a function of the environmental conditions. The life expectancy in Karakalpakstan is the lowest in Uzbekistan and does not extend beyond middle age for men or women. It is about 10 years shorter than the national average for both men and women. What is more disturbing is that death is coming early to the youngest generations. Field research has shown that the teen age group is the most sensitive population to the environmental changes. Their bodies are still in maturation and thus this age group is under ongoing careful observation.

Common Diseases

The lower life expectancy is but one aspect of the total picture. When we look at general health and well-being and what brings people into the hospitals, there are several obvious trends. The general health of the population is compromised by the ecological changes to the surrounding agricultural land, as my colleagues describe in greater detail in other chapters. Scientists and medical professionals from all over the world have realized that for the 35.8 million people who live in the greater Aral Sea region, which in Uzbekistan includes Karakalpakstan and Khorezm, sickness is prevalent, and has been for over 10 years.

Why is sickness pervasive among the children, women, and men who live in the region? There is very little industry or mining activity. Most people have been engaged in farming and small town occupations. The negative influences are brought on by ecological changes at a scale the local people continue to have very little control over. These are changes in the water quality, soil quality, air quality, and food quality, all of which have been negative. Life has become more difficult for everyone and much of the population is rural, with limited access to services (Tulchinsky & Varavikova, 2002).

Second, there has been a demonstrated increase in air pollution resulting from blowing sands, salt, and dust rising in the desiccated lakebed. The winds pick up the salt, chemical residues and mineral deposits, which combine into billowing dust clouds that rise into the atmosphere and blow south and east. The particulate composition has changed both for rain and for snow. The mineral content has been shown to have increased by a factor of 6–7 times. Particulate matter suspended in the air has generally increased by 10–12 times the accepted norm (Lisitsina & Yu, 1987).

In effect, this has caused regular and ongoing air pollution that is showing up as an increase in respiratory ailments and hospital admissions.

The air is no longer pure and clean, and the records from hospitals would suggest that the air is no longer even safe to breathe in some places. Many diseases are on the rise, including respiratory ailments like bronchitis. People also suffer from deeper allergic reactions in their bodies resulting from inhaling the air and the food they have eaten. What is striking about the available health statistics (Statistical Bureau, 2006, 2008, 2010) is how Karakalpakstan compares to the national average. For example, when looking at incidence of respiratory illness, the national average for Uzbekistan is 76 per 100,000 people in 2005 and 68 in 2007; this even reflects a decrease. For Karakalpakstan in 2007, the number was 136 per 100,000 in 2007 and 138 in 2008, about double the national average. No other region has counts this high even though some of them have much more industry. Blood diseases such as anemia and leukemia are widespread among the population of the ecological disaster regions. Karakalpakstan has had some of the highest rates of blood diseases, when compared to other regions and to the national average. However, some progress is being made. We are seeing what we hope is a steady decline in blood disease rates in Karakalpakstan since 2008.

Then there is the matter of birth defects. The number of different birth defects has grown throughout in the ecological disaster zone, however, our concern with the information that we are able to review is with Khorezm, where we are seeing a negative trend for birth defects as environmental conditions are becoming more unstable. Incidence of birth defects documented there show an increase from 138 cases out of 100,000 in 2005 to 185 cases in 2010. When we compare Khorezm to the neighboring area to the east, Bukhara, it had four times the number of birth defects in 2012 (185 vs. 45) (Statistical Bureau, 2006, 2008, 2010).

Khorezm also leads in maternal mortality rates. In the last few years, there have been reports of substantial impacts to maternal health and also child mortality in the international media, particularly through such organizations as UNDP and Doctors without Borders, which are active in the region. Complications during pregnancy, childbirth, and during post-natal care have increased in the Khorezm region, where an incidence of

1,323/100,000 was calculated in 2009 and, the next year, an incidence of 1,813 cases was reported. (Statistical Bureau, 2006, 2008, 2009, 2010, 2011).

CONCLUSIONS

The loss of the Aral Sea is, first of all, an ecological disaster which has had direct impacts on the population living in the surrounding regions. Not only that, there are ongoing challenges that may only get worse in the near future. The region is showing clear signs, as described prognostically in the international scientific literature, of climate change. The meteorological conditions are experienced regularly by the local people especially as geographical effects. These changes include, for example, an increase in the number of dust storm days. As measured by the Muynak meteorological station, these have increased from 29 to 90 days in a year. The temperature in the summer stays at 25–32°C, reaching a maximum of 41°C in the daytime. Winter temperatures reach −5 to −8°C in daytime. Official data on temperature change in the region during the past 10 years is not available, but the changes tending toward more extreme summer highs and winter lows have been consistently observed.

The role of cotton as a crop is changing, and less is being grown in the region currently. However, lingering effects from past pesticide use, and ongoing pesticide use, will manifest in the health portfolio of the people. The changes in agricultural productivity have led to an increase in poverty, which affects the nutrition and caloric intake particularly for women of child-bearing age and young children.

The mortality rate for each demographic including children, women and men, young and old is higher in the regions of Karakalpakstan and Khorezm, which are closest to the lakebed of the former Aral Sea.

Second, there have been improvements in the last two to three years, especially as a result of the Health I and Health II projects supported by international donor agencies, which have led to a stabilization of the mortality rate. This is attributable to ground level improvements made in the region both within the health sector and to the ecology of the Aral region. Some of the donors involved included the World Bank, the Asian Development Bank, and the government of the Republic of Uzbekistan.

The health interventions made included the provision of higher quality food supplies and clean drinking water to the population of the Northern Bukhara, Karakalpakstan, and Khorezm regions. As a part of long-term planning, the organizations created the conditions for decreasing local unemployment. Preventative care has improved through the provision of higher quality medical service to the local population. For example, prophylactic medical observation of the residents and early detection of any health problems is now available at a wider scale. Furthermore, the availability of more medical services has created the conditions for the patients to have access to necessary in-patient medical treatment.

A third and final point is that while the improvements in recent years are significant, there is much more that needs to be done at every level of preventative and curative care to meet the needs of Karakalpakstan as the zone of direct impact of the loss of the Aral Sea, and also the other affected regions of Khorezm, Navoiy, and Bukhara.

REFERENCES

Alibekov, L. (2010). Ecological problems of Uzbekistan. *Journal of problems reclamation deserts* (in Russian). Samarkand University Press.

Karimov, I. (1997). *Uzbekistan on the threshold of the twenty-first century: Tradition and survival.* Tashkent: Uzbekistan Press.

Lisitsina, Yu. P. (Ed.). (1987). *Management of social hygiene and organization of public health services.* Moscow: Medical Press.

Statistical Bureau. (2006). *Public health services of Uzbekistan in 2005.* The Statistical Collection of State Committee on Statistics of the Republic of Uzbekistan. Tashkent: Uzstat.

Statistical Bureau. (2009). *Public health services of Uzbekistan in 2008.* The Statistical Collection of State Committee on Statistics of the Republic of Uzbekistan. Tashkent: Uzstat.

Statistical Bureau. (2011). *Public health services of Uzbekistan in 2010.* The Statistical Collection of State Committee on Statistics of the Republic of Uzbekistan. Tashkent: Uzstat.

Tulchinsky T. H., & Varavikova E. A. (2002). *The new public health.* Jerusalem, Israel (in Uzbek) (Translated from English), Second Tashkent Medical Institute press.

CHAPTER 18

ECOLOGICAL CHANGE IN THE ARAL REGION: ADAPTATIONS BY THE SPOONBILL AND BLACK-CROWNED NIGHT HERON

Mukhtor Turaev

ABSTRACT

*An indicator of the ecological damage done by the Aral Sea disaster is the fate of species that made use of this lost ecosystem. In this paper, two water birds, the spoonbill (*Platalea leucorogia *L) and the black-crowned night heron (*Nycticorax nycticorax*) are discussed. Primary research by the author shows that both species adapted to changed conditions by extending into regions not previously within range and using human-formed rather than natural environments. The spoonbill has made extensive use of new ponds and lakes formed by rising water tables in the Bukhara region. The heron has taken advantage of urban and human occupied settings for expanded settlement. Both are thriving in new environments but their old range in the Aral region is no longer suitable for their survival.*

Disaster by Design: The Aral Sea and its Lessons for Sustainability
Research in Social Problems and Public Policy, Volume 20, 283–290

ISSN: 0196-1152/doi:10.1108/S0196-1152(2012)0000020029

Human activity in nature frequently adversely affects the biological diversity of the area and leads to gradual changes in the structure of the bio-system and eventually to a change in the composition of the region. At the end of the last century, this situation was observed in the Aral Sea. The drying up and increasing salinity of the sea caused a regional change of habitat. Surface water was dramatically diminished and, as a result, wading bird habitat was lost. Wading birds were, in fact, major victims of the Aral Sea disaster.

Since the 1970s, at the same time as the Aral region dried up, human activity was creating new artificial habitats in the central part of the country in the forms of reservoirs, fisheries, movable ponds, and artificial lakes. These water bodies have played an important compensatory role in the life of migratory birds and waterfowl by replacing lost water habitat.

Of particular interest has been the Amu Bukhara channel in the Bukhara region, greatly expanded during this time. This irrigation channel contributed to the rapid decline of the Aral Sea even as it created replacement habitat. The ecological mechanisms are interesting to examine. First, the Amu Bukhara channel contributed to the Aral disaster because it diverted massive amounts of water from the Amu Darya river that would have otherwise reached the Aral Sea. Second, because such irrigation channels in Uzbekistan are unlined, they not only lose considerable water to the arid atmosphere through evaporation, but they also leak into the ground through the sandy soil. Third, the extent of this leakage becomes evident as the water table is engorged. Finally, as the extent of irrigated land increased year by year, groundwater levels in the region rose, forming a number of temporary water bodies, including lakes Karakyr, Khadicha, Zikr, Devhana, Chuchkahana, Hadzhamsayat, and Kumsultan. Since 2000, these new habitats have attracted aquatic and coastal birds, including such species as the White-headed Duck, Glossy Ibis, Little Egret, Pygmy Cormorant, mute swan, and other rare and red-listed species. These are now routinely seen in the waters of Bukhara region.

In this paper, I discuss the adaptive response of two affected species, the rare spoonbill (*Platalea leucorogia* L) and the black-crowned night heron (*Nycticorax nycticorax*). Fig. 1 provides the location for the observations to follow.

SPOONBILL

Spoonbill is one of the species affected by recent changes in the region. Much is known about the ecological distribution of spoonbill in Uzbekistan

Fig. 1. Observation Points for the Spoonbill (*Platalea leucorogia* L) and the Black-Crowned Night Heron (*Nycticorax nycticorax*) Within the Zerafshan Valley. Map: Leslie Raucher. *Source*: United States Geological Survey (USGS), National Aeronautics and Space Administration (NASA), Esri Inc. (2010).

over the time that the Aral Sea declined (Kenzhegul, 2006; Mitropol'skii, 2007; Sagitov, 1987a, 1987b; Shernazarov, 1996).

Until the middle of last century, the main habitat for spoonbill were waters of the Aral Sea and the delta of the lower reaches of the Amu Darya and Syr Darya rivers. As these waters receded, nesting spoonbill expanded to the southeastern areas of the country. Spoonbill nests were recorded in the middle reaches of the Syr Darya and in Lake Aydarkul (Shernazarov, 1996) and a bit later in the southern part of the country, at Lake Alan (Shernazarov, 1996).

Historically the lower reaches of the spoonbill migration was found in the middle Amu Darya and Zarafshan rivers, where the species was sparse (Maslov, 1947; Mitropol'skii, 2007; Sagitov, 1987a, 1987b). Recent observations, however, place spoonbill nests on the lower reaches of the

Zarafshan river, much further east than the historic range. Since 2002, we regularly observed their breeding ponds in the lower reaches of the river Zarafshan. As an indication of this rapid spread into new habitat, my team found an inhabited spoonbill nest on March 27, 2002 in the Gulf of Tudakul. The nest was plunk in the middle of the colony of gray herons and great cormorants (Turaev, Shernazarov, Bakaev, & Rakhmanov, 2003). That day we found six nests in all. We returned after 12 days to find 18 nests. By late April, we calculated 29 nests with clutches. We were witnessing the eastern spread of the spoonbill on a major scale for this rare bird.

That trend continued in the following years. In the same pond in the Gulf of Tudakul, we counted 100 nests in 2003 and 120 in 2004. Continued expansion into new areas has also been witnessed. Since 2005, the birds have changed their place of breeding. For example, we found occupied nests in the fish farm at Kagan, in Lake Hadicha. We know from the spoonbill literature that where bodies of water are widely spread, nests will reappear in the same location year after year. But where multiple bodies of proximate water are found, nest locations will vary from year to year. Thus, breeding spoonbills were observed over time on the same island on lake Aydarkul (Shernazarov, 1996). However, in the Bukhara region, where the newly emergent water bodies are located closer to each other and are beginning to form lake systems, the location of spoonbill nests varies from year to year.

The reason for nesting place relocation is likely to be human impacts that interfere with spoonbill reproduction. This appeared to be the case in our observations. In 2005 nesting spoonbills were reported as follows: 112 nests on Lake Karakir, 65 on Lake Zamonbobo, 12 on Lake Zikr, 6 nests on Lake Hadicha, and 32 nests in the fish farm ponds at Kagan. By the beginning of 2009, the birds had been pushed into yet a new habitat, nesting on lakes (29 nests and clutches) and Chuchkahana, Hadzhamsayat (14 nests with clutches) located along the Amu Bukhara canal.

These nesting spoonbills tend to join multi-species colonies including large and small cormorants, gray herons, great and small white herons, and Glossy Ibis. While these other species tend to mingle, spoonbill tend to nest in a distinct part of the colony, often in the middle. The triangular shaped nests are made from year old reed stalk and placed on cane. The nests are placed at a height of about one and a half to three feet (44–86 cm) above water level. Distance between nests ranges from one and a half to ten feet (44 cm to 2–3 m) depending on the state of security from predators.

Timing of migration and the spring term of nesting for spoonbill depend on the progress of spring. Birds were observed to appear on the waters at the end of March and beginning of April. After a short period of time the birds begin to nest, although the appearance of nests was found to range between the end of March and the end of June, depending on the location. The extremes were found, respectively, at Lake Tudakul reservoir and at the Kagan fish farm.

The earliest eggs were laid in nests in early April and continued to appear through mid-May, corresponding to norms for this species (Shernazarov, 1995, 1996). Birds gradually build their colonies; therefore, they do not begin to nest simultaneously. On the May 18, 2009 observations in the colony at Kagan fish farm, we found clutches in different stages of incubation as well as newly hatched chicks. We made similar observations on May 31, 2009 at Lake Chuchkahana. We found chicks in nests as late as the end of August.

These variations reflect spoonbill adaptation to biotic, abiotic, and human influences. As an example, in 2008, at the Kagan fish farm, the pond where the colonies were formed dried up completely. As a result, the colony became prey to stray dogs, jackals, and cats. We also observed in various reservoirs the effects of strong winds and poaching, hunting, egg collection, and destruction of nests. Under such conditions, the birds are forced to re-nest, accounting for observed nesting as late as the end of June. We observed that birds forced to re-nest were likely to place two and even three eggs per nest, rather than the average of about four eggs of 7 cm × 3.4 cm inches in size.

This period of late breeding in artificial ponds in the steppe zone is risky, particularly during early stages, during laying or incubating, or at the time when young chicks need to be intensively fed. When fishermen approach a colony, the parents temporarily leave their nests. Sometimes nests are abandoned for hours, placing the chicks at great stress and vulnerability. Uncared for chicks often die of sunstroke. Nests can also be lost during rain events with similar outcomes. It is interesting to observe the similar success of nests established in fish farms and in lakes, averaging 2.8–2.9 chicks. Both parents become involved at hatching.

The timing of migration depends on the course of autumn and tends to occur between early September and late October. Once they can fly, chicks completely abandoned the colony and join a new flock in another location. By September, bird populations on the reservoirs begin to shrink. Nests are abandoned as birds move to other bodies of water. We observed 60 spoonbill that remained during late September to early October on the eastern part of Tudakul reservoir in a flock of pink pelican, black stork, and small and great white herons. They did not depart for water bodies further south until the second half of October.

BLACK-CROWNED NIGHT HERON

Another species of interest is the black-crowned night heron. The behavior of this large bird has also changed significantly in the post-Aral period. Traditionally, the heron did not nest in Bukhara region (Maslov, 1947; Sagitov, 1987a, 1987b). In recent years, the status of the species has changed. At present, herons are constantly nesting, with small numbers even staying in mild winters. They have also expanded their range into the lower lands of the Zarafshan River to the east. Data was collected from five districts between 2000 and 2011, including Bukhara city.

Herons often form mixed species colonies with other water birds, although they also live in segregated colonies on islands in lakes, rivers, and Tugai habitats. Nesting herons in the Bukhara region have only been identified over the past decade. In this short period, the number of nesting birds within identified colonies has grown significantly and expansion into new territories in Bukhara region has been observed. In Bukhara oasis, heron colonies have been identified in the reeds of Karakir and Zamonbobo lakes, where 47 nests were counted on April 23, 2003, as well as on the shores of Zarafshan River in Vobkent district, where 11 nests of heron were found on Odok Lake near Khargush village on 14 April, 2002. New colonies were found in Alat city, center of Alat district, in May 2000.

Unlike the spoonbill, the heron tends to use the same nesting places from year to year, as confirmed by our observations. At the same time, the number of observed nests has steadily increased, from 12 in 2001, 24 in 2003, 28 in 2006, 100 in 2008, and 300 in 2009.

In 2004, a new colony in Romitan city, in Romitan district, was found to contain 22 nesting pair of herons. By 2009, this colony contained 861 nests, two-thirds belonging to rooks and the other third to heron. That year, another colony of 56 heron nests, again shared with rooks, was found on the central avenue of Bogiturkon. In early May 2008, a colony consisting of 12 nests was identified in Jondor city, Jondor District. Also in 2008, two colonies were registered in Bukhara City: 21 nests in the Sharq district near the airport and 45 nests opposite the main bus station. The two Bukhara colonies had dramatically expanded the next year, doubling at Sharq and tripling near the bus station. These colonies were all shared with rooks. Expansion within the city next brought colonies to parks, groves, and kindergartens in Bukhara City. An interesting adaptation found in these urban colonies was that nests in the city were located higher than in natural conditions, from 5 to 14 m or more above the ground to ensure safety from teens and terrestrial predators. Like rooks, herons preferred tall trees, such as willow and plane trees at the territory near to the city airport.

A comparison of habitats reveals that birds living in humanized environments are starting to nest earlier than breeding species in natural terrains and recorded breeding more than 20 years ago. A factor may be that, in the reeds, nests are crushed by winter conditions, forcing birds to build new nests every year. In one 2009 field observation at Romitan, herons repaired 34 old nests, took over 17 old rook nests, and built 217 new nests. Heron had to begin this process of nest building as early as late February, contemporary to pairing, in order to be ready for egg laying as early as March 9. In the Bukhara colony, where weather was less of a variable on nest viability, eggs were laid later, on March 23 in 2009 and April 2 in 2008. These dates are considerably earlier than in the past, when laying eggs in Bukhara oasis at Karakir Lake was found to occur between late April and the first half of May (Sagitov, Pukinsky, & Pukinskaya, 1989).

At all of the observed sites, the average fertility of herons ranged between 2.5–3 eggs per laying. In the Romitan colony, mass hatching was observed in late March, with 66 of 86 nests having produced broods by the March 28 visit there. In Bukhara city, chicks did not appear until April 22, almost a month later. At the same time, some couples were continuing to build new nests.

Urbanized terrain includes numerous people, buildings, paved streets, traffic, and a lack of open areas that would normally supply habitat in wet areas. As a result, adult birds cannot forage for themselves, leaving no opportunity to collect food for chicks near the colony. Birds are forced to fly out of the city for food, found in canals, reservoirs, and shallow lakes. Such territories are located at a distance of half mile to a mile (800 m–1.5 km) in Romitan, two-thirds of a mile to one and two-thirds miles (1–2.5 km) in Alat and two-thirds to two and one-third miles (1–3.5 km) in Bukhara. City birds become stressed by other human behavior; we have seen that herons abandoned one Bukhara colony after their nesting trees were felled.

What do heron eat? In Khorezm oasis, two-thirds of the herons' diet is made up of fish (Turaev, 1995). A more detailed study was done of 27 chicks in Romitan colony which had fallen from their nests due to such forces as strong winds. The findings reveal a diet of 38% fish, 27% lake frog, 15% water snake, 7% rodent, 3% terrestrial invertebrates (ants, wasps, mole, Coleoptera), and 9% aquatic invertebrates (shrimp, leeches, water beetles, dragonfly larvae).

In conclusion, heron have adapted to climatic changes in the Aral region by colonizing in several towns and even a big city like Bukhara. They have expanded their range and bred successfully, showing adaptation to the conditions of urbanized terrains of southwestern Uzbekistan. Joint nesting with rooks apparently contributes to the safety in the reproductive period. Nesting and hatching of broods was able to occur much later in urban areas than in rural settings.

Our discussion of both the Spoonbill and the Black Heron indicate that these two bird species have adapted to new range and habitat in the wake of the Aral disaster. Changed environments result in behavioral adjustments. Although these findings offer an indicator of how much the Aral environment has been degraded and species driven off by loss of habitat there, these two species have adapted to new environments even as their native setting was destroyed.

REFERENCES

Kenzhegul, K. (2006). *Ecology and the value of fish-eating birds in the lower reaches of the Amu Darya River and south of the Aral Sea*. PhD thesis, Tashkent, p. 20.

Maslov, N. M. (1947). *Ptitsy Bukharskoy oblasti* (pp. 45–70). Scientific works of Bukhara State Pedagogical Institute, Bukhara.

Mitropol'skii, O. (2007). *Spoonbill – Birds of Central Asia* (pp. 81–87). T.1. Almaty.

Sagitov, A. K. (1987a). *Aistoobraznie – Birds of Uzbekistan* (Vol. 1, pp. 17–54). Tashkent.

Sagitov, A. K. (1987b). *Ciconiiformes – Birds of Uzbekistan* (Vol. 1, pp. S44–S46). Tashkent.

Sagitov, R. A., Pukinsky, Y. U. B., & Pukinskaya, M. V. (1989). Today's settlement copepods and wading birds on the lakes Karakyr Bukhara region. *Fauna and ecology of birds in Uzbekistan* (pp. S193–S205). Samarkand.

Shernazarov, E. (1996). *Antropogennaya transformatsiya fauny, naselenii i ekologii vodnyh I okolovodnyh ptitsi Uzbekistana*. Dissertation, National University of Uzbekistan, Tashkent.

Turaev, M. M. (1995). *Fauna i naselenie ptits vodoyomnih nizoviy rek Amudaryi*. Dissertation, Bukhara.

Turaev, M. M., Shernazarov, E., Bakaev, S. B., & Rakhmanov, R. (2003). New materials for breeding of some species of birds on the hydrophilic reservoirs Bukhara. Contemporary problems of ornithology of Siberia and the Central Asia. *2nd International Ornithology Conference*, Ulan Ude (Vol. 2, pp. 101–104).

United States Geological Survey (USGS), National Aeronautics and Space Administration (NASA), Esri Inc. (2010). Global Land Survey, 2010, Landsat ETM+, 30m scene GLS2010_Enhanced, USGS, Sioux Falls, South Dakota.

CHAPTER 19

ENVIRONMENTAL CHANGE AS A THREAT TO THE KHOREZM HERITAGE

Gavkhar Salaevna Durdieva

ABSTRACT

Khiva is a UNESCO World Heritage site with ancient mudbrick buildings and a complete city wall. These are threatened from below by water and salt, and we seek to understand how the deterioration is happening and can be stopped. We have done investigations in the subsoil over several years to determine that destruction is caused by waterlogging, shifts in the ground, termites, and other damage. The damage has been caused by agricultural practices, especially leaky irrigation systems. Some of Khiva's buildings are becoming structurally unsound from water damage. Our studies at several sites confirm that water diversion will be necessary to prevent further infiltration of salt and water. Khiva's legacy is contained in the ancient buildings which are an important part of Silk Road history. Our research contributes to the understanding how to protect mudbrick buildings from environmental factors and deformation processes.

Disaster by Design: The Aral Sea and its Lessons for Sustainability
Research in Social Problems and Public Policy, Volume 20, 291–304

ISSN: 0196-1152/doi:10.1108/S0196-1152(2012)0000020030

INTRODUCTION: ARCHITECTURAL LEGACY OF THE ARAL REGION

Emerging threats to our historic architectural heritage come at a time when there is growing international interest in our cultural resources. After a considerable amount of time when Central Asia existed within the confines of the Soviet Union, Uzbekistan, and its neighbors have emerged as diverse nations sharing common waterways and cultural heritage. That cultural heritage is alive in the customs, dance, music, arts, and crafts of the region. The cultural heritage is also found in a built environment created during the thousand-year history of the Silk Road.

The region south of Karakalpakstan, site of the Aral Sea, is known as Khorezm after the ethnicity and historical lineage of its people. From the 12th to the 20th centuries, Khorezm was under Khorezm's shahdom. Although there were many great cities in the region over this sweep of time, the best representation of the period's architectural achievements remain in the city of Khiva, located south of the Amu Darya not far from Uzbekistan's current southern border with Turkmenistan. Khiva's walled inner city fortress, the Ichan Kala, continues today in a state of remarkable preservation (see Fig. 1).

Fig. 1. General View of the Complex "Ichan Kala" Khiva. *Note*: Photo is taken by author Gavkhar Salaevna Durdieva.

Khiva is one of the most ancient cities of Central Asia. The architectural monuments of Khiva are concentrated within the limits of Ichan Kala. Here the historical record merges with the living culture directly because so many of the ancient buildings are intact and inhabited. At 26 square hectares, the "Khiva State historical architectural museum-reserve Ichan Kala" is one of the largest in the CIS, a Commonwealth of Independent States formed after the breakup of the former Soviet Union. Khiva has been granted status as a World Heritage site by UNESCO (Mankovskaja & Bulatova, 1978). Interest in tourism is increasingly evident.

There is every reason to believe that Khiva, the former capital of Khorezmia, has retained the classical flavor and feeling of our old historic cities more than any other city in Uzbekistan. Khorezm has more than 100 world famous masterpieces of architecture in its territory, but the massive structure of Ichan Kala ("Inner city") and the oldest part of Khiva, is the most important. Founded in the first centuries of the Common Era (CE), Ichan Kala is officially preserved both as a state museum and as the first historic city in Central Asia on the register of the UNESCO World Heritage Cities. With its dedication on December 12, 1990 it was given the byname "A museum under the open sky."

What is preserved today has a long history as a bustling city in the desert. From the 10th to 20th centuries, the Khorezmian shahs kept their residence here, and later the Khiva khans their palaces of feudal nobility. The inner city also housed the mosques, madrassas, living quarters, and bathhouses of the common people, even mausoleums. The buildings and alleys alternate in a contrasting rhythm. The Ichan Kala museum reserve still has its surrounding city wall, which measures 2,200 meters. Partially restored and fully approachable on foot, the wall's foundation is more than 2,500 years old. There are many architectural monuments like Ichan Kala in the world, but only the Ichan Kala museum reserve has retained its integrity in quite this way. One of the challenges for us today is thus to study the construction techniques with the scientific methods available to us, so that we may preserve the understanding of the ancient builders for our future generations.

ENVIRONMENTAL THREATS

Some architectural monuments in this ancient region have weathered hundreds of years exposure to repeated natural calamity, including flood, drought, and storm. They have also suffered from innumerable acts of human folly, including war, theft, and poor water management. What is unique about our time is that this irreplaceable legacy is today exposed more and more to "technogenic" influences. Humans, not nature, are the source of the current crisis and their impact has been magnified by the scale and energy amassed in their actions. Impacts to architecturally and historically significant buildings and settings are, therefore, secondary impacts of human-caused environmental disaster. These threats are intensified by environmental changes tied to the whole Aral disaster.

The roots of today's environmental problems can be found in historical changes to landscape and water. Of particular importance is salinization. This threat to our buildings is greater today than ever before. The decision to intensively cultivate huge territories of desert created unprecedented conditions on a massive scale. Irrigation ditches bringing water for crops were not insulated and thus seeped water into the ground. As moisture accumulated in the ground, the level of subsoil waters rose gradually. When water is forced up through the mineral-rich subsoil, sulphates and chlorides naturally present in the soils are pushed to the surface. These salts become highly concentrated in the soil. Such saline soils, found throughout low parts of Central Asia, are responsible for secondary salting and declining harvests.

In other chapters, this phenomenon is discussed for its relevance to agriculture, drinking water, and health. Here my concern is the effect of salt intrusion on architectural monuments and historically important buildings, specifically how it deforms structures and intensively erodes walls and foundations. This problem has emerged as a significant threat to the historical legacy throughout the cities located in low areas of Central Asia. In Uzbekistan, the threat is most evident in the regions of Bukhara, Khorezm, and Karakalpakstan.

The most dramatic of the regional changes is, of course, the decimation of the Aral Sea. Superseded only by the larger North American Great Lakes and the Caspian Sea, the Aral Sea ranked fourth in size among the inland water bodies of our planet. The threat to the sea and the threat to our buildings has been the same, excessive diversion of waters from our nearby Amu Darya and the more northerly Syr Darya.

Up until the twentieth century, the Amu Darya used to reach all the way to the Aral Sea.[1] It has allowed civilizations to thrive on its banks and replenish the water lost to evaporation in the Aral. Today its flow has been diminished to a range of fluctuation from 45 to 95 km^3/year, with a mid-annual debit of 79 km^3/year. The next river in importance, the Syr Darya, originates from mountain ridges of the Tian Shan range and constitutes a volume of some 38 km^3/year. Historically, an average of 50 km (fluctuating between 65 and 35 km) water from these two sources discharged into the

Aral Sea Basin, the rest were used for irrigation, or were lost to filtration or evaporation. Just from natural evaporation, the surface of the Aral Sea decreased by 1 m annually (Holmuradov & Khasanov, 2003).

BUILDINGS DEFORMED BY SALT

One can easily perceive how the changing surface waters affect the natural environment, but the effects on the built environment are no less dramatic. There is a clear boomerang effect resulting from the ecological changes instigated by irrigation practices. More and more of our ancient buildings suffer the cost and direct damages of modern practices. As noted, the visible impact in ancient cities is the deformation of historical, architectural monuments because of salt intrusion through the subsoil. The salt rapidly degrades the building substrata at the foundations and bases of walls. Salt intrusion leaching up into the higher brickwork is more than a minor concern.

However, other threats combine with salt intrusion. As the waterlogging of the ground occurs and salts are shifted upward, the structural capacity of the ground is reduced. The increasing instability of the soils deforms the surface and also causes the buildings to sag. Although all buildings made of traditional building materials – local clays and mudbricks – are affected, our concern here is with historical monuments of architecture. The structures, especially in the Ichan Kala, tend to be massive in size, heavy, and inherently prone to structural damage as the ground underneath shifts its shape and firmness. Some other mudbrick buildings beset by salt intrusion brought on by waterlogged soils in Karakalpakstan and Khorezm have collapsed.

Then there are the winds. In just a year, between 15 and 75 million tons of dust blow off the drained and exposed bottom of the former Aral Sea. These dusts are a mixture of sand and salt. The wind comes from the north and deposits dust across the southern Aral region. This mixture can settle onto the land near the Aral Sea at the astonishing amount of 520 kg per hectare (Holmuradov & Khasanov, 2003).

Mineralized subsoil waters have had substantial effects in all of the historical centers located in low areas. Most notable are the impacts to the sizable oasis of Bukhara to the east, as well as smaller settlements in Karakalpakstan and Khorezm. These cities share one attribute: they are all located in the extensive valley of the Amu Darya. The large-scale proliferation of irrigation and the insufficient insulation of the canals noted above have led to a sharp increase of subsoil waters and the concomitant degradation of the region, departing sharply from long-standing historical patterns of sensible use. Magnificent buildings that have lasted for many hundreds of years in this arid region may soon crumble if no action is taken.

ARCHITECTURAL LEGACY THREATENED

Over the past decade, it has become increasingly clear that these environmental forces are causing extensive damage to world famous masterpieces of architecture in the historical cities near the Aral Sea, especially in Khiva (Mankovskaja & Bulatova, 1978; Rashidov & Sirojiddinov, 2003; Tamirshunoslik Open Company, 1998). An example of how human actions have direct consequences for the region can be found only 12 km south from the city of Khiva. There, the neighboring country of Turkmenistan has since 1982 constructed an irrigation canal, the Tashauz branch channel, that at full length is 180 km long and has a capacity of 400 m/second (Holmuradov & Khasanov, 2003). The seepage does not recognize the international border, but already the channel has deteriorated the condition of 175 thousand hectares of irrigated arable lands in Khorezm. Moreover, it has further aggravated the hydrogeology of the ancient city of Khiva.

In particular, the salted subsoil waters have proved to be physically and chemically aggressive to building materials. As a result, intensive erosion has been occurring both underground and aboveground to monuments, causing deformation and even in some cases full destruction. These effects are evident and threaten catastrophic destruction to some clay-walled structures in Karakalpakstan. The repercussions for Khiva's Ichan Kala are also increasingly apparent. Striking deformities are evident in some monuments of the Ichan Kala, including inclination of minarets in the Saidniyoz Sholikorbai complex, the Djuma mosque, and the madrassa Kutlug Murad Inak. The ancient, walled inner city of Khiva is pockmarked with signs of structural decay, insect infiltration, and water damage. The salt lines above human height are reminiscent of high water marks in flood-prone areas.

Termites Moving In

The process of bogging and secondary salting has also affected biota. An example of such a secondary impact is seen in the behavior of termites. In response to the changes, termite activity, and aggression has increased. Termites have started to intensively destroy structures. Their attack on the wooden portions of historical monuments is quite astonishing to behold.

Fig. 2. General View of the Wooden Constructions Damaged by Termites at the Yusuf Yasavulboshi Madrassa in Khiva. *Note*: Photo by Gavkhar Salaevna Durdieva.

Recent research carried out by the Institute of Zoology of the Academy of Sciences of the Republic of Uzbekistan has shown that out of 56 structures within Khiva's fortress Ichan Kala, termite damage was evident in 31 of them, with severe damage in five structures. The damaged structures include such historical monuments as Kunya Ark, Palvon Kari, the Djuma mosque, the minarets of Khorezm's shahs, and the mausoleum of Pahlavon Makhmud. Fig. 2 illustrates the damage.

THREAT TO FOUNDATIONS

Of particular concern is termite damage to the underground foundations of monuments. It is impossible to preserve structures unless one can assure the reliability of their bases and foundations. Buildings undermined by termites, by subsidence in shifting soil, by salt intrusion, and by waterlogging can become structurally unsound. It is therefore necessary to monitor changing levels of subsoil waters, rot, termite damage, and other physical factors that deplete the building's resilience. At the same time, it is necessary to monitor human activity and use of buildings because this is another factor affecting the stress and strain on foundations that may have been physically weakened from environmental causes.

But how can one see problems buried in the ground in time to take action? My work has employed an imaging technology called "multi-horizontal complex electro-exploring." The technique allows the scientist to see what is occurring underground around the foundation of buildings.

In a longitudinal study from 2003 to 2010, seismic and electromagnetic data was collected at historical monuments in Ichan Kala. Bore pits were then dug to confirm and sharpen the accuracy of the method. Special attention was paid to foundations of buildings already showing evidence of deformed structures and foundations, including the Djuma mosque, Kutlug Murad Inak, and Saidniyoz-Shalikorbai. I gained confidence that the approach I employed was reliable. What emerged from the assessment was a basic finding – deformation occurs in formerly dry areas now inundated with water, so-called "rehumidifying zones."

The focus of the current research is to determine the durability and habitability of these monuments. While the research continues over time, the ecological conditions worsen every year. We know with certainty that human activities of water management are causing these problems and that the "technogenic load" continues to increase on the soils and clays. Two loads of new damage are caused for every load of improvement we can achieve.

A better portrait for the approach can be gained by examining two case studies.

Case One: The madrassa Kutlug Murad Inak. Remote testing revealed a "rehumidifying zone" was located beneath the angular eastern column of the endangered Kutlug Murad Inak madrassa, shown in Fig. 3.

An investigation was conducted to identify the source. Bore pits revealed the water table at a depth of 3.5 meters, ruling this out as a likely culprit. Instead, cracks were discovered in sewage pipes installed during the Middle Ages but still in use, allowing sewage to wet an area between 2 and 3.5 meters wide. The usual cause of rehumidifying zones was not present, but the problem was.

Case Two: The Minaret of the Saidniyoz-Shalikorbai Mosque (see Fig. 4).

A visual inspection of the minaret readily reveals that it has been significantly deformed. Moisture on the walls of the minaret was found 5 meters above the ground along with white stains, indicating corrosion of the clay-surface coat. An even more serious issue was an evident tilting of the minaret. The minaret is 30.5 meters high and 5.3 meters wide at the base.

Fig. 3. Two Views of the Endangered Minaret at the Madrassa Murad Inak in Khiva. *Note*: Photo by Gavkhar Salaevna Durdieva.

The trunk of the minaret is noticeably inclined toward the southwest. Measurements between 1998 and 2010 revealed that the tilt increased 152 millimeters over that 12 year interval. But what was the source of these problems? Electromagnetic testing revealed foundation degradation and a rehumidified zone. Bore pits were made in 2002 and 2009 (see Fig. 5) to confirm this finding. The water table was found to have increased 70 cm during those seven years (Fig. 5a and 5b).

Observations of a rising water table over multiple year periods do not clarify whether change is progressive or whether water levels fluctuate over shorter intervals. To address this question, a survey of water tables was undertaken for different sites within Ichan Kala at intervals between January 10 and June 21, 2004 (see Fig. 6).

Fig. 4. A General View of the Minaret of the Saidniyoz Shalikorbai Mosque. *Note*: Photo by Gavkhar Salaevna Durdieva.

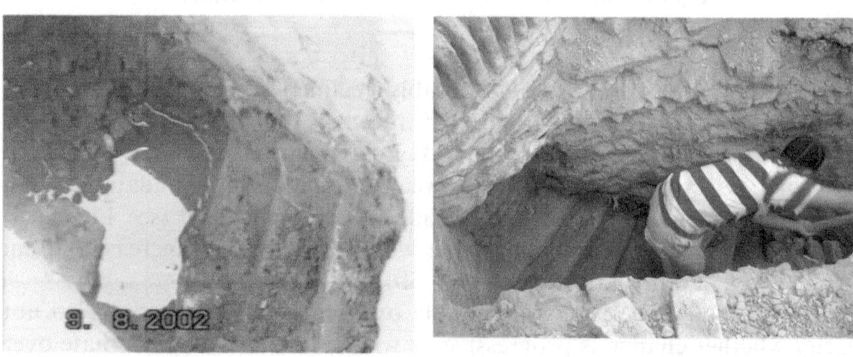

1-Shurf (2002) **2-Shurf (2009)**

Fig. 5. Installed Bore Pits at the Endangered Minaret in the Complex of Saidniyoz-Shalikorbai Mosque, Khiva. *Note:* Photo by Gavkhar S. Durdieva.

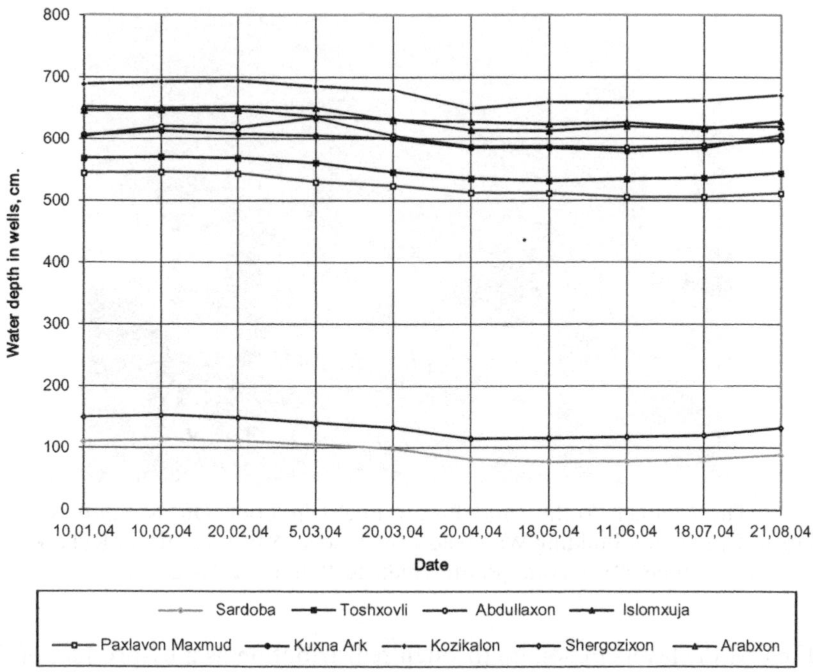

Fig. 6. Change in Level of Subsoil Waters over Time (2004). *Source*: Gavkhar Salaevna Durdieva.

Measurements during this half a year reveal some variation over time. We draw several conclusions from this. First, the water is confined within the contours and density of the soil and surface bedrock. In Ichan Kala the depth fluctuates between 2 and 7 meters. Fluctuation also occurs according to the month. The water table sinks to its lowest level during drought periods in July. The maximum water table level in Ichan Kala was observed in February–March, when substantial rain falls. During this period, the water table reaches between 1 ½ and 6 ½ meters above the bedrock. As this water pushes upward through the ground, it inundates the foundations, flooding the surface. It is a strange concept to think about the surface level being flooded by underground water. But this condition, as discussed above, is the most dangerous for the ancient buildings, soaking the foundations of structures both above and below ground and bringing salt not only to the surface of the land, but as far up building walls as the flood level can take it. One does not need to mark the high flood here; it is marked by the salt line.

Fig. 7. The Chapter Author and Editor Edelstein Look On as Editor Gadaev Points to Cracks in a Building Wall below the Evident Salt Line (Ichan Kala, Khiva, June 2011). *Note*: Photo credit to Ilkhomjon Niyazov.

Finally, the layers of soil go through redistribution as a by-product of the water level rising and falling. As the water moved up and down through the subsoil (with a range of 0.5–0.8 m in a year), the soil particles in the top part of the aeration zone become salinized and more compacted, while the particles in the lower layers of the aeration zone become separated and the salts wash out of them (Fig. 7).

A DEADLY COMBINATION OF WATER AND SALT

It is evident from these results that, when the water table rises sharply, underground waters soak the Ichan Kala. Several undesirable processes are then set in motion that, literally, undermine the architectural treasures clustered there (Rashidov & Sirojiddinov, 2003). First, humidity increases above the saturated area. Humid conditions, in turn, weaken the ground, resulting in additional subsidence below foundations. The resulting inconsistent ground thickness means that a building is now resting on an uneven base. As a result, it shifts and the structural integrity of the entire building becomes threatened (where the hand is pointing in the photo, see Fig. 7).

Second, as the water table rises and falls and rises again, salts are forced upwards into air pockets and accumulate in the top margins. This, in turn, unleashes further deterioration of the environment. Soil compaction occurs on the site. The soil becomes less porous and less water filters through. As a result, when it rains, water pools on the surface rather than draining; moisture accumulates under the bases of the foundations. Buildings waterlog above and below ground simultaneously.

Finally, as the water table lifts and then lowers, moving within the air spaces in the upper soil, the ground level landscape is altered as soluble salts dissolve and are washed out, leaving the ground weakened and uneven. Instead of sitting on firm ground, foundations now sit on inconsistent and structurally diverse soils with the predictable result that buildings are prone to settling, cracking, and shifting. Such conditions are not uniform, so that one building may be little affected while the next is severely deformed to the point of being threatened.

Although efforts can be made to protect and restore each threatened building, a much more appropriate mitigation would be to address the cause of the problem. To avoid inundation of building foundations from rising water tables and surface accumulation of water and salt, drainage needs to be dramatically improved for the entire region surrounding the Ichan Kala complex. Repair of existing drainage systems inside the walls of this ancient fortress must be undertaken so as to not harm its historic integrity. Often keeping drainage systems cleaned out is all that is required.

CONCLUSION

Modern human land use has unleashed a destructive force into the environment south of the Aral Sea. Magnifying problems caused by earlier residents a millionfold, these new demands on the environment drive water up through the ground to flood the surface and salt up from the underlying soil and rock. The process degrades and destabilizes buildings that have withstood prior tests of time. Unless water can be diverted away from these architectural monuments, our historical legacy remains at grave risk. Where damage is already evident, as with the leaning of minarets, expensive restoration efforts are required. If no actions are taken, an even more costly result will occur, the loss of a priceless architectural legacy. Ichan Kala today remains a living museum, both an ancient and a modern community set in the midst of these endangered architectural gems from the past. Will its life be extended or terminated?

NOTE

1. Some sources indicate that the Amu changed course over time and drained into the Caspian first, even had several outlets, then more recently drained only into the Aral, but now the Amu does not reach its former destination in the Aral (*Ed.*).

REFERENCES

Holmuradov, R. I., & Khasanov, A. Z. (2003). Design of monuments of architecture of Samarkand. Samarkand.

Mankovskaja, A., & Bulatova, V. (1978). Monuments of architecture of Khorezm. Tashkent: FafurFulam.

Rashidov, T. R. & Sirojiddinov, Z. S. (2003). Problemy obespecheniya soxrannosti pamyatnikov arxitektury (In Uzbek). *Problemy mexaniki.* [The problem of maintenance of safety of monuments of architecture, *Mechanical problems.*], *5*, 7–13.

Tamirshunoslik Open Company. (1998). *The conclusion on engineering-geological conditions of minaret Djuma in Khiva.* Unpublished report. Tashkent.

CHAPTER 20

WHOSE DISASTER IS IT ANYWAY? ROMANCING THE WORLD HERITAGE STATUS IN UZBEKISTAN

Flavia Alaya

ABSTRACT

Over the past two decades, four Uzbek historic cities and four "intangible cultural heritage" traditions have been raised to the World Heritage List. Yet none of these is in Karakalpakstan, and a painful disconnect appears between the zeal to protect the cultures and monuments of southern Uzbekistan and inaction in identifying and addressing the huge cultural as well as environmental losses most directly associated with the death of the Aral Sea. In this chapter, a U.S.-based cultural historian and conservation-preservation practitioner offers impressions and cultural and spatial material analysis of some of the historic places included on a recent study team tour of Uzbekistan to explore the impacts of the Aral Sea disaster. It is apparent that three-term president Islam Karimov has made culture a linchpin of his program of Uzbek growth and security, cultivated a higher and higher profile for his regime within UNESCO, and

Disaster by Design: The Aral Sea and its Lessons for Sustainability
Research in Social Problems and Public Policy, Volume 20, 305–322

ISSN: 0196-1152/doi:10.1108/S0196-1152(2012)0000020031

focused intensely on interpretation and material conservation of historic sites favoring the themes of his own regime. The focus of this chapter is the disconnect between the emerging national and achieved Aral Sea narratives.

> The scientific courage and unrivalled scholarship of the great Amir Temur's grandson Mirzo Ulughbek demonstrated in the conditions of the medieval ages cannot but astonish the contemporary scientists. The life and scientific activity of Ulughbek stands as one of the founding stones of our people's spirituality and shows what a profound significance was paid to develop the fundamental sciences in our land, so many centuries ago.
>
> — Islam Karimov, President of the Uzbek Republic, quoted on the national monument at the Ulugh Beg observatory in Samarkand

INTRODUCTION

In spring of 2011, an eclectic set of American disciplinary specialists made its way to the far corners of Uzbekistan, drawn together by a common interest, not just in seeking out the state of the Aral Sea in its slow but apparently inexorable disappearance, but also in teasing out the environmental and economic policies leading to the "disaster" and any prospective policy shifts that might alter future scenarios. It was a group also uniquely (if in some sense hubristically) attuned to looking beyond the Sea to the other ecosystems – economic, agricultural, medical, and cultural – that have impacted its degradation and are secondarily impacted by it.

On balance, our visit was intended as a learning experience for us rather than an Uzbek consultancy with "outside experts." Under the generalist umbrella of the Trust for Mutual Understanding and as guests of a group of Uzbek university scholars, no one except our group leader, perhaps, needed to be an "expert" even in matters relevant to our disciplines in Uzbekistan or Central Asia. My team role arose from a fortuitous multi-layered perspective in cultural history that combines theory and scholarship with experience as a "cultural worker" – that is, as a practitioner of cultural resource analysis and protection necessarily crossing several cultures and disciplines (history, architecture, archaeology). It helped that this ensemble of qualifications included some familiarity with the functions and processes of UNESCO, the organization that talks about, thinks about, refines, and defines cultural resources considered World Heritage sites, and sets the cutting-edge standards for their protection.

WORLD HERITAGE SITES AND THE DEFINITION OF UZBEK IDENTITY

Uzbekistan – at least southern Uzbekistan – is replete with World Heritage sites. Or it is now, that is, since independence. The *Convention Concerning the Protection of the World Cultural and Natural Heritage,* adopted by the General Conference of UNESCO in 1972, was not ratified by Moscow until 1988, and the first application for an Uzbek listing (for Ichan Kala – the ancient fortress city of Khiva) did not occur until October 1989, with official inscription into the list in 1990. The nominations of sites in Samarkand and Bukhara appear to have been delayed over an extended declaration period correlated with the process by which the newly independent state worked to develop a cultural policy of its own.[1]

Like every other aspect of Uzbekistan's national identity formation, this process of *cultural* self-globalization is bound up with the country's *political* self-redefinition. This is not just because the Soviets, until *glasnost* and *perestroika,* may have lacked enthusiasm for endorsing the "universalist" themes of Western culture embedded in the concept of World Heritage (Red Square itself was not named to the list until 1990, the same year as Ichan Kala). More is owed to Uzbekistan's need to create a new post-Soviet identity, a process complicated by its having inherited a national self-concept and even national boundaries designed more for Soviet convenience than as an organic response to cultural or political imperatives of its own. In one instance, even the final listing for a World Heritage site as prominent as Samarkand was postponed until Moscow's influence in defining it was moot, giving the new Uzbekistan (and the World Heritage Board) discretion to include only "Muslim" sites within its complex of historic resources.[2]

From a theoretical standpoint, then, it may be safe to say that Uzbekistan's existence as a post-Soviet nation-state was not, as with some more classic historical revolutions, the *result* of a complex indigenous process of national identity formation – of previously "imagined community" – but only the beginning of such a process. The early and relatively quiet removal of Soviet megamonuments represented only the most obvious symbolic self-reinvention.[3] Since then, Uzbekistan's single two-decade president, Islam Karimov, has evidenced a very clear hand in "re-imagining" this cultural identity, and shaping Uzbek national culture in the image of his claims to continued power.

As with other registers of Uzbek national self-expression, this has been a top-down, not a bottom-up process. What has emerged as a cultural policy is a deliberate and hegemonic reaching back through time, elaborated in several tomes of his *Ideology of National Independence* and enunciated, and against all subsequent ethnic and political differences, to a traditional if somewhat imaginative Muslim past of high art, high literature and high science, and the slow, methodical introduction, again top-down, of symbolic memorials to heroes of traditional "Uzbek" culture.[4] What has been called the "cult" of Amir Temur that initiated this process of replacing former Soviet or Soviet-allied political culture heroes has now yielded to a more varied pantheon, with poet Alishir Navoiy, for instance, and the da Vinci-like Timurid genius Ulugh Beg[5] prominently ushered in to replace former Soviet or Soviet-allied political culture heroes as the subjects of normatively gigantic public sculpture and monuments. Mosques and madrassas – awe-inspiring both in size and complex engineering as well as decorative craftsmanship – have been rebuilt, in some instances from the ground up, largely to interpret that same "founding" period.

But it is also clear that the nation's cultural "branding" through the World Heritage program, centered at its UNESCO headquarters in Paris, has had a distinctive part to play in this carefully manipulated process of symbolic embrace and implicit repudiation.[6] And in relation to the essential subject of our inquiry, it is also fairly clear that this process has clearly *not* included anything remotely touching the "world heritage" significance of the Aral Sea, let alone its other significances, and doubtless never will.

Why, may be even more of a politically vexed question than getting to the root of the Aral Sea disaster itself. It is a question I can explore only through the limited evidence now at my disposal, but I believe even this offers clues to implicit as well as explicit Uzbek standards for the definition or exclusion of other cultural monuments.

WHY *NOT* THE ARAL SEA?

Perhaps the first thing the reader not steeped in World Heritage lore needs to know is that under the wider embrace of UNESCO's World *Natural* Heritage list, the Aral Sea might, in theory, have been eligible for World Heritage listing. The standard normally applied to structures was even amplified in 1993 by the articulation of the category called "cultural landscape," which offered a protective umbrella for sites that may combine natural and cultural environments.[7]

Michael Glantz, the senior scientist who has done so much, so singlehandedly, to call attention to this issue, floated such a World Heritage notion as long ago as 2007. "Providing the Aral Sea with World Heritage Status," he said, "can serve to encourage governments in the region to seek ways to restore the sea to a usable level." He did not spell out how this might actually work, beyond creating one more incentive for a restoration justified for many other reasons, e.g., to restore biodiversity, economic activity, and livelihoods. What he added is instructive, however, if only between the lines:

> It would provide a modicum of hope for the future for inhabitants near the sea, the Karakalpak people, who have been left with little hope and few options short of migrating to other parts of Uzbekistan.[8]

Teasing out Glantz's cultural message from between the lines is not difficult, but still leaves the premise utopian. Moving such an national, even international agenda would require some expressed commitment not just to the economic benefits of restored industry and biodiversity in this region, but to the preservation and, to a degree essential to any effective nomination, even *celebration* of the culture and contribution of the Karakalpak people. I will have more to say about this in detailing the experience of our visit to Muynak, it became apparent to me in the course of our visit that such a heightened profile for Karakalpakstan, which already has a certain dubiously sovereign status within the national consciousness, is not on the Uzbek agenda.[9] Five years onward since Glantz wrote this plea, it may be fair to call his proposed project impractical. The near complete draining of the portion of the Aral Sea, its so-called southern "lung," which lies, or once lay, fully within Uzbek national boundaries, has fairly assured that UNESCO will leave this culturally and (apparently) politically distasteful route unexplored.[10]

One wants to say that this disaster cannot possibly be permanent, that the Aral Sea "problem" must look less dire behind the scenes, at the level of policy formation, than on its face. But nothing even hints at an offer of hope, and the official Uzbek response, practical or theoretical, seems thoroughly evasive at best, calculated at worst. The country is a signatory to IFAS or EC-IFAS (The International Fund for Saving the Aral Sea, with the EC for "Executive Committee"), a Central Asian task force created in response to international outcry, whose Uzbek branch centered in Nukus, the Karakalpakstan capital.[11] But false expectations of its capacity to bring an international voice, perspective, and resources to the problem may only have resulted in a sense of deeper frustration and impotency. In one highly memorable moment of our visit, its white-haired leading official, waving his arm in a gesture theatrically reminiscent of something out of Dickens's Circumlocution Office, told us candidly that studies of the Aral Sea problem, stacked together, would fill his spacious meeting room from floor to ceiling. At the level of private non-profit activism, Yusup Kamalov, also based in Nukus and representing the Union for Defense of the Aral Sea & Amudarya (UDASA) – the only apparent Uzbek NGO with an active interest in the problem – has for years engineered a passionate, near one-man campaign against official inaction. Even he appears to have been rendered ineffectual.[12]

Outsiders offering assistance find a similar Catch-22, a point driven home at a seminar arranged by the faculty at Mamun Institute at Khiva, where presenters outlined with what appeared repetitious clarity how much of the Aral water loss and concomitant soil degradation could be attributed to inefficient methods of agricultural hydration, that might be addressed with simple solutions like concrete or other impermeable liners in the irrigation canals.[13] Advice like this appears to be politely entertained, and routinely ignored, conveying an atmosphere of deathless Soviet *apparatchik* logic: Let us spend all our resources on meetings and studies until the Sea drains away, hopefully taking the problem, or at least this particular form of the problem, with it.

OTHER CULTURAL IMPACTS

As we know, the impacts of this seeming heads-in-sand policy on the rest of Uzbekistan are quite real, meaning actual, visible, physical – and Uzbeks know what they are. Some of their and our contributions to this collection will have spelled out just how such impacts affect the present and future well-being of the Uzbek people – and not just those in Karakalpakstan.[14]

Yet even in terms of secondary impacts, this policy of apparent inanition has also allowed featured World Heritage sites, otherwise so clearly a part of Karimov's Uzbek "cultural revolution," to be measurably endangered. Cultural impact specialists have not been silent, with concern expressed at every level from the local to international that the same agricultural groundwater seepage problem that has slowly but inexorably drained the Aral Sea and weakened the irrigation-delivery power of the Amu and Syr Darya Rivers is also infiltrating the subsoils of cities and undermining the same treasured structures so painstakingly identified, glorified and, by all expectation, *protected* by their World Heritage listings.

Both the historic city of Bukhara and Ichan Kala at Khiva are two of the urban aggregations surrounded by agricultural territory exhibiting clear evidence that rising water tables infused with corrosive salts have not only degraded the water supply and measurably impacted human life and health, but also created a distinctly unhealthy infrastructure environment for ancient buildings. These conditions have resulted in spalling stone and clay above ground, with salt and water weakened, termite infested foundations below. The fine grit of desert sands, rendered poisonous by high concentrations of chemicals in the seabed as the Aral has retreated, blows westward and southward to pollute the air all Uzbeks breathe, but it is also silently sanding the brilliant blues and gold glazed-tile finishes on some of the most exquisite relics of Islamic culture.

These effects, documented by heritage experts elsewhere, are far from lost on the Uzbek cultural specialists we met with at the Mamun Academy, located virtually at the gate of Ichan Kala, the country's first and northernmost World Heritage location.[15] But though these specialists can and do convey their own sense of urgency, and make no explicit effort to hide the origin of these problems in the degradation of the Aral Sea region, they do not appear to have had an impact on policy. It is impossible to know whether such homegrown critique will receive any better hearing as a result of cooperative efforts like this one.

SEEING THE SEA

Still, as Michael Glantz hinted in his provocative World Heritage solution, it is the most obvious of the cultural impacts of the Aral Sea disaster that may actually be the most officially ignored: the loss of the Karakalpak village-based fishing and fish processing communities around the Aral Sea as it existed until the 1960s.

Perhaps the highest profile of these is Muynak, a small city at the southwestern edge of the sea, which we visited on the last day of the northward leg of our journey. It took some work to convince our hosts that seeing the Sea – or what it has left behind in its inexorable retreat – was an essential piece of our mission. But once committed, we left our less-than-comfortable hotel in Nukus at dawn for the long morning ride over rough road to Muynak. After hours of bleak, sweeping vistas of salt-infested Karakalpakstan desert, we were not entirely surprised to find it a near empty shell of a place. Notwithstanding a bit of crumbling industrial infrastructure and a dusty, untended post-Soviet "cultural palace" to interpret it, it might have been a lost city of the desert.

We had no plan to stay, and we had sufficiently tested the state of the Karakalpakstan tourism industry at Nukus the night before to not even consider it. But even in the register of "hospitality" the contrast with the South was dramatic. It seemed as if the only people *not* discouraged from visiting were disaster day-trippers like us.[16]

Indeed, to the extent that Muynak has been recently interpreted as a cultural site at all, *absence* itself has been made the attraction. A massive viewing platform featuring the rigid white sail of a phantom ship, pointing to the sky, stands at the old Aral Sea, shoreline looking out over a parched seabed of wind-streaked yellow sand and brush stretching to the horizon. Around the platform, superimposed on this desolate landscape, a series of billboard-size maps display the dramatic half-century tale of shrinking blue water in heartbreaking 10-year intervals. We had seen such maps elsewhere, but nowhere so effectively, or perhaps with such unintended irony, as in this desolate place. The viewer's eye is drawn to the hulks of some half-dozen corroding fishing ships sprouting from the dry shoreline, dark silhouettes against the bright sand, surreally backlit by brilliant blue desert sky. The scene is so defiantly staged, so dominated by a sense of ghoulish destination theater, that it is difficult to tell whether it is meant to express shame or celebration.[17]

Scholars might disagree whether Muynak alone, or even some combination of cultural resources around the Aral Sea, might have met the canonical targets of World Heritage. But the reality remains that by the standards of cultural resource identification developed over the past half-century this loss, not just of the Aral Sea, but of an entire Aral Sea culture, is its own disaster. Even at UNESCO, among those who cultivate, develop, and protect these standards, there is agreement that cultural sites falling below World Heritage standards need not be regarded as expendable, and are owed an even richer register of appreciation when they are threatened.[18]

Yet for all the cultivation of relations with UNESCO, the option of acknowledging this fading heritage, and perhaps bringing resources to interpret it before it has vanished, has been ignored. Nothing seems clearer, here, in the world of Uzbekistan cultural messaging, than the utter marginalization of this loss as against the officially pursued, highly charged southern Uzbek World Heritage agenda.

ROMANCING UNESCO

It is no secret, as I have said, that the cultural self-globalization of Uzbekistan as part of its work of national self-identification is essentially a top-down process, ruled heavily by a president who understands the potency of a controlled cultural hegemony in the international public sphere. His apparent assumption is unassailable: that in this sphere, wealth (or the appearance of it) attracts more wealth, and power more power. Thus the Uzbek project: to use the glory days of cultural monumentality to "brand" Uzbekistan as rich, as rich now as it ever was, and dissociate it from the subaltern status of its recent Soviet past. The message, in this political environment controllable and deployable at will, makes it a minor matter that the nation's real "riches" may be far from well-distributed, its cultural riches making a kind of smokescreen for the continued economic marginality of many of its parts.

There is also usefulness, indeed a certain political magic, in giving an aura to the present regime as a renaissance of an imaginary past that embraces the formidable power of Amir Temur (the warrior king the English Renaissance made famous as the cruelly unstoppable Tamburlaine) and Ulugh Beg, the generous-spirited scientific intellectual of the period the West knows as the Renaissance. These two culture heroes combine the force of apparent opposites: ruthless conquistador-style power and formidable intellectual leadership, recalling a Muslim world that appeared to flourish economically at the same time that it dominated the universe of science, art, and letters. Invoking this prior imaginary Golden Age (somewhat as Mussolini once did with Rome) has the ancillary function of obscuring unpleasant distinctions – or facts – about the uses of power, or revising them into an apparently seamless discourse.[19] Finally, tying them together (however ahistorically) with the long period associated with the evocative concept of the "Silk Road" – rooted in economics as well as a rich and complex cultural history – opens a further register of cultural "personality." Like a carpet intricately woven on both sides, it becomes a useful message turned inward or outward, affirming the "openness" to foreign economic investment, eastward and westward, that the present leadership cultivates, and an intolerance of the closed-minded local "Islamist fanaticism" that might interfere with it.

Only an examination of the Uzbek educational system might begin to answer deeper questions about how this cultural strategy works in any popular sense, and whether, in a world of global internet fact-checking, it suppresses or provokes critical thinking. But it is perhaps a safe guess that for young Uzbeks these dualisms in tension create a complex and powerful cultural ego ideal, even as they mask certain inequalities, especially ethnic and financial ones, and blur the line between religiosity and religious submissiveness, power and strength, mother wit and real intellectual, artistic, scientific, and political knowledge.

In any case, an example of this controlled mixed message in a culturally interpretive site – something like a time capsule of the national character – can be found in both the structural symbolism and public-space language of the recently opened Ulugh Beg observatory in Samarkand. The visitor is first greeted at the initial stair-level by a modernistically heroic, post-Soviet representation of this intellectual giant, scroll in lap, gazing outward in contemplation of the universe. This is followed by a long stairclimb – real and symbolic, of course – which arrives at a wide, high deck effectively above the city. Here stand twin versions of the towering blue-gold entry archway that has become the signature awe-inspiring architectural feature of almost every official Uzbek monument.

One doorway leads to a small viewing platform for the gigantic astrolabe housed in the observatory, where we are told the gifted astronomer mapped the movement of the stars and sun.[20] Like so many Timurid-period monuments it has been painstakingly reconstructed so as to partially replicate the largely destroyed original. Via the second archway, one enters an interpretive structure offering to tell the Ulugh Beg story in more detail. Inside, we enjoyed the guidance of a youthful docent, impressive in her command of English (like all the young Uzbeks we met). Skilled enough in fielding questions, she was also vaguely annoying in her almost humanoid zeal to rattle off everything she wanted us to know in an almost uninterruptible memorized flow.

Yet even here, what a contrast to play against recent history at Muynak. One senses that this observatory complex may be only the leading edge of something larger planned for this site, but it is already highly congenial to tourism: the handsomely tiled exterior offers a powerful ceremony of entrance that yields to a modernistically bright interior filled with intriguing (if somewhat scantly interpreted) artwork, manuscripts and cultural objects. The displays make no secret of tying together the period of Silk Road openness and the Timurid period of great scientific quests for new knowledge, nor do they hide the desire to create a single unbroken national narrative that is a *bona fides* of cultural power.

It is here, on the main entry wall, that we see the presidential quotation that opened this essay:

> The scientific courage and unrivalled scholarship of the great Amir Temur's grandson Mirzo Ulughbek demonstrated in the conditions of the medieval ages cannot but astonish the contemporary scientists. The life and scientific activity of Ulughbek stands as one of the founding stones of our people's spirituality and shows what a profound significance was paid to develop the fundamental sciences in our land, so many centuries ago.

It is a kind of mission statement for Karimov's view of this complex as a time capsule of the national character, and a perfect medium for his message. Here he can put a primarily Western spin on what is world-historical about Central Asia and at the same time highlight a source of "difference" in a proud heritage which he only hints at (and soft-pedals) in the phrase: "One of the founding stones of our people's spirituality." It is a phrase that, without mention of the word Muslim, subtly ties together not just East and West, culture and power, but science and religion. Clearly an essential political component of Karimov's national cultural project is to bracket together scientific macho and Muslim religious identity without so-called "extremism."

There is certainly nothing essentially "wrong" with this – every nation ultimately seeks to define itself via a master narrative that elevates the past and points the way to its future achievements and national integrity – all the better if that narrative appears to resolve opposites, the more contradictory the opposites, the better. And it is certainly no surprise that, despite the obvious survival of many of the political and organizational structures of power that obtained in the Soviet period (not to mention the use of Russian as the official language), the Uzbek narrative elides the nation's more recent history as a Soviet satellite. Less obvious and perhaps more sinister is that, as with so many "master" cultural narratives, this one also calculatedly sidesteps the present (or recent) ethnic or cultural composition of what now constitutes the Uzbek people. As any anthropologist will tell you, and as southern Uzbek attitudes toward northern Karakalpakstan often inadvertently express, this uneasy mix may be much more complex than can be represented in any Silk Road narrative, and much more of a cauldron of old animosities than a model of Muslim cultural maturity.[21]

Culture in Uzbekistan is clearly a process largely orchestrated by the president and designed to empower his regime. But it is also a family affair, with his own iron hand wrapped in the silk-and-velvet gloves of his two beautiful and worldly daughters, Gulnara Karimova and Lola Karimova-Tillyaeva. Gulnara, or "Googoosha," as she is called by her father (a name she has just adopted for her iconic pop music persona), is considered an artist in her own right, and manages an empire of complex cultural, political, and (decidedly) financial influence from her home in Geneva (Karimova, UzNews.net). Her younger sister, Lola Karimova-Tillyaeva, who lives in Paris, was named Uzbek permanent ambassador to UNESCO in 2004, and since 2009 has been a member of its executive board responsible for formally elevating nominated sites to the World Heritage List.

Lola Karimova-Tillyaeva has managed an impressive portfolio related to culture since the affirmation of Samarkand as "Crossroads of Cultures" in 2001, including the addition of four so-called "intangible" traditions to UNESCO's "Representative List of Intangible Cultural Heritage of Humanity": the "cultural space of the Boysun District" south of Samarkand, in Surkhandarya, and the "classical music tradition of Shashmaqom" in Bukhara, both in 2008, plus "Katta Ashula," characterized by UNESCO as "a type of traditional song that forms part of the identity of various peoples of the Ferghana Valley"; and Navruz, "a tradition of seasonal celebration" in the Syrdarya District, near Tashkent, in 2009.

Also in 2009, she set up the Uzbekistan 2020 foundation "to facilitate cultural and intellectual exchange between the countries of Western Europe and Uzbekistan, as well as promote cultural tourism to Uzbekistan." This process may not have many vocal objectors, or those willing to be vocal. But as a sales program for heritage tourism investment, it promises to be highly effective. These cultural choices also appear to encourage the cross-border and cross-continental collaborations so beloved by the UN. It is clear from the tourism literature of the UN World Tourism Organization (UNWTO) and elsewhere that the collaboration of countries "along the Silk Road" as world heritage tourism destinations is seen as a marketing advantage for economic development.[22]

The importance of these cultural sites to the official image-building of Uzbekistan can be measured by the clear dedication of vast sums to their restoration. The city of Khiva is preserved largely from its seventeenth century core, now a well-inhabited city, actually more of a large village, an exercise in what the western world calls "living history." Life within its walls is in some sense a "reenactment" of medieval Uzbek culture. Religious sites, like Shah-i-Bandi and other mausolea have been completely rebuilt from the ground up. Some have been wonderfully done, actually, if you like that sort of reconstruction. They are not just impressive, they are visited and enjoyed and respected. The Ulugh Beg observatory may be a shameless and expensive display of national pride, but surely much of that pride is justified by any standards of cultural self-interpretation. The display and the narrative are beautiful, and engage smart, talented young people in the show-business of cultural tourism.

Is it fair, then, to say that the Uzbek world sadly divides, between Karakalpakstan and the rest? Some of the most poignant cultural "monuments" in Uzbekistan, not really monuments but anti-monuments, are there, in that big and desert state. Not just the rusting ship hulks outside of Muynak, but the withering potted grasses and curling pictures and posters in its little museum, remembering each worker of the fishing industry in a motley photograph, exhibiting, if dustily, the splendid, vivid oils that vacationing painters must have painted of the fishing scenes at sunrise, at sunset, again and again, half a century and more ago – a collection without curator or interpreters that is unlikely to survive the passage of much more time.

How curious is it that this "cultural palace" of Muynak, actually in some sense a derelict vestige of a passing Soviet culture, and the last articulated human memorial to the Aral Sea, may be in some sense one of the most important and perhaps least visited cultural sites in the country? How much more curious that a similar Soviet era vestige, the artwork collected by Igor Savitsky in the State Art Museum of the Republic of Karakalpakstan, stands similarly disregarded? What international message is conveyed when one of the most extraordinarily representative collections of early modernist painting anywhere in the world – work often repudiated by the Soviets and now buried by the Uzbeks – has been summarily moved to a single floor of a museum that was already not exceptionally large, and the curators forced to crowd some of the most beautiful artwork of the twentieth century into a string of small rooms, four or five deep, virtually on top of each other?[23]

SO, WHAT ABOUT THE ARAL SEA?

Surely it is no accident that the Silk Road cities are about as far as they can be from the scene of the Aral Sea catastrophe and still be in the same country. It is a biggish country, but bad roads and deserts make it bigger. Even from Khiva (the northernmost Silk Road city) in Khorezm to Nukus in Karakalpakstan is a great stretch, let alone from Khiva to Muynak. From the perspective of ordinary national political life in Tashkent, the Aral Sea is far away, a bit like a repressed crime – a repressed crime for which the Silk Road has been made both political and cultural compensation.

It may be appropriate, then, to give Mickey Glantz the last word of hope in spite of experience:

> There are examples of heritage sites that serve as memorials to sad experiences in human history. The Aral Sea, once the world's fourth largest inland sea and now not even on the

> list, deserves Heritage Status as well as restoration. It will take a long time to accomplish this task. (Glantz, 2007)

So it will, Mickey. So it will.

NOTES

1. President Karimov, essentially a former Soviet official who transitioned into autocratic power at home, has made defining Uzbek culture a leading edge of his policy. Four cultural sites that may be described as cities have now been inscribed on the World Heritage List: the fortress city of Ichan [or Itchan] Kala at Khiva in 1990; the Historic Centre of Bukhara in 1993; the Historic Centre of Shakhrisabz, home of Amir Temur (grandfather of Ulugh Beg), in 2000; and Samarkand – Crossroads of Cultures – in 2001. In addition, four intangible "sites" from various Uzbek cultural traditions were inscribed on the *Representative List of the Intangible Cultural Heritage of Humanity* in 2008 and 2009. Reference is made to these later in this essay.

2. The UNESCO Advisory Bodies evaluation for Uzbekistan's re-nomination of Samarkand in 2001 states, "The site of Uzbekistan was first proposed for nomination in 1990, but it was deferred because some major monuments had not been included in the proposal. The site was again proposed in 1991, and the Bureau again deferred consideration, proposing that only Muslim monuments be included in the proposal." A well-informed contributor to the World Heritage Forum (online) suggests, "It would very be interesting to know which 'non Muslim' monuments were present in that earlier nomination!! Perhaps the Soviet opera house or statues of Lenin?" Retrieved from http://www.worldheritagesite.org/forums/index.php?action = vthread& forum = 8&topic = 1581. Accessed on June 30, 2012.

3. See Bell (1999), which addresses the emergent symbolic meaning of replacing the statue of Lenin with the statue of Amir Temur as hegemonic national "father."

4. See March (2002). March describes this series of works as "required reading throughout all levels of education in Uzbekistan and … heavily propagated through mass media, state institutions, cultural associations, and organs of local … administration" (p. 371).

5. This name is variously spelled in English as one word or two, and as Ulug Beg or Uluk Bek or as I have spelled it here, in addition to the spelling (Ulughbek) used in the heading quote from President Karimov.

6. An official presentation at UNESCO's international conference in 2003, at the time of the adoption of the "intangible heritage" convention, ringingly highlighted Uzbekistan's wealth of cultural property for that audience: "the Great Silk Road which provided the region with early links to world trade," and "over 7000 items of cultural property … under state protection, of which approximately 2500 are buildings, 2700 are archaeological objects, and more than 1800 are monuments. Ten cities are classified as historic, including Samarkand, which is also one of four items of Uzbek cultural property to figure on the UNESCO World Heritage List." The only caveat also has the ring of officialdom: "Cultural property is subject to the conflicting demands of research, tourism, town planning and protection of the landscape. However, according to President Karimov, cultural property is above all

national property to be kept as intact as possible for future generations to enjoy." I have not yet determined which "ten cities" are included in Karimov's list.

7. UNESCO officially adopted the convention for "cultural landscape" in 1992 to recognize "Combined works of nature and humankind [that] ... express a long and intimate relationship between peoples and their natural environment." It also adopted a convention for "intangible culture" in 2006, opening another possible option, but one made rather more ironic than they may have intended, now that the Aral Sea heritage has been disappeared.

8. Glantz's argument first appeared in the journal *Ambio*, and was picked up online by *Ogmius*, Newsletter of the Center for Science and Technology Research: "Providing the Aral Sea with World Heritage Status can serve to encourage governments in the region to seek ways to restore the sea to a usable level." Mickey Glantz, National Center for Atmospheric Research. Retrieved from http://sciencepolicy.colorado.edu/ogmius/archives/issue_19/ogmius.pdf. Accessed July 5, 2012. Glantz's (1999) book, which was the go-to source for our study trip, is *Creeping Environmental Problems and Sustainable Development in the Aral Sea Basin* (Boston, MA: Cambridge University Press).

9. Karakalpakstan, something like an Indian reservation in the United States (to borrow a comparison from my study-team colleague, Astrid Cerny), is a "republic" within the republic – in the language of the Constitution: "formally sovereign and sharing veto power over decisions concerning it." But the serious social and economic impacts of this status, which seems to underscore only the limitations of sovereignty, derive from its being effectively unable to exploit its own presumably vast resources, including the oil and minerals now said to lie under the Aral seabed.

10. Recent stories have claimed an emergent rescue of the Sea's northern "lung" owing to an upstream damming strategy. But because this project lies largely within neighboring Kazakhstan and actually diverts water away from Uzbekistan, further depleting the flow to the southern sectors of the Amu and Syr Darya Rivers, the impact of this dam-based recovery may not be felt in Uzbekistan except negatively.

11. "The main purpose for creating IFAS was to attract funds for Aral Sea related projects under the Aral Sea Basin Programme 1 (ASBP 1) and later ASBP 2. A few ASBP 1 projects were funded in the early years but it was generally considered to be poorly managed. The number of Aral Sea projects has been diminishing in recent years and there appears to be no projects funded at present. Trust in IFAS, ICWC, and the whole associated structure has suffered over the years for several reasons: (1) limited progress on regional cooperation on water resources management, (2) frequent and continued conflict within the IFAS/[ICWC/SIC structure], (3) very poor project management under ASBP 1 and others in which projects were not completed and money disappeared, (4) the questionable value of continuing the ASBP in the current, changing water management climate, which is the main role of IFAS. IFAS and ICWC realize that questions have arisen about the continuing value of IFAS, and recognize that international organizations, including UNDP, are skeptical. The IFAS and ICWC people agree that some changes are due if regional cooperation and coordination on water management is to progress. Some stakeholders also raise the point that there may be an important role for the UN to be played in supporting progress and the changes that need to be made to achieve it. ... It is argued that IFAS remains valuable because it is a forum in which the presidents can meet on water issues. This has validity for now and for the short term

because it unfortunately remains necessary for presidents to intervene on water issues and IFAS can play a diplomatic role. *However, the long-term goal must be to depoliticise the management of water, which means removing presidential forums from the water management sector.* [italics mine] ... IFAS is not just a part of, but the official head of, the whole water management structure. Discontinuing IFAS requires a revision of the governing statutes. ... **IFAS is headed by the five presidents and discontinuing it would require discussion and agreement at the presidential level to do so.** It should be noted that the President of Turkmenistan and the Minister of Foreign Affairs of Kazakhstan have both made the suggestion of discontinuing IFAS." Retrieved from http://waterwiki.net/index.php/IFAS_-_International_Fund_for_Saving_the_Aral_Sea. Accessed on July 8, 2012. Emphasis added.

12. Retrieved from http://www.springerlink.com/content/p4u5m8610885350g/. Accessed on July 8, 2012. Kamalov's chapter: "The Aral Sea: A matter of mutual trust" (2010) forms part of the NATO Science for Peace and Security Series C: Environmental Security, and appears to be a stinging indictment of the system that failed to resolve this problem.

13. John Lamers is a senior researcher in the Department of Ecology and Natural Resources Management of the Center for Development Research at the University of Bonn, Germany.

14. Karimov himself, writing in 1998, less than a decade after his accession to power, is unsparing in his evocation of the disastrous consequences of Aral Sea loss, including its health and food impacts. Yet his observation that "the roots ... date back far into the past, but the scope of it has expanded over the last decades" is unspecific enough to lack that reassuring ring of accountability in the present. (1998, 76ff).

15. "For the last decade there have been serious problems connected with preservation of world famous masterpieces of architecture. In particular, the salted subsoil waters, owing to its aggression in relation to building materials as a result of difficult physical and chemical processes, begin intensively to erode underground and aboveground constructions of monuments and as a result deformations often occurs and in some cases their full destruction. As an example, such deformations may bring catastrophic destruction to some clay-walled structures of Karakalpakstan to strong deformations in some monuments of the complex "Ichang-Kala" in Khiva (inclination of minarets of the complex Saidniyoz Sholikorbai, the mosque Djuma, the medrese Kutlug Murad-inak, etc.). The influence of mineralized subsoil waters in the historical centers located in low territories of such oasis as Bukhara, Karakalpakstan, and Khoresm is the most notable. All these cities are located in the valley of the Amu Darya. Large-scale irrigational work described above have led to a sharp increase of subsoil waters that has accordingly worsened historically established geological situation in the region"; see for example the preceding chapter by Durdieva.

16. But see http://www.visituzbekistan.travel/tours/nukus-moynaq-aral-sea-khiva-tour/, where visitors are actually encouraged in hyperbolic terms to make this trip – though of course *not* to stay over. Accessed on July 9, 2012.

17. It is unclear who is responsible for the maps, possibly IFAS? But an echo of this same unintended irony obtains in the words of President Karimov himself, addressing the "challenges to stability and progress" facing Uzbekistan in the twenty-first century in his 1998 book on that theme. With unrelenting use of the

words "ecological disaster" and "global catastrophe" over several pages, and characterizing water taken from the Aral Sea as "drained blood," he elicits the greatest possible drama from the situation: "The Aral crisis is one of the biggest ecological and human catastrophes in recorded history .... Never before has such a case been witnessed in history: in the lifetime of only one generation, the death of a whole sea" (Karimov, 1998).

18. My own 1997 experience at ICCROM, the educational wing of UNESCO in Rome, is telling, since I was invited there on the strength of what Europeans call my "conservation" work on behalf of the Landmark Historic District in Paterson, New Jersey, an industrial founding site dating back only to 1792 – quite a stretch from some others who represented sites thousands of years old, or indeed, from Venice, which we visited together as part of an historic urban planning exercise. I was surrounded in our seminar by seasoned cultural operatives representing very different historic places and cultures from all over the world, and not necessarily on the world heritage list. Everyone shared the conviction that it was our task to interpret almost any local cultural resource, if it was resonant enough, in ways that might be held in international esteem.

19. See, again, Andrew March, who underscores Karimov's approach as an "abuse of history," and explores and argues its ideological and public sphere impacts in detail (March, 2002, pp. 380–382).

20. "In *Zij-i Sultani*, his astronomical table and star catalogue, Beg determined the length of the sidereal year as $365^{d}\ 5^{h}\ 49^{m}\ 15^{s}$, which has an error of $+25^{s}$, making it more accurate than Copernicus's estimate, which had an error of $+30^{s}$. Beg also determined the Earth's axial tilt as 23.52 degrees, which remains the most accurate measurement to date. It was more accurate than later measurements by Copernicus and Tycho Brahe, and it matches the currently accepted value precisely." Retrieved from http://en.wikipedia.org/wiki/Ulugh_Beg_Observatory. Accessed July 9, 2012.

21. But note, below, the direction taken to "embrace" cultural diversity in recent World Heritage listings.

22. Lola Karimova-Tillyaeva's UNESCO website provides an elaborate press kit for President Karimov's bond with UNESCO, citing his visit to UNESCO's headquarters in April 1996 as "a watershed moment in the development of cooperation between Uzbekistan and UNESCO ..., when President Karimov addressed the members of the Executive Board as well as attended the ceremonies to mark the 660th anniversary of Amir Temur, 14th-century statesman." She also marks his attending "an opening ceremony of the international exhibition: 'Dawn of Science and Culture in the Era of Timurids,'" and celebrates "the final meeting of the 155th session of the Executive Board of UNESCO, held in Uzbekistan on November 6 and 7, 1998, which resulted in the adoption of the Tashkent declaration: 'Culture of the world and UNESCO activities in the member states.'" The site names 10 commemorative dates in Uzbekistan's history "marked under the aegis of UNESCO" and alleges the backing of UNESCO for the International Institute of Central Asian Studies in Samarkand, plus eleven UNESCO departments, 10 community learning centers, as well as 46 associated schools, and says that a "significant number of projects have been implemented in Uzbekistan since the country joined the international organization." Although I have been unable to verify this fact from any

other source, the site also alleges that "President Karimov was awarded UNESCO's Avicenna Gold Medal for "his contribution to the development of national culture, the preservation of historic monuments as well as his work in boosting ties with UNESCO." Retrieved from http://lolakarimova.com/en/uzbekistan_unesco. Accessed on July 15, 2012.

23. Opened in 1966, the Karakalpakstan State Museum of Art, also known as the Nukus Museum of Art, houses a collection of over 82,000 items, ranging from antiquities from Khorezm to Karakalpak folk art, Uzbek fine art and, uniquely, the second largest collection of Russian avant-garde art in the world (after the Russian Museum in St. Petersburg). See http://en.wikipedia.org/wiki/Nukus_Museum_of_Art, as well as the museum's own very good website, which describes it as "home to one of the largest collections of archeological objects and folk, applied and contemporary art originating from Central Asia."

REFERENCES

Bell, J. (1999). Redefining national identity in Uzbekistan: Symbolic tensions in Tashkent's official public landscape. *Ecumene*, *6*(2).

Glantz, M. (1999). *Creeping environmental problems and sustainable development in the Aral Sea basin*. Boston, MA: Cambridge University Press.

Glantz, M. (2007). Can a concept (World Heritage Status) save a sea (the Aral Sea)? *Ogmius: Newsletter of the Center for Science and Technology Policy Research, 19*, 1 ff. Retrieved from http://sciencepolicy.colorado.edu/ogmius/archives/issue_19/ogmius.pdf. Accessed on July 5, 2012.

Kamalov, Y. (2010). The Aral Sea: A matter of mutual trust. *NATO Science for Peace and Security Series C: Environmental Security*, *VI*, 5–42, 367–369. Retrieved from http://www.springerlink.com/content/p4u5m8610885350g/. Accessed on July 8, 2012.

Karimov, I. A. (1998). *Uzbekistan on the threshold of the twenty-first century: Challenges to stability and progress*. New York, NY: Palgrave-Macmillan.

Karimova, G. Retrieved from http://realgoogoosha.com/

Karimova-Tillyaeva, L. Retrieved from www.lolakarimova.com.

March, A. (2002). The use and abuse of history: 'National ideology' as transcendental object in Islam Karimov's 'Ideology of national independence'. *Central Asian Survey*, *21*(4), 371–384.

UzNews.net. (2012, February 12). Where does UNESCO Ambassador Lola Karimova actually work? Retrieved from http://www.uznews.net/news_single.php?lng = en&sub = top&cid = 31&nid = 19146. Accessed on June 30, 2012.

SECTION FIVE
DESIGNING SOLUTIONS: SOCIAL, ECOLOGICAL AND TECHNOLOGICAL APPROACHES

During our visit to the Aral Region, our exchange team had a lengthy meeting with a former Karakalpak leader, Ubbiniyaz A. Ashirbekov. Director of the Executive Committee, Nukus Branch, International Fund for Saving the Aral Sea. As we sat in his cramped office in Nukus, he pointed in despair to the bookshelves on his wall, filled to the ceiling with multiple millions of dollars of studies of how to address the Aral disaster. The sea level had fallen and in its place there rose a pile of books. If only they could have been employed to dam the remaining waters they might have served some utility. Much as the Soviets before them, those designing solutions to the region's problems think big. A big disaster requires a big solution with big impacts and a very big budget. But, perhaps a big disaster requires many small solutions instead, each designed to mesh with the others but not so interdependent that the failure of one brings down the lot.

If we are capable of creating disaster by design, then we also capable of designing sustainable solutions to avoid mistakes or try to undue them. The notion of ecological design is a fundamental concept to offset the hopelessness that disaster on the scale of the Aral Sea easily inspires.[1] The simple question is where do we start?

In the prior discussion of agriculture, numerous suggestions were made for shifting to approaches that will reduce the environmental impacts that have had such a negative impact on health and environmental quality in Uzbekistan. These ranged from abandoning pesticide intensive approaches, diversifying crops, improving the efficiency of irrigation channels and employing drip irrigation. Other avenues also come into play to mitigate problems and offer more sustainable solutions. They include efforts to address the problems of potable water affecting populations in the Aral region, shifting to the use of renewable forms of energy and employing permaculture design principles.

This section begins with another contribution from Yusup Kamalov, the scientist and activist who has spearheaded the defense of the Aral Sea for decades. Kamalov's thinking has always been about the sea and the region, but also about the larger issue of ecological integrity, environmental justice and rights of nature.

In Chapter 21, "New Thinking and New Approaches: A Bioregional Response to the Lost Aral Sea," Kamalov works on two levels. First of all, he states the philosophical position for the rights of the Aral Sea and Amu Darya. Then, he offers a detailed proposal for how to manage water in the river system in a way that unites the bioregional system (the river) while addressing the geopolitical reality of five Central Asian states. The combination of these levels makes for a provocative paper, indeed.

The success of a concept such as proposed by Kamalov, as he well knows, requires the development of civil society so that there are forces outside government to push for sustainable alternatives and avert future Aral scale disasters. Western-style organizations such as The Union for Defense of the Aral Sea and Amu Darya are one option. But are there indigenous alternatives? In Chapter 22, "Renewable Social Energy: Mahalla as a Traditional Institution for Sustainability," Tashkent-based population geographer Lola Gulyamova offers an Uzbek perspective on a way forward. She answers the question, "Where is the Uzbek voice?" Describing the traditional mechanisms for participation in decision-making and promoting collective interests, she explains how "mahalla" have served community interests. The point is that societies open up only in ways that they are culturally ready to become open. Modern Uzbekistan is an amalgam of remnants of the Soviet way mixed with forces of globalization and modernization and a strong element of Uzbek tradition. It will be interesting to see how the intrinsic nature of the Uzbek people can work to make the public into a sustainability implementing force.

There are social approaches to sustainability and technical ones. Engineering plays as much a role as does public engagement. The importance of engineering solutions is illustrated as we turn to the question of potable water. Throughout the greater Aral region there is a major health crisis around water and sanitation. In Karakalpakstan and the larger Aral region, drinking water has too high a salinity to be healthful. Pesticides and other pollutants add to the problem. Poor sanitation adds a pathogenic component to the pollution problem as well, fanning disease. Groundwater from deep wells is plentiful in the region. But the long-term reliability of these wells and the cost and inadvisability of drilling new wells hampers access to freshwater. In Chapter 23, "Clearing the Pipes: Providing Potable Water through Well Restoration," Abror Gadaev and Gulmira Boboeva offer an engineering solution to well clogging, the major factor in shortening the lifespan of existing wells.

We next turn to the issues of renewable energy. By relying on the sun for energy directly, rather than on fossil fuels comprised of stored solar energy extracted from the earth, people can meet their needs through a renewable form of energy that comes with comparatively little environmental impact. Solar is an obvious good choice for Uzbekistan, with its ideal 300 sunny days a year and consistent weather patterns. As developing countries like Uzbekistan make choices for the future, bypassing a core dependence on fossil fuels is one way to leapfrog past developed counties with misdirected infrastructure while addressing human needs directly (Sachs et al. 2002).

In Chapter 24, "Renewing with Renewables: Direct Solar Energy Use in Developing Countries," William Makofske presents a primer for how this can be achieved. Makofske illustrates his chapter with applications from the northeastern United States, where he has designed a near-zero energy home for a somewhat more challenging climate than that found in Uzbekistan. Modified to fit the context, his model is completely applicable. What is important as a takeaway from Makofske is that his view of solar and wind energy is a distributed or decentralized view, with energy being generated at the point of use rather than transmitted. He, therefore, does not fall into the scale trap seen with large solar and wind installations which may or may not be appropriate for all applications and which engender greater impacts due to size and number.

In any case, the application to small scale, decentralized locations is what is required in rural and remote areas of Uzbekistan, including the Aral region. We were delighted to find a physicist as a member of our core Uzbek team who had been working on just this context. In Chapter 25, "Getting the Salt Out: Innovative Solar Technologies for Rural Clean Water," Eshkuvat Arzikulov, collaborating here with Qamariddin Srojev, sets forth a plan to use solar collectors for remote water wells both for pumping and water treatment. A close collaborator of Abror Gadaev's, we see them as the perfect team to lead new efforts to provide reliable potable water in the Aral region.

How does a renewable agenda get set for Uzbekistan? In our travels, we encountered considerable enthusiasm for oil and gas extraction activities in Uzbekistan related to its emerging role as a significant regional energy producer. Exploration for fossil fuels is occurring throughout the country, from east to west, including in the bed of the former Aral Sea. In working on our project, we had the good fortune to meet an accomplished young Uzbek whose doctoral dissertation addressed precisely this question.

In Chapter 26, "Renewable Energy as a Key Factor for Sustainable Development in Uzbekistan," Ashraf Khodjaev looks at the policy choices facing his nation and charts a course toward renewable energy choices. Rather than becoming enamored of fossil fuel extraction, he soberly recognizes that at best such activities buy some time for Uzbekistan to develop, in the very near future, the renewable energy infrastructure it will need as a growing country.

If we were to look for an integrative methodology that would bring together the solutions discussed in this chapter, we could do no better than with the permacultural approach espoused by Andrew Jones in Chapter 27, "Permaculture Restoration of the Aral Sea Watershed." Used around the globe with great success, the application of the core principles of this system of design is restorative while capable of creating resilience. Its application to desertified, salinized and barren landscapes has yielded new paths to ecological integrity. One can read this chapter and imagine an approach to the Aral Sea that requires no huge dams or river diversions, but instead invites hundreds, perhaps thousands, of small projects for restoration and environmental health, eventually linked in a healthy ecosystem.

Michael R. Edelstein
Editor

NOTE

1. My favorite book on the subject is Van der Ryn and Cowan (1996).

REFERENCE

Sachs, W., Acselrad, H., Akhter, F., Amon, A., Egziabher, T. B. G., French, H., Haavisto, P., … , Agrawal, H. (Eds.). (2002). Fairness in a fragile world: A memo on sustainable development. *New Perspectives Quarterly*, *19*(3), 8–42. Blackwell Publishing. Retrieved from http://dx.doi.org/10.1111/j.1540-5842.2002.tb00082.x

Van der Ryn, S., & Cowan, S. (1996). *Ecological design*. Washington DC: Island Press.

CHAPTER 21

NEW THINKING AND NEW APPROACHES: A BIOREGIONAL RESPONSE TO THE LOST ARAL SEA

Yusup S. Kamalov

ABSTRACT

In these comments, I share my long-term thoughts on the Aral disaster. I begin with my philosophical position. For the restoration of the Aral Sea to even become possible, we must change the current human relationship to the Earth. Then, within that first context, I turn to a concrete proposal to address water management affecting the Aral region.

PHILOSOPHY

At each stage of development, humankind discovers more about the properties of the biotic and abiotic world that surrounds it. Whether gradually or by revolutionary change, the perceptions of humans about their place in the universe are changing. Such a shift is occurring from an

Disaster by Design: The Aral Sea and its Lessons for Sustainability
Research in Social Problems and Public Policy, Volume 20, 327–333

ISSN: 0196-1152/doi:10.1108/S0196-1152(2012)0000020033

anthropocentric paradigm to a nature-centric one. The anthropocentric doctrine is not accepted formally by any authority, but it appears in the structure of states, in economics, in policies and in the behavior of the majority of the Earth's population. While humankind exploits nature as if it were a slave and they the master, some of us attempt to save Nature from this bondage. The ideas of Copernicus and Galileo were implanted into the mind of civilization during hundreds of years; today we are forced to scramble to effect an immediate change in paradigm. And here is the choice. Either we salvage Nature and rescue ourselves or we destroy Nature and us along with it. Our challenge is to change all systems of mutual relations between man and Nature in order to replace the system that has led to ecological catastrophe with a new one that takes into consideration the interests of natural ecosystems. Our reason is twofold: that these systems provide life to humankind and that Nature has a "right" to be allowed to function. The fact that Nature cannot express itself in terms of human language and protect its interests through courts makes the mission of the present day humanist even more important and honorable. Our task is not merely to improve human understanding but to make possible survival.

NATURE AS A LIFE-PROVIDING SYSTEM FOR HUMANS

Given current levels of technological development, humans can delude themselves easily into thinking that they can live in artificial environments while Nature perishes. This really is a delusion. Alternatively, we can recognize nature as our life support system and vow to maintain it. Even from a narrow economic view, the advantages of this alternative approach are obvious.

To construct a life-sustaining artificial environment would require solving a nearly endless numbers of uncertain problems. In contrast, returning to the Earth as a natural basis for life support simply requires that we adopt, or in many cases re-adopt, known approaches. Our understanding of ecology is sufficiently advanced to amplify conditions necessary for our lives and avoid harmful properties. Technocrats should equally proclaim the Earth as our common home, an indivisible life-providing system for all humankind. Once proclaimed, this recognition allows us to refocus on real obstacles. Most of all, the existence of state boundaries disregard the laws of Nature. They are intended to support the existing socioeconomic conditions of human populations, but they fail to maintain the integrity of natural objects.

ROOTS OF ECOLOGICAL CATASTROPHE

The division of the Earth into territories, belonging to different states and disregarding the unity of the biosphere, is the most serious barrier to solving global ecological problems. Today we are confronted with a long list of states formed throughout history almost entirely through conquest and the ambitions of a few aggressive conquerors. The internal configuration of states has been shaped over time by different cultures. *But never in history have states been formed to address the threat of humankind's assault on Nature!*

State boundaries cut the flesh of the biosphere for illegitimate reasons. Natural resources, even the air, are declared to be a state property! Nonsense is obvious, yet no country admits constitutionally that Nature cannot be the property of any state. Government's concern is only with keeping the spatial and social environment within which that government derives authority to act firmly in its control. Governments are capable of directing citizens to solve some problems effectively. In fact, states remain the only force capable of somehow confronting cataclysms at present, although their power may be used for good or bad.

There are optimistic signs that the state system is evolving. Countries are becoming increasingly open, sharing cross flows of ideas, information, and capital. Boundaries between states become more transparent. The European Union is opening them fully. One begins to see conditions that allow ecological problems to be addressed in a wider scope than in the framework of one single state. In fact, if one considers the existing states as "political-economic units of society," then at the continental scale, it is possible to single out a defining feature of Nature as an "ecological-economic unit of the biosphere" (EEUB). Then it is possible to comply with the law on indivisibility of the biosphere.

The most convenient example of integrated and crucial natural features is the rivers. Even when located within the boundaries of a single state, they are currently much divided between municipalities and private property owners and do not appear as integrated systems. Thousands of firms and companies use the rivers for different needs, garnering enormous profit, but little benefit is returned to the river as it passes through these intermediary contexts. No payments are made to the river for the water taken, for the pollution caused, nor are dividends of added quality given back to the river.

Rivers are "units of the biosphere." Such units require an "owner" to keep the river healthy as an ecosystem and, simultaneously, managing its contributions as an economical enterprise. This owner, acting in the capacity of a "steward," maintains the efficiency of the river's ecosystem by managing the profits generated by the river as a supplier of food, water, transport, and vacation spot. The steward acts as an intermediary between the human socioeconomical environment and Nature, allowing the river to become a full and equal partner in the economy.

THE CONCEPT: NATURE AS AN EQUAL PARTNER

To address the growing ecological crisis, we must grant economic independence to the "ecologic-economic units of the biosphere," beginning with rivers. All countries would agree in concept that a river belongs to Nature. Nature "owns" the human life support system. The river is a part of this system. Once this understanding is ratified by governments sharing a river, we can move to the creation of such "ecologic-economic units." For want of a better term, I will refer here to these units as "stewardship enterprises," charged with serving as an intermediary between the river and society. The stewardship enterprise of that river then sustainably makes use of the river's services for generating income in return for a fee while fully maintaining the health and productivity of the river. The scale, budget, and operations of the steward enterprise would be determined based upon the quality and quantity of water in the delta of the river and inflow to the sea. The enterprise would be monitored by the government or governments sharing the river's drainage to assure, on one hand, that it is fully utilizing the river's services for profit while, on the other, protecting the river fully. Funding of the stewardship enterprisers (e.g., salaries, operations) would be dependent on carrying out both parts of this mission. The stewardship enterprise would further be responsible for protecting downstream populations and systems, and liable for damages caused by the river resulting from such occurrences as floods, droughts, snow, and mudslides. Monitoring would occur along the river, from headwaters to delta. The stewardship enterprise would be motivated to promote innovation for clean facilities along the river that would add profit while protecting or enhancing water quality. Success would add to the income of the stewardship enterprise, but as a public entity, it would not retain its profits beyond certain levels of incentive. Rather, surplus profit would go into a fund controlled by the authorizing states to be spent on such things as access to alternative and backup water and river systems, as might be required.

APPLYING THIS TOOL IN THE ARAL SEA BASIN

Of course, my inspiration for creating the stewardship enterprise concept is its application to the Aral Sea region, where human-made ecological crisis is several decades old and has become a protracted disaster of great severity.

At root in the disaster is a fundamental misunderstanding of the situation. Despite the generally accepted opinion that this region lacks water, actually the main source of misfortune is the overabundance of water! The old technology of irrigation requires the presence of thousands of artificial pools catching runoff of "superfluous" water from the irrigation channels. Ninety percent of these channels are not lined with any sort of waterproof material. The watering of crops is carried out by simply flooding the fields. All of this results in a raised groundwater table, which transports salts to the surface of soils and compromises their fertility. The salted soil then gets washed with more fresh water, which then drains off more impure water into yet more saline pools around the fields, which have in recent years begun to perpetuate as a vicious circle.

One can take the model of Venice, the precious but precarious lowland city of canals in northern Italy, sitting on artificial islands on the Adriatic Sea. All agricultural territories of Central Asia resemble Venice, but on a larger scale. When the Aral Sea was displaced on the upstream and middle stream of the rivers Amu and Syr, the water stayed there in the form of thousands of different-sized artificial pools. Such a movement of huge masses of water over the past 30 years has changed the local climate. More rain falls in winter, increasing land and mudslides in the mountains. The sizes of glaciers in the Pamir and Tian Shan mountains are reduced. The stage is being set for a real deficiency of fresh water in Central Asia to occur.

At a time of such ecological crisis, and when there is a current shift to free market experiments in the region, the time is ripe to implement the stewardship enterprise as a new form of "ecologic-capitalist" enterprise serving as a transition between old and new systems. Since it will be transboundary in administration, it will reduce the likelihood of conflicts between states. This "ecologic-economic unit of biosphere" would protect the glaciers, clean riverbeds, maintain forests along the rivers, and strengthen riverbanks. The stewardship enterprise for this river would care for this part of the biosphere.

The governments of Central Asian states realize the necessity of solving the problem of the Aral Sea and undertake efficient steps in this direction. Thus, the International Fund for Aral Sea (IFAS) has been created, to which the Central Asian states contribute a part of their gross national product. IFAS, in turn, has created the "Interstates Coordinating Water Commission" to oversee water allocation in the region. Also formed were the two "Unions of Water Economies" for the Amu Darya and Syr Darya to take immediate charge of maintaining the state of the rivers in all their length. From the perspective of bioregional management, these steps get us a big part of the way we need to go.

My concept of the stewardship enterprise would unite all of these functions of IFAS as an "ecologic-economic unit of biosphere." A fundamental change would be in the financing of the new enterprise, which would take payments from states proportionately to the quantity and quality of water consumed. In other words, states would pay for water drawn from a market formed around the water in the basin of the Aral Sea. This market would protect the rivers from the impacts of commodification by recognizing the economic interests of Nature while integrating all of the Central Asian states. This new arrangement will not be decreed in a day, but incremental shifts in policy over several decades will still yield immediate benefits to Nature and to humanity.

The creation of this new form of biosphere services market would consist of three components: the commodity, the owner of the commodity and the consumer of the commodity. The commodity in question here are the waters and services performed by the healthy Amu and Syr river systems. To have long-term value, the rivers must be preserved both in terms of quality and quantity. The "owner" must be, before any other attribute, responsible for the quality and quantity of the commodity. The consumer must have the option "to buy or not to buy" rights of access to the service of the river. Other details must be put in place as well, including infrastructure, a capable banking system, convertibility of currency, and an insurance system. Implementation would not be rushed and would follow several stages.

At the first stage, the Council of the Central Asian States Presidents would establish two new "ecologic-economic units of the biosphere," a stewardship enterprise each for the "Amu Darya" and "Syr Darya," building on the existing IFAS entities. These stewardship enterprises would immediately begin collecting fines for river pollution, with the understanding that those fines would increase yearly every 5 years until they covered the cost of cleaning the contaminated waters. These funds would be used to start maintaining the rivers, building a system of both hydrological and ecological monitoring, and to create an insurance fund. Next, the stewardship enterprise will gradually implement a system for users to pay for the water they use.

The second stage involves the introduction of a minimum payment for water to rise over time until it covers the cost of obtaining clean water. The rate of payment increase will be determined by the Presidents' Council according to the ability of participating countries to pay.

The third stage would not arrive for 30 years, after full implementation of the first two stages. The stewardship enterprises will work independently but under the general control of the Presidents' Council, which will monitor whether they are successfully maintaining the quantity and quality of the water reaching the Aral Sea. The stewardship enterprises will need to remain transparent enough for the ongoing attention of the public, the media, and international auditors. In this way, the stewardship enterprises can manage the river resource for the populations of Central Asia, taking steps to acquire additional resources from outside Central Asia as may be necessary. And the Aral Sea will be restored!

The peoples of Central Asia have the richest history and cultures but also have an historical perspective on the rise and fall of civilizations. The unique Aral crisis invites a revaluation, the chance to pioneer a new civilization based upon a mutual relationship with Nature and equality of biosphere and society. We have a chance to become an example for the entire world in solving a complex philosophical, economic, and political problem in a new way informed by the loss of the Aral Sea.

THE ARAL MUST BE SAVED AND RESTORED!

CHAPTER 22

RENEWABLE SOCIAL ENERGY: *MAHALLA* AS A TRADITIONAL INSTITUTION FOR SUSTAINABILITY

Lola Gulyamova

ABSTRACT

This chapter discusses population issues in Uzbekistan as regards current trends in urban social organization. The majority of the population is young and confronted with challenging problems both global and local in nature. Uzbekistan is also experiencing a demographic shift as changes in the national economy prompt people to move to urban areas from the countryside. This shifts the numbers and sizes of cities, but also places new demands on natural resources like water. Uzbekistan's geography has advantages and limitations for urban sprawl. Its culture offers a major advantage described in the author's original research: a form of spatial and social organization for civil society found in Uzbek cities. Problems of urbanization can be addressed through this feature in Uzbek society called the mahalla. Mahalla *promote stability and continuity within a community and a means whereby common interests can be represented outside the community. This chapter investigates the contributing factors to that stability and* mahalla's *ability to handle the dynamics of rural to urban migration.*

Disaster by Design: The Aral Sea and its Lessons for Sustainability
Research in Social Problems and Public Policy, Volume 20, 335–348

ISSN: 0196-1152/doi:10.1108/S0196-1152(2012)0000020034

INTRODUCTION

In the globalized world we live in today, there is no disaster that is just local in nature, and the consequences of any environmental catastrophe echo all over the world. Uzbekistan's environmental challenges are epitomized by the disaster of the Aral Sea, which has had a severe impact on the surrounding areas, and put over 2.5 million people at risk. The Aral Sea, fed by rivers now used for irrigation, has shrunk to a fraction of its size in the 1960s. From 1960 to 1992, the sea's area was halved and its volume quartered. Today it contains just 30% of its former volume (UN Nukus conference). Its shoreline has receded by 120 kilometers and the sea level has declined more than 16 meters. By 2018, the once 16.1 meters deep sea may become a desert. The implications are not only a radical restructuring of the surrounding environment, but carry significant social impacts.

BRIEF INTRODUCTION TO THE POPULATION GEOGRAPHY OF UZBEKISTAN

Uzbekistan is a landlocked country that has abundant resources including gold, copper, natural gas, oil, and uranium. Some 80% of the country is a flat desert, with mountain ranges rising in the far southeast and northeast. The Ferghana Valley, located in the northeast, is the most fertile region, and has many urban settlements and industries.

Strategically located in the heart of Central Asia and with 29.6 million inhabitants in 2012, Uzbekistan has about one-third of the entire Central Asian region's population. This has the advantage of giving us huge human capital, but also the disadvantage of needing to feed and provide livelihoods for the largest population in our region.

The country has a very young and rapidly growing population with an annual growth rate of 1.3% in 2007, currently about 1%. About 35% of the population is younger than 16 years, and more than 62% of the population is younger than 30 years (State Committee, 2012a). One of the main challenges is that this country is facing potentially serious job shortages and integrating young adults into the workforce is a serious problem for economic development.

The Human Development Index (HDI) is a composite index measuring average achievement in three basic dimensions of human development – a long and healthy life, knowledge, and a decent standard of living. This index is used to define progress for health, education, and income. It is also useful for assessing factors such as environmental risk and inequality, thereby stressing their reduction. And the main indicators for calculation of HDI are life expectancy at birth, mean, and expected years of schooling and gross national income (GNI) per capita. The index ranks countries relative to each other, so it is a helpful measure for understanding the current state of human development in Uzbekistan. Using data from 2009, the United Nations' Human Development Report (UNDP, 2011) ranks Uzbekistan 115 out of 187 on the HDI and 99th out of 155 countries on the Gender-related Development Index (GDI). During the past 5 years (2005–2011), the HDI in Uzbekistan achieved a medium rank, improving from 0.611 to 0.641. Improvements resulted from efforts by the national government to improve lifestyle standards through structural reforms and programs. By way of comparison, the regional HDI for Europe and Central Asia (they share the same region as classified by UNDP and the World Bank) increased from 0.644 in 1980 to 0.751 in 2011, placing Uzbekistan below the regional average of 0.682. The United States' HDI score improved during the same period from 0.837 in 1980 to 0.910 in 2011. According to the HDI and GDI of other Central Asian countries, Uzbekistan ranks in the mid range on the list with the highest HDI for Kazakhstan (0.745) and the lowest for Tajikistan (0.607).

Basic demographics for Uzbekistan are also worthy of note. Life expectancy at birth for males is 70.6 years and 75.2 years for females (2009), although this is highly variable by region. For comparison, in 2011, the highest value of life expectancy in the world was 81.1 years in Norway and the lowest one was 48.4 years in the Democratic Republic of Congo. Uzbekistan is on the higher end for life expectancy. This is an accomplishment of recent decades, now challenged by the many changes related to climate change that lie ahead for our desertified country. The lowest life expectancy is in Karakalpakstan, the westernmost and very rural region located near the former Aral Sea. The loss of the sea has had a measurable negative impact on life expectancy in this region.

Uzbekistan has good human capital overall, although some regions, including Karakalpakstan, have experienced setbacks in recent years as the local economy experiences major shifts. We have a young population with significant potential for continued growth. However, the average annual growth of the Uzbek population has decreased from 2.2% to 1.1% in the past 20 years as the result of changes in social and cultural life (UNDP, 2011). More than half the population today, about 15 million people, lives in urban areas. This trend is likely to increase as the country modernizes and develops economically. The standard of living continues to improve in accordance with government policies. The government is investing heavily in human development including poverty alleviation and education (Government of the Republic of Uzbekistan, 2005). In the past year, more than 230,000 students studied in 59 universities. The GDP per capita is $2,875 in 2009.

URBANIZATION AND CLASSIFICATION OF RURAL AND URBAN POPULATIONS

As we saw above, Uzbekistan's young and growing population is changing the composition of Uzbekistan. Table 1 shows that during the 30 year period from 1974 to 2004, which includes both the Soviet period and the time since independence, urban population has grown faster in medium-sized and large cities. In fact, the number of cities of medium to large size doubled. During the same period, the number of small cities with population of less than 10,000 citizens changed only slightly. This finding heralds the beginning of a process of rapid urbanization, where life in big cities is viewed as more attractive than in smaller ones. Rural populations leaving the countryside are not going to the nearest small urban area, but are moving toward the larger cities. This is significant because Uzbekistan is a predominantly agricultural economy with a population that until now has been 70% rural. As found in other countries, urbanization alters peoples' social, economic, and cultural life. We can thus expect major changes over the next few decades.

In 2009, the Uzbek government reclassified 951 communities from rural to urban, increasing the total number of urban settlements in the country to 1,184. Tashkent, Uzbekistan's capital, is the only large city. A metropolis with over 2 million residents, it is the largest city in Central Asia by

Table 1. Number of Settlements Based on Population, Years 1974 and 2004.

	Number of small cities Population <10,000	Number of cities Population 10,000<50,000	Number of medium-sized cities Population 50,000<100,000	Number of large cities Population >100,000
1974	85	54	9	8
2004	71	128	17	16

Source: UNDP in Uzbekistan (2007).

population. Several other population hubs historically attract people, notably Samarkand, Bukhara, Andijan, and Ferghana, all ancient cradles of culture and civilization dating back to the Silk Road. In addition, there has been substantial in-migration to the newer towns of Navoiy, Chirchik, Almalyk, and Angren, built after 1950. Today there are 17 large cities and 208 medium-sized cities in Uzbekistan.

The rural population is spread throughout Uzbekistan, often in villages that have less than 2,000 people each. The desert areas are inhabited, but at very low density, mostly by pastoralists. The villages are mostly located in the long Amu Darya corridor and Syr Darya and Zerafshan Rivers and along the roads. The predominant occupation in the villages remains farming. Interestingly, though agriculture has a massive footprint in the economy, tying up 90% of the freshwater use and enabling a majority of the population to live in rural areas, its importance to the economy is waning. In recent years, agriculture has shrunk from 31.1% of the economy in 2004 to 24.0% in 2007 (State Committee, 2012b). This has at least two explanations. One is that more of the rural population is idle, no longer employed in agriculture because there is less to do, and related to that, the shifting role of cotton in the agricultural yields gained annually. Agriculture is no longer the dominant contributor to GDP, but rather is on par with industry and manufacturing. The rural labor force has become more mobile in its search for work, which in turn explains the expansion in urban population.

CITIES BUILT ON OASES

There is a unique feature to Uzbekistan that makes it different to talk about "cities" here than in most other places in the world. The historical settlements were all founded on oases, and over time these have, as in the case of Samarkand or Bukhara, grown into cities. Population has historically been unevenly distributed due to the extremely dry climate, which made areas with little or no water uninhabitable. The population density is much higher in water-rich regions, and it is an important reason why Uzbekistan does not have a history of many large cities. In the past, settlements could only grow as large as there was water available to support them. The Silk Road passed through Uzbekistan and the settlements benefited from the trade caravans passing through as much as the trade caravans needed the oases at intervals to get food and water.

This has literally meant a big divide between urban and rural populations. In the past, the size of a settlement was determined by the availability of food and land suitable for cultivation. Today the divide is the lack of opportunities for employment in rural areas and the big difference between lifestyles in various parts of the country. Seeking better opportunities is the main reason that people move out of the rural areas and away from such places as the ecological disaster zone in Karakalpakstan. It is considered a normal trend in many parts of the world for there to be a rural to urban migration pattern, as a region becomes less agrarian and more industrial. In Uzbekistan this is not without risk. The population shift leads to urban expansion and a subsequent imbalance of demands on natural resources, particularly fresh water. Oases are supported by aquifers, but the water tables do not readily support populations past a few hundreds of thousands. The Samarkand water supply system for example has been stretched thin by the number of users and neighborhoods. It is in the process of being upgraded and modernized to meet the demands on it.

On the positive side, people in Uzbekistan do not live in the same rural isolation as in historical times. Migration rates and flows demonstrate the preferences of local people to change their lifestyle and be more integrated with global society. Furthermore, as many of the chapters in this book have illustrated, we have severe environmental challenges today, which are also complex. They cannot be considered local. Cotton is grown over a wide area, so the pesticide and irrigation related damage is also over a wide area. The Aral Sea disaster inevitably influences the lives of the population in the region and beyond. In recent years, moving from rural to urban areas reflects people's dream to improve their living standards, as much as it is a move away from the ecological problems.

CHALLENGES FOR URBANIZATION: IS SMART GROWTH POSSIBLE IN UZBEKISTAN?

Uzbekistan is facing the same issues as other arid countries as their populations grow. The limits on water and land resources are quickly felt when decision-makers, planners, and micro-economists try to build with urban expansion in mind. Balancing this with the needs of irrigated agriculture is a big challenge for the country. What is the best balance between urban and rural areas? How can valuable land be protected so that it can feed the growing population?

When we look at growth and population changes in Uzbekistan, we can recognize patterns in the statistical movement of people, and we need to consider redefining cities and living spaces. Many countries today are facing the problem of rapidly growing urban areas. The rapidly growing urban population worldwide is a major challenge from environmental, social, and political perspectives. Today the total population on the Earth is 7 billion, and there is no sign of it slowing down. Cities will take up more and more of the available land area. This means that as cities continue to expand more quickly, some researchers believe they may no longer be the concrete jungles of the past, but may grow around natural features such as forests or coastlines, enveloping them (Yale 2011). In Uzbekistan we have already enveloped our Amu River and expanded the habitable areas with irrigation, but a lot of country is desert. Our spatial organization of the growing population is necessarily going to have to adapt to this constraint.

In Uzbekistan, the urban population increased from 8.3 to 9.7 million between 1991 and 2008. This trend heralds a massive change from a predominantly rural, agricultural nation to an urbanized nation. The urban population is expected to reach 20 million by 2030. Fig. 1 shows the total

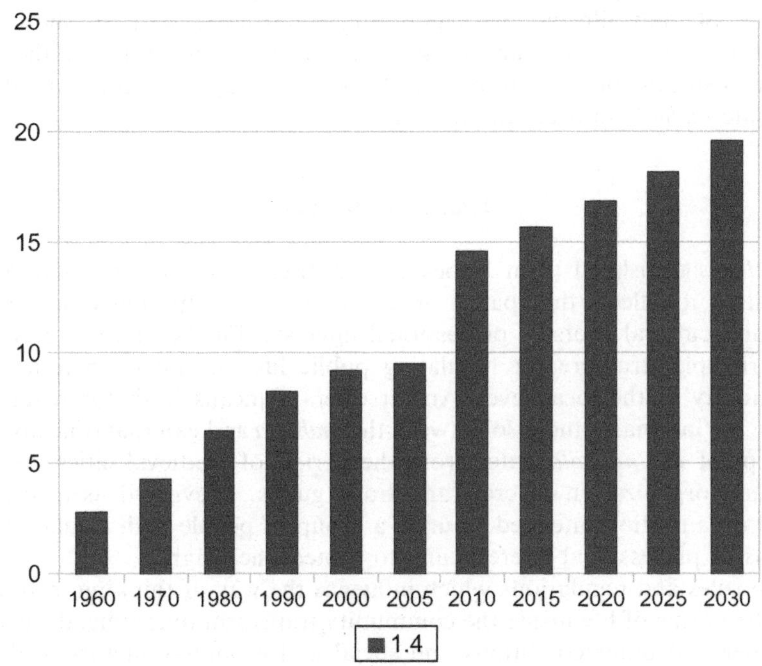

Fig. 1. Projection of Increasing Urban Population (in millions) Between 1950 and 2030. By permission of the author.

population growth. This rapid transformation from rural to urban population is evident. As we contemplate these changes, we must consider carefully what kind of society we want, even as we plan to accommodate the influx of people to cities despite the aforementioned geographical constraints on water and land.

MAHALLA: A SOCIO-CULTURAL ORGANIZATION FOR URBAN POPULATIONS

I would like to make the case in this chapter that Uzbekistan is unique when it comes to urbanization. Above, I have discussed population growth patterns, and how an oasis does not allow for typical urban sprawl patterns due to water constraints and the surrounding desert. For the remainder of the chapter, I am going to discuss another distinctly Central Asian concept, the *mahalla*. This word has different meanings among the Central Asian ethnic and culture groups and is integrated to a greater or lesser extent in the realities of daily life in our neighboring countries. I want to show how important *mahalla* is for our Uzbek culture. Then I want to suggest the ways in which sustainable urbanization in Uzbekistan ought to, in fact needs to, integrate *mahalla* into urban planning.

What Is a Mahalla*?*

Mahalla is translated from Uzbek as a district or a neighborhood. More generally, it reflects the spatial organization of a community defined by common cultural roots or professional interests. Furthermore, it serves as the principle structure for regulating public life. *Mahalla* functions as a community at the local level. And it offers a means both for regulating public life internally and also between the *mahalla* and external officials. The concept of the *mahalla* dates from the period of medieval cities and the common organization of craft and trade guilds. It evolved as a form of social organization intended to unite a group of people with similar ethnic origins or professional interests and to protect their rights.

The rules and regulations which influence the way of thinking, behavior, and the norms of life inside the community transform over time, depending on inner and outer conditions and social and economic factors. Cultural traditions intrinsic to the people of Central Asia before the conversion to Islam (in the eighth century), as well as religious traditions acquired after the conversion to Islam, were the main support structure for the development of *mahalla*. Traditions that appeared in the eighteenth and nineteenth centuries as the result of the conquest of Central Asia by the Russian Empire and the traditions that arrived with the Soviet times have also had their influence on *mahalla* development. As a steadfast element of society, *mahalla* pursued and pursues the interests of a definite group of citizens, where the guiding principles are mutual aid, consent, and respect in their struggle for survival. It is a mutable, spatial form of population distribution and a form of public involvement in decision-making. Collectively, *mahalla* make up a unified civil society, representing an interesting and distinctive example of self-organization and self-governance.

The development of civil society in Uzbekistan is a "top-down" process. Our leadership is involved in rethinking the role of existing civil institutions in public and social life. The president of the Republic of Uzbekistan, Islam Karimov, has expressed a vision of transforming our nation "From Strong State to Strong Civil Society." The president proposes that the *mahalla* serve as the model for designing our approach to public involvement in decision-making (Karimov, 2012). Recently, the government of Uzbekistan has begun to define the legal basis for *mahalla* and other existing and newly established civil institutions. Strengthening the *mahalla* as a civil institution continues to be one of the priorities for further development.

TRIBES AND SETTLEMENTS

The spatial distribution of population in our country and the formation of *mahalla* result from the assimilation of our nomads. People have traditionally formed groups around blood relationships (kith and kin) and shared membership in a tribe. This basis for grouping still exists. Even in modern cities, one may find that relatives or people sharing a common place of origin all live in the same districts. This principle of association remains dominant throughout society, guiding domestic public relations.

The names of settlements and their subsections give a sense of how people settled according to ethnic and tribal rules. For example, in Samarkand and Bukhara provinces, many settlements have such names as Arabkhona, Arablar, and Eronkhona. These indicate early settlement by Arabians and Iranians. Before World War II, there was little inter-marriage between representatives of different tribes and ethnic groups. *Mahalla* in this case pursued the principle of ethnic and tribal isolation. This tendency still prevails, and people think about *mahalla* in regard to marriage. The tribal principle is influential both inside and outside the *mahalla*. In their everyday life, people prefer to associate with those who are close or familiar to them, belonging to a common *mahalla* helps people.

The institution of family is a basis for *mahalla* and even for the further development of society. A new law, "About Family Entrepreneurship," signed by the presidents of the Republic of Uzbekistan on April 26, 2012, is evidence of this approach. Its main idea is the strengthening of the family and the legal regulation of family-based entrepreneurship.

OCCUPATION AND *MAHALLA* IN CITIES

In cities *mahalla* are also formed by groups of people with shared professional interests. Such *mahalla* then pursue these common interests in addition to tribal or relational ones. Some districts today even bear the names of a specific trade or craft because craftsmen settled in that location in old times. A *mahalla* served as a place for people of specific trades to live near each other, offering competitive services in the marketplace. To be a member of a craftsman's *mahalla* was considered respectable and provided social status. Today, in many cities, one still finds *mahalla* named after the main profession of its inhabitants, such as *mahalla* named after the gunsmith (*O'kchi*) or blacksmith (*Taqachi*). The guiding principle for regulating public relations in such instances is the protection of incorporated professional interests.

MAHALLA ADAPT TO MODERN LIFE

Today's *mahalla* is the symbiosis of cultural and social traditions with significant social changes that occurred in the twentieth century. Religious traditions were repressed during the Soviet era, but *mahalla* served as a principal means for saving these traditions. Today *mahalla* are regaining dominance and some are restoring the cultural traditions suppressed during the Soviet period. The modern form of governance and *mahalla* management is based on a mixture of traditions, and depends on the professional, cultural, and religious orientation of its inhabitants.

Three types of *mahalla* exist today. Each is characterized by self-governance at the local level, agreement among the majority of community inhabitants about decision-making, and adherence to shared cultural and religious traditions.

In older districts of cities and also in rural areas, inhabitants are usually of the same ethnicity and religious traditions; regulation of public life is strictly based on adhering to Islamic law. In these *mahalla,* public consensus is more conservative and is not open to innovation.

At the same time, in newer sections of the cities, inhabitants are likely to be of diverse ethnic origin, religion, and place of birth. There, *mahalla* has a more flexible form of self-governance and regulation of public life. Public opinion is more tolerant and progressive in adopting changes in cultural and social life.

The third type of *mahalla* reflects a state of transition either to the more conservative or more progressive forms. The process of the transition depends on such factors as literacy rate and the social orientation of inhabitants.

In all types of *mahalla*, inhabitants elect leaders, and their preferences are reflected in formation of committees within each *mahalla*. The inhabitants of a *mahalla* pay close attention to public opinion, which is reflected in the deference accorded to respected senior members according to their status and age.

STATUS IN THE *MAHALLA*

In most cases the Chair of the Mahalla Committee is a respected gentleman. A significant subcommittee of the *mahalla* is the Council of Women, chaired by a respected lady. The financial support from government enables Mahalla Committees to manage everyday costs in administering the *mahalla*, to carry out cultural and social events (e.g., celebrating Independence Day, the Navruz holiday, Veterans' Day, and others) and to collect donations by wealthy people whose philanthropy helps to support the poor, orphans, the indigent, and aged, also, *mahalla* improvement. This support is respected and encouraged.

The national government is consistently improving the legal status of civil institutions including *mahalla* in order to expand self-governance. According to national law, the self-governance of citizens is guaranteed by the Constitution in the case of local government institutions. Furthermore, according to this law "The self-governance is the independent activity of citizens for solving matters of importance to the local community with consideration to is own interests, historical heritage, local habits, and traditions."

Basic principles of citizens' self-governance institutions are based on social justice, humanism, transparency, and independence of decision-making on matters important to local community, public mutual aid, and democracy. Democracy is exercised within the *mahalla* as an equal right for all inhabitants to take part in decision-making and independent conflict resolution of matters important to local community. Fig. 2 shows *mahalla* members attending a meeting.

In order to better support the *mahalla's* efforts nationwide, a charitable Mahalla Fund of the Republic of Uzbekistan was established. This fund carries out activities in accordance with its mission "to provide financial support to the indigent and poor, and social aid during transition to a market economy." On October 8, 1992, the president of the Republic of Uzbekistan signed a decree regarding the proper allocation of funds to the national charitable Mahalla Fund. The acceptable sources of funding include state government resources, donations from enterprises, associations, and philanthropic donations made by citizens. This fund offers significant direct funding to local governments to supports the activities of *mahalla*.

Today *mahalla* is a robust pillar for all community members, which fulfills its mission to offer support and aid. Usually people come together to offer

Fig. 2. Representatives of *Mahalla* in a Rural Area of Uzbekistan Attend a Meeting. Photo by Lola Gulyamova. October 27, 2011.

mutual assistance, such as to provide aid for weddings and funerals, a traditional habit. A *mahalla* has its own equipment, buildings, and supplies and all of these are publically owned. In other words, all inhabitants have equal rights and access to use them.

CULTURE AND VALUES OF THE ENTIRE NATION

Uzbekistan is a country with a rich history and cultural traditions of several ethnic groups, also called nationalities from the Soviet days. In small- and medium-sized cities, the population is usually mono-ethnic, and it strives to preserve this culture and its traditions. Multi-ethnicity is a noteworthy element of social life in big cities, however. All ethnic groups have their own traditions and practices for maintaining them. A culture seeks to preserve meaning and its values, so regulating its public life and public opinion are an important matter. At the same time some values of the entire nation (i.e., Uzbekistan) may be described as the guiding principles of *mahalla*.

One of these shared values is seniority. It is carried down from the nomadic past and has in many cases been a positive feature of how *mahalla* came to develop. The life experience and wisdom of elders and their issued opinions are considered for decision-making. Respected elders are often the members of the Mahalla Committee. In many cases their opinion is authoritative and moreover, the Council of Elders regulates the norms and standards of public life and the policy of the Mahalla Committee.

Another shared value is parental orientation. We have a proverb "A *mahalla* is a father and a mother," which shows the importance of this institution in the social life of people. The institution plays a significant role in raising children and their education. Community life inculcates to its youth the values of patriotism, devotion to community, selflessness, and readiness to provide mutual aid. For adults, it reinforces these values through joint activities, mutual aid, and support.

The third principle is mutual aid. It has been the basis for development of *mahalla* from the beginning of their existence. Everyday life is connected to the development of relations between neighbors and all inhabitants of the *mahalla*. This form of contributing to each other's well-being through public effort is the best example of living a group's values. Many generations have respected and lived by these traditions.

People in the past established the rules for living in *mahalla* and for regulating matters of daily life in the community with respect to the interests of subgroups and individuals. Today social life is changing and modern technologies provide opportunities to learn about other traditions and cultures. *Mahalla* are adapting and accepting new trends in social life, in accordance with its grassroots connections and ethical norms and standards.

In sum, the form and function of the *mahalla* reflects the approach of the Uzbek ethnic majority culture; other ethnic groups accept the positive attributes of this form of self-governance while preserving their own cultural traditions. Similarly, in other eastern countries, the spatial and social organization of the population is founded upon a principle of tolerance between communities of different origin and religion. Today this form of self-governance dominates all over the country, and it helps to bring people together for solving important issues for the community. As Uzbekistan's people work toward a mode of social organization that can promote a sustainable future, in stark contrast to the approach that led to the tragedy of the Aral Sea, the *mahalla* represents an important cultural asset.

REFERENCES

Government of the Republic of Uzbekistan. (2005). *Welfare improvement strategy paper of the Republic of Uzbekistan for 2005–2010*, March 2005. Tashkent.

Karimov, I. (2012). *Development of educated and intellectually rich generation as the main condition for sustainable development and modernization of the country*. Opening Speech by the president of Republic of Uzbekistan Islam Karimov at the International Conference, Tashkent. February 16.

State Committee of Statistics of the Republic of Uzbekistan. (2012a). *Demographic indicators*. Retrieved from http://www.stat.uz/uz/demographic/

State Committee of Statistics of the Republic of Uzbekistan. (2012b). *Macro- economical indicators*. Retrieved from http://www.stat.uz/en/economy/

UNDP. (2011). *Human development report. Sustainability and equity: A better future for all*. Retrieved from http://www.undp.org/content/dam/undp/library/corporate/HDR/2011%20Global%20HDR/English/HDR_2011_EN_Complete.pdf

UNDP in Uzbekistan. (2007). *Uzbekistan in figures*. Retrieved from http://statistics.design.uz/data_finder/2585/pdf/population_dynamic.pdf

Yale. (2011, October). *Growth of cities poses significant global environmental risks*. Yale School of Forestry and Environmental News. Retrieved from http://environment.yale.edu/news/article/growth-of-cities-poses-significant-global-environmental-risks/

CHAPTER 23

CLEARING THE PIPES: PROVIDING POTABLE WATER THROUGH WELL RESTORATION

Abror Gadaev and Gulmira Boboeva

ABSTRACT

This chapter addresses the need for groundwater resource management in the arid and semiarid zones of Uzbekistan, broadly and specifically the potential to prolong the functional life of water wells by rehabilitating capacity and developing new methods and devices for cleaning and protecting well filter screens. The urgency of this concern is felt most deeply in areas lacking potable water, most notably the larger Aral region. If achieved, the result will facilitate access to adequate supplies of potable water in a region where the surface waters are seriously depleted and polluted. Springs that once bubbled to the surface to form oases were surrounded by life in the desert. Perhaps reliable water wells are the basis of tomorrow's new oases.

INTRODUCTION

The Aral Sea zone is beset by adverse environmental conditions and poor quality of drinking water. The only contact with potable water is to reach

Disaster by Design: The Aral Sea and its Lessons for Sustainability
Research in Social Problems and Public Policy, Volume 20, 349–356

ISSN: 0196-1152/doi:10.1108/S0196-1152(2012)0000020035

water stored safely in deep aquifers well below the polluted water table. Existing deep wells tend to lose productivity due to clogging over time and often fail during peak demand. A method for water well reclamation, "combined deep water well rehabilitation technology," has been developed that will restore existing well capability, avoiding the need to dig new wells. This process introduces no hazardous materials into the groundwater or environment, addresses the range of clogging materials likely to be encountered, and avoids damage to the well itself. It has wide applicability throughout Uzbekistan and Central Asia.

THE PROBLEM

Central Asia is an agricultural region with a hot and dry climate and only two main surface water sources, the overtaxed Amu Darya and Syr Darya. Fortunately, another option exists for the region; existing artesian groundwaters 40–100 meters deep that have been naturally protected and can provide ecologically safe and naturally protected sources for drinking water, as well as for industrial and municipal needs. To tap groundwater, wells are drilled into the water saturated sand layers encased within the underlying bedrock. Gravel stone is placed into the well to stop sand clogging. Properly installed, water wells are efficient and reliable. However, in some parts of Uzbekistan, most notably the Aral region, water tends to be "hard," namely full of mineralized salts of calcium, iron, and magnesium. As a result, wells lose productivity over time as filters become clogged by a mortar formed of sand and gravel chemically bonded by salts of calcium and iron oxides and corrosion from the metal screens, for which hard groundwater, with its high mineralization, is a cause. The problem is invisible below ground, but is readily evident when screens and pipes are unearthed. Fig. 1 shows a filter screen caked with such deposits. Fig. 2 displays results for calcium testing in a water pipe, showing the buildup of deposits at various intervals. It can easily be seen how such conditions restrict water flow through the well, reducing efficiency and eventually total function.

In Uzbekistan, groundwater is the primary source of quality drinking water, representing 85–90% of the general water budget and 93% of drinking, industrial, and municipal supply. In some regions, as in the Aral region, groundwater is the only safe source. There are problems with the current situation. Water quality commonly does not meet drinking water standards. Given Uzbekistan's hot and dry climate, the long summer (from May until October) results in sharp increases in water consumption that

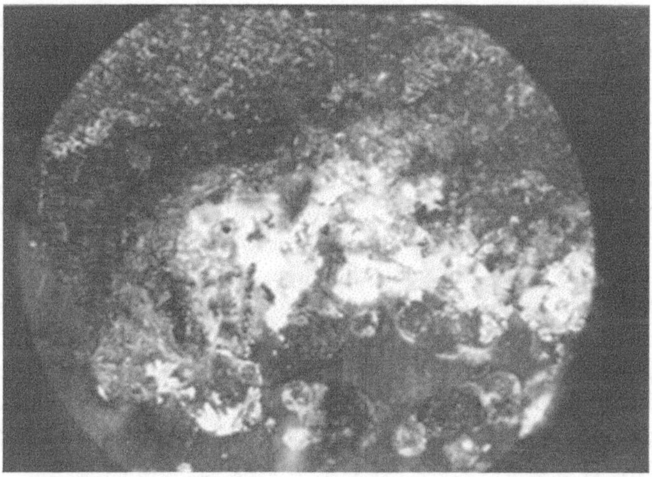

Fig. 1. The Clogging Salt Deposit on the Well Screen (Kyziltepa Water Well, Uzbekistan). *Source:* Photo by Abror Gadaev (Photo Is Taken from Microscope Analysis of Well Clogging Deposits in Kiziltepa, Bukhara Region).

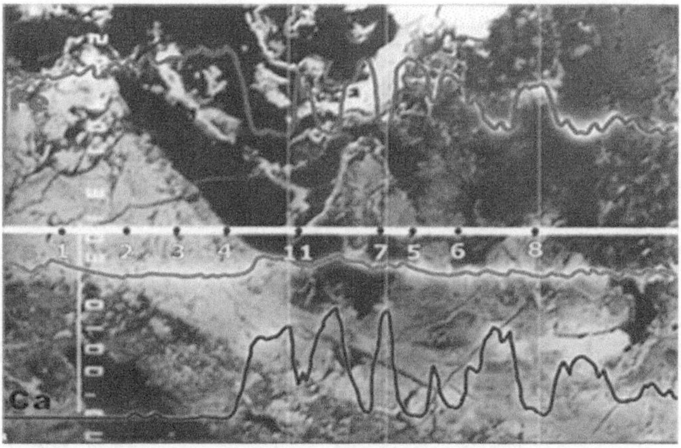

Fig. 2. Results of X-ray Analyses of Salt Clogging Samples Taken from Kyziltepa Water Well Filter Screen. *Source:* Photo by Abror Gadaev (Photo Is Taken from a Microanalysis of a Well Clogging Deposit in Kiziltepa, Bukhara Region. The Picture Shows Long-Term Formation of Clogging Deposits).

exceeds what wells are able to supply. The inability to meet this peak demand means that water shortages result. The urgency of the situation is particularly noticeable in the Aral Sea disaster zone, where the quality of the surface waters and environmental conditions are clearly unsafe and, therefore, there is near total dependency on unreliable groundwater systems.

The problem encountered is fortunately not the draining of overtaxed aquifers pushed beyond their sustainable yield, but rather wells that suffer declining productivity. The existing regime of water management does not plan on, and thus, plan for, such shortages. There is no observable way to predict when wells are about to experience problems. A given well may have many years of operation prior to experiencing a sudden loss of capacity. Of course, age of the well correlates with incidence, but there are other factors, as well, including invisible natural conditions encountered underground. Deficiencies are common in the construction of wells and the design, maintenance, and operation of the entire pumping system. A frequent cause of clogging is the debris left from prior rehabilitation efforts, when there is incomplete removal of plugging deposits. As a result, the original pore volume of the gravel pack, well screen, and fractured formation are impaired by the very steps intended to fix the problem.

The decline of wells does vary dramatically, however, from region to region. In the hard waters of the Aral region, well decline is dramatically more rapid than in other areas. For example, in Samarkand, where water is comparatively "soft," wells work properly on the order of 40–50 years without intervention. Of the approximately 120 wells found there, a few (3 or 4) require rehabilitation. In the Aral region, in contrast, the situation is different. Our tests were conducted in Kiziltepa in the Bukhara region, part of the greater disaster zone. There, wells required restoration in only 4–6 years! Of 25 wells, half encountered during the research were deficient.

As a general rule, a well that has lost more than 40% productivity is a candidate for "rehabilitation," the term used for cleaning filters and clearing out the gravel filter zone. Effective water well rehabilitation requires the removal of all deposits in order to restore pore volume and thus well capacity. A variety of physical and chemical approaches have been developed for well rehabilitation, but their successful application depends on the prior identification of the plugging material. For example, a chemical treatment to dissolve material requires knowledge of just what needs to be dissolved, a daunting puzzle due to the variability of the complex geochemical and microbiological underground environment and also the need to fully remove the plugging while not also affecting well construction.

THE SOLUTION

The authors have been engaged in a long-term project intended to vastly improve the life and effectiveness of water wells. A new water well rehabilitation technology was developed called "combined deep water well rehabilitation technology." This process combines carbon dioxide in a mix of states (solid, gaseous, and liquid) with acid applied cyclically in strong impulses to remove calcification from water well screens and the proximate gravel zone.

Since 1991, "combined deep water well rehabilitation technology" has been tested on more than 200 wells in Ukraine and Uzbekistan with excellent results. Success has been found for a wide variety of well conditions, including vertical wells, drainage wells, both shallow and very deep wells, small diameter to large diameter pipes, and screened and injection wells. Restored wells have deposits removed from both inside the well screen and the formation surrounding the well.

The effectiveness of the technique rests on the combination of four factors. First, the injection of liquid into the well is impulsive and cyclic to maximize controlled mechanical force. Second, injected acid plays the role of chemical solvent. Third, a liquid is injected capable of causing extreme agitation through phase change to gas. Finally, all of this occurs in a tightly confined space, allowing for the injected liquid to deliver the coup de grace. All but the first step requires further explanation.

A solvent effect is created because an active component in the liquid injected into a clogged well is carbonic acid. Carbonic acid is only a mild acid; under atmospheric conditions it produces a pH of ± 6.0, near the neutral boundary between acid and base. However, under the pressure of an aquifer or a sealed well, the pH is reduced to 4.0 or less, more acidic but still relatively mild. As the pressure is released, and the gas escapes, the pH returns to above pH 6.0 and thus is self-neutralizing, requiring no disposal concern.

Gaseous and liquid carbon dioxide is injected into the well at a temperature of approximately 0°F. The injection contains tremendous potential energy, described as the energy of dissolution, which when released becomes the energy of detachment and energy of agitation. The release of this energy is capable of causing the detachment, dissolution, and removal of sediments and encrustation from the surfaces within the well screen and its surround.

Then there is the final ingredient. The injection of carbonic acid with liquid and gaseous carbon dioxide has the desired effect because it occurs in the small, localized space of the well. Extreme agitation is created not only by the mechanical pulsing and force of the injection, but by the chemical behavior of the liquid carbon dioxide. Without being too aggressive for newer plastic wells constructed using PVC or HDPE, the combination of carbonic acid, agitation, and localized space delivers the necessary energy to detach the encrusted plugging material. The bulk of the energy actually comes from the phase changes unleashed by this combination. The expansion rate from a liquid to a gaseous state of the carbon dioxide is almost 700 times in volume. This agitation achieved with liquid carbon dioxide is similar to that seen when dry ice is placed in water using a special container and the carbon dioxide changes from a solid to a gaseous state.

There is an art to the process as well. The feed rate and volume of the injected carbon dioxide must be controlled while the injection and downhole pressure is carefully monitored to make certain that pressures can dissipate

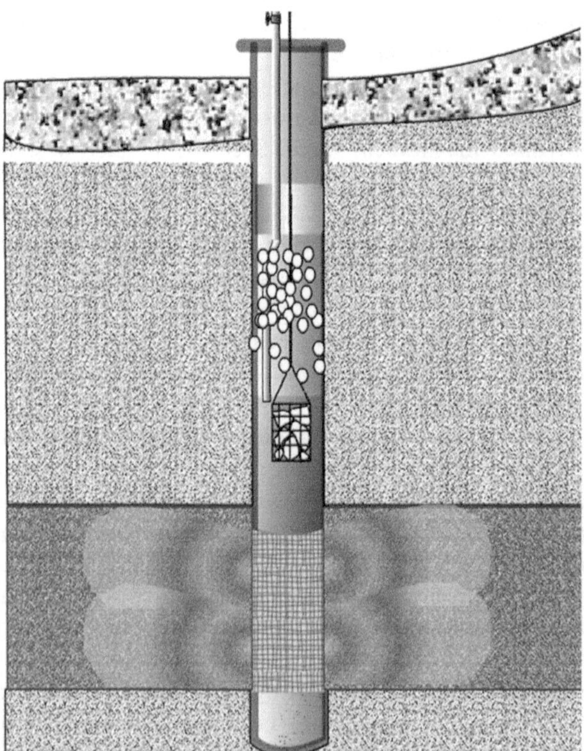

Fig. 3. Deep Water Well Rehabilitation Process. *Source:* Created by Abror Gadaev (Picture Illustrates the New Water Well Rehabilitation Technology).

into the surrounding rock and the well itself is not damaged. Recent developments allow for use of a more compact container for introducing chemicals and solid "dry ice" to the localized space in the well near the filter area. Fig. 3 depicts this process.

In our trials, we have found "combined deep water well rehabilitation technology" to have the ability to effectively disrupt and remove a wide variety of plugging deposits, in a wide variety of well construction types and materials, offering superior deposit removal and more complete pore volume recovery, without adversely effecting well construction. All of the test wells continue to perform although their use varies with other factors, including physical condition of the well and existing pollution or potential pollution of the water catchments area. Our data suggests that the recovery approach returns the productivity of treated water wells to 90% of their original capacity. At a cost of approximately $1,350 dollars per well, restoration of water well capacity offers a savings of approximately 15–20% over construction of new wells. This represents an important savings for the farmers and local governments footing the bill. However, more importantly, because every well into the groundwater represents a potential avenue for surface pollution to contaminate the otherwise shielded artesian aquifer, abandoning old wells in favor of sinking new ones is undesirable. Although in the polluted Aral region, care is taken to double case wells to minimize pollution, avoiding unnecessary penetrations of the aquifer and risking the proliferation of abandoned and uncared for wells is essential.

CONCLUSION

Water is both an essential of life and increasingly also a limited resource. Safe and clean water resources are already in deficit; increasingly people are deprived of a stable drinking water supply. Sustainable development involves addressing the availability of this necessity for all, now and in the future. Achieving this goal is a tall order. In this paper, we illustrate one of the engineering scale steps that can be taken toward the desired goal. There is a need for many such steps.

ADDITIONAL SOURCES

Gadaev, A. N. (1994). *Methods for regenerating the productivity of artesian wells.* Young scientists' works, Ministry of Higher Education, Samarkand, Uzbekistan (pp. 44–47).

Gadaev, A. N. (2006). *Regeneration of clogged water wells* (in Russian) (157 pp.). Samarkand, Uzbekistan: Samarkand State Architectural and Civil Engineering Institute.

Gadaev, A. N., & Boboeva, G. S. (2002). *Geochemical hydrodinamics of water well rehabilitation. Problems of architecture and construction* (Vol. 2, pp. 37–41). Samarkand, Uzbekistan.

Gadaev, A. N., & Niyazov, I. (2010). Uzbekistan: Water resources management and health issues. *Ecology and Noospherology Journal, 8*, 16–20.

Gadaev, A. N., Tugay, A. M., & Prokopchuk, I. T. (1991). Using complex reagents to remove salt deposits from artesian well filters. *Journal of Science and Technology in Urban Public Works, 77*, 38–41.

Gadaev, A. N., Tugay, A. M., Ternovzhev, V. E., & Kochetov, G. M. (2007). Forecasting of water well filter's clogging process. *Problems of Water Supply, Wastewater and Hydraulics, 8*, 16–20.

Gadaev, A. N., & Boboeva, G. S. (2005). *Construction of wells & intake structures* (160 pp.) (in Uzbek). Samarkand, Uzbekistan: Samarkand State Architectural and Civil Engineering Institute.

Tugay, A. M., Emelyanov, B. M., Gadaev, A. N., & Apanasenko, V. E. (1991). Complex acid treatment of water wells. *Water Chemistry and Technology Journal, 10*, 52–54.

CHAPTER 24

RENEWING WITH RENEWABLES: DIRECT SOLAR ENERGY USE IN DEVELOPING COUNTRIES

William J. Makofske

ABSTRACT

The use of fossil fuels in developing countries places increasing economic, health, and environmental costs on the population. In decentralized and rural communities without existing grid systems, direct solar technologies provide the basis for electricity production, for water pumping and hot water, and for heating of houses. Examples and case studies for each of these direct solar technologies are presented which may be directly applicable or potentially modified for rural development in countries such as Uzbekistan and Turkmenistan, which have ample direct solar resources. Related design involving both daylighting and passive cooling are described as part of the incorporation of passive solar heating techniques.

INTRODUCTION

In the last 20 years, as the price of generating solar electricity from photovoltaic (PV) cells became cheaper, nonprofit groups and for-profit

Disaster by Design: The Aral Sea and its Lessons for Sustainability
Research in Social Problems and Public Policy, Volume 20, 357–371

ISSN: 0196-1152/doi:10.1108/S0196-1152(2012)0000020036

companies have proliferated in order to provide electricity for rural areas in developing countries. For example, the Solar Electric Light Fund, a nonprofit group formed in 1990, installed 110,000 solar home lighting systems in India within 7 years (Williams, 2010). Today, many countries and development experts see the advantages of bringing PV electricity to poor people in rural or decentralized environments such as sub-Saharan Africa (Nygaard, 2009) or China (Jungfeng, Signeng, Miaji, & Linguan, 2007). Such development will only become more important as cheap and more accessible fossil fuels diminish and prices rise. Along with the price rise, there is a rush to develop less accessible, expensive, and much more environmentally damaging fossil fuel energy sources through tar sands development, deep water oil and gas drilling, fracking for natural gas in shale rock, and other exploitive and expensive technologies. It is clear that many developing countries will not be able to afford more expensive fossil fuel energy, and certainly, the planet's climate cannot afford it either. Fortunately, there are many relatively inexpensive solar-based technologies for electricity, heating, and hot water that can provide much of the energy needed in developing countries, with positive impacts on the environment, peoples' health, and the climate. This chapter explores some of these options.

Renewable (solar-based) technologies can be broken into several categories. There are those that are based on direct solar energy, including solar thermal power plants (California Energy Commission, 2012a; Wikipedia, 2012a), solar PVs, passive and active solar heating, daylighting, solar hot water, solar water pumping, and others. Indirect solar technologies include such options as hydropower, wind power, wave and current power, and biomass. Many countries have several possibilities for renewable energy development. The economics of each choice varies considerably depending on resource availability, technology and implementation costs. However, the energy resource with the most energy potential and widest distribution is direct solar. For example, Uzbekistan has an extraordinary solar resource, with 300 days of sunshine per year.

For both developed and developing countries, increasing use of solar technologies is becoming a necessity. As a widely available energy source with essentially zero fuel cost and an infinite lifetime, solar energy reduces or eliminates dependence on ever more expensive fossil fuel energy. Along with less fossil fuel use comes environmental and health benefits to the population as air, water and land pollution is drastically reduced. Around the globe, climate change is having increasing impacts on agriculture and water supplies, including flooding and drought frequency. These impacts can be lessened over time by using renewable energy technologies, which significantly reduce greenhouse gas emissions. Renewable technologies can also provide more security and resilience since they are locally based. They also can have significant positive economic impact since they provide local long-lasting jobs through manufacturing, installation, and repair, and, in addition, reduce the outflow of money to import fossil fuels. For most renewable technologies, operating and maintenance costs are low.

Along with the use of energy technologies of any type, it is increasingly clear that both energy efficiency and energy conservation need to be considered. Energy efficiency involves technology that performs a given task with less energy, and it is typically applied once. Examples include insulating a house or using a more efficient heating device; for efficiency, energy savings occur over time without further intervention. Energy conservation, on the other hand, involves changes in behavior that save energy and these actions must be repeated to realize on-going savings. For example, car pooling, reducing thermostats at night, and wearing warmer clothing indoors in winter are examples of energy conservation, and are all repeated actions.

In spite of the vast abundance of solar resources, it is absolutely essential to couple solar technologies of all types with energy efficiency and energy conservation. The most obvious reason is that solar technologies cost money and resources to build. It makes no sense to invest more money in expensive energy systems than you need. In addition, all solar technologies provide power or energy proportional to their size, or equivalently, the collector area exposed to the sun. Since collector size must be matched to the needed power or energy demand, to minimize the initial cost of a solar technology, demand must be minimized. Once the initial cost of a solar technology is paid for with energy savings, future energy harvested from the device is essentially free and consists only of maintenance costs.

In the contribution by Eshkuvat Arzikulov discusses the potential for solar water purification and solar water pumping in this volume (Arzikulov, Chapter 25, this volume). The focus in this chapter is on several of the other technologies associated with direct solar, including solar PVs, passive solar heating, daylighting, passive cooling, and solar hot water. These applications can be used in any setting, but will be particularly effective in small-scale decentralized or rural communities with sunny climates.

SOLAR ELECTRICITY

While electricity can be produced by large-scale solar power plants, particularly in a dry desert climate with ample direct solar resources like

Uzbekistan, such systems are typically large MW (megawatt) power systems that require associated electrical grid systems to distribute the energy (California Energy Commission, 2012b; Wikipedia, 2012a). Solar PVs (National Renewable Energy Laboratory, 2012a), able to be used at any scale, would be more directly applicable to the power needs of decentralized villages. Solar PVs, also called solar cells, provide direct DC electricity from solid state devices that are connected together to form panels (Fig. 1). Panels in turn can be connected to form arrays of any size which can be matched to particular electrical power needs, such as water pumping, or running electrical appliances, lighting, electronic equipment, etc. Electrical power can be stored as electrical energy through the use of such devices as batteries, capacitors, and pumped hydro storage systems, although batteries are commonly used today.

Solar cells provide power with no moving parts, no noise, and lifetimes on the order of 30 years. There are many types of solar cells that are manufactured today, but the market is still dominated by semi-crystalline and crystalline silicon cells whose efficiencies have increased to around 15–18%. There is also newer thin film PV that uses less material and promises further reduced cost. Even so, PV prices have dropped considerably over time as increasing demand has allowed mass production methods to develop, and it is expected that prices will continue to drop over the next decade, perhaps to 50 cents a watt, a cost which is highly competitive with all other means of producing electricity (U.S. Department of Energy, 2012a).

Over the past 30 years, PV applications have moved from powering remote devices, to off-grid houses, to grid-connected houses using net metering, and

Fig. 1. Photovoltaic Panels Being Installed. (Credit W. Makofske)

now to large-scale utility-connected arrays that may be 200 MW or even larger. PV arrays greater than 500 MW are in construction (Wikipedia, 2012b). In countries or areas with a limited or nonexistent electrical grid, PV arrays represent a viable method of providing electricity to decentralized or remote locations, particularly those with limited or nonexistent hydropower or wind resources which are currently somewhat cheaper to develop than PV. This situation currently exists in parts of Uzbekistan.

Direct current PV systems consist of the panels, and other components which may include charge controllers, batteries, and wiring. However, using a DC to AC inverter can provide AC power. In addition, the AC power can be grid connected through net metering, and excess electricity production can enter the grid while turning the meter backwards. The grid then supplies power to the house at night or during cloudy weather. This procedure eliminates the need for battery storage, and is now widely used.

In a remote setting without an existing electrical grid, PV is most economical if the power can be used directly at the time it is collected, eliminating the need for electrical storage. For example, PV power can freely run equipment such as water pumps where the water is then stored in tanks, or run fans to provide immediate cooling, eliminating the need for electrical control or electrical storage equipment (see Fig. 6). PV can also be used to run a pump to circulate hot water from solar collectors to tanks, again without the need for any other electrical backup. However, there are many uses of electricity that don't use much power, particularly for lighting and communication equipment, where modestly sized batteries can provide electrical storage at an acceptable price. For this reason, the use of PV in remote villages around the world has been growing rapidly (Jungfeng et al., 2007; Nygaard, 2009; Williams, 2010). While manufacturing PV panels is a high tech industry, connecting PV panels, control equipment and batteries together to run various devices is relatively simple, and has been found to work well in many different societies around the world. This allows people in remote locations to have some essential refrigeration, reliable lighting at night, and basic communication equipment, thereby allowing connection to the vast knowledge and resources available through the Internet.

PASSIVE SOLAR HEATING AND DAYLIGHTING

The use of direct sunlight to heat dwellings has been around for many centuries and has been used by many cultures (Butti & Perlin, 1980). In the past 30 years or so, this approach has been rediscovered in the form of passive solar design, which takes advantage of the sun to collect and store heat for buildings. There are numerous approaches to solar heating houses, including simple south-facing windows, sunspaces, attached greenhouses, and Trombe walls (National Renewable Energy Laboratory, 2012b). While passive design can be optimized in new buildings, in many cases, retrofits can also be made to existing buildings.

The simplest design of a passive solar house includes the following two critical elements: a south-facing window orientation with solar access (i.e., no or little shading during desirable collection times over the heating season), and overhangs or shading that blocks sunlight during times when solar heat is not desired. The sizing of the window system matches the dwelling's heating needs; it should not substantially overheat the dwelling. Other desirable components for solar passive heating include the use of thermal mass for heat storage at night and during cloudy weather, and the use of movable insulation to reduce the heat loss back through the windows at night. Passive south window design is remarkably effective (a square foot of window area in most climates can displace a gallon of oil burned at 70% efficiency, approximately 100,000 Btu), simple, and long-lasting. Sunny climates will have even better performance. Of course, exact south orientation is not required; angles as much as 30–45 degrees from due south can work reasonably well for solar collection.

In a sunny climate such as found in Uzbekistan where clay and concrete materials are used for houses, the use of these materials as thermal storage mass can help provide a very high percentage of heating needs. This approach allows the sun to hit a mass wall or a concrete floor which rises in temperature during the day, and subsequently releases its heat at night. This technique has been used extensively in the southwestern United States, where passive solar design using adobe and concrete has flourished over the past several decades (Arizona Solar Center, 2012a; National Renewable Energy Laboratory, 2012b; New Mexico Solar Energy Association, 2012).

Attached greenhouses or sunspaces (California Energy Commission, 2012b) can also be an effective way to provide passive solar heat to a dwelling. These south-facing structures gain solar energy and then can passively transfer heat into the house via open doors or windows. Overall, heat gain is similar to that obtained by using south-facing windows, although there typically is more heat loss through the structure's glazing. Some advantages of this method are that greenhouses can be retrofitted to existing structures fairly easily, and they can be used to grow food in the winter. Appropriately sized thermal mass in the form of water barrels or concrete can also prevent the greenhouse from freezing in most climates.

In Uzbekistan, a simple approach might be to minimally modify existing high mass housing design to provide the right amount of solar gain and storage through either south-window direct gain designs which heat thermal mass in the building with sunlight directly or indirect gain designs where the sun first heats a south wall, called the Trombe wall (Torcelini & Pless, 2004; Wikipedia, 2012c) which subsequently releases heat to the house. Of course, the amount of solar gain and storage that is needed depends on the thermal heat loss from the house, and on climate temperature variations, so some method of reducing heat loss at night will likely be desirable. Another method which may work is the use of mass-produced affordable attached solar greenhouse structures that provide both heat and food. Fig. 2 shows an integrated solar house in New York State in a climate with 6500 heating degree days which includes south window and greenhouse solar heating, and both a PV and solar hot water heater on the roof. The house is well insulated and airtight, using an air-to-air heat exchanger. The 3.9 kW PV system provides for 100% of appliance and lighting use, and backs up the solar hot water heater. The south windows and greenhouse provide the heat equivalent of burning 300 gallons of oil for heating. The near-zero net energy house uses a total of 70 gallons of oil over the year. Clearly, for Uzbekistan and Turkmenistan, some experimentation will be needed to find the right combination of design and construction modifications to make climate-appropriate passive solar heating both economically and culturally

Fig. 2. Integrated Near Net-Zero Energy House. *Note:* Location is New York State; House Incorporates Solar PV, Solar Hot Water, Passive South Windows, and Attached Greenhouse. (Credit W. Makofske)

desirable, and simple. Luckily, there is much solar design information and data that is available in similar climates, like the U.S. southwest.

A secondary, but important co-benefit of passive solar design for heating is also providing good daylighting. This is usually accomplished by good

Fig. 3a. Overhangs in Solar Design. *Note:* Shading Provided by Overhangs Protect Against Unwanted Solar Gains in Summer. (Credit W. Makofske)

Fig. 3b. Overhangs in Solar Design. *Note:* Overhangs Allow Full Sun in Winter When Passive Solar Heat Gain is Needed. (Credit W. Makofske)

window placement and sizing, and should be incorporated into the building design. Another benefit of good solar design is passive cooling (Arizona Solar Center, 2012b), a design approach that reduces overheating of buildings in warm or hot weather without the use of fans or air conditioning. This is often achieved in conjunction with solar passive heating by incorporating simple overhangs or other methods that allow the sun to penetrate the dwelling during the heating season and block sunlight during the warm weather. This is illustrated in Figs. 3a and 3b. Overhang design is latitude dependent and can easily be applied to most houses. Passive cooling could also involve the use of Trombe walls that vent to the outside in the summer and also drive air flow through cool tubes buried in the earth to help cool the house.

SOLAR HOT WATER

Another direct solar technology that can be easily incorporated into remote and rural villages, particularly if integrated with solar water pumping, is solar hot water. There are two basic approaches to solar water heating, passive (without the use of pumps and electricity) or active (using pumps and electricity) (Florida Solar Energy Center, 2012; U.S. Department of Energy, 2012b). Both could be feasible in remote locations.

Passive solar hot water could be as simple as placing a black hose in a glass covered box. However, a typical passive solar hot water approach places a tank with a dark surface in an insulated glazed box. This is called a breadbox or integral collector, and is widely used in various nonfreezing climates around the world (Fig. 4). As hot water is used, water flows into the tank, either by pumping or gravity-feed, to replace it. This system could be modified to work in a non-severe but freezing climate.

Another widely used passive solar hot water approach is called a thermosyphoning system. In this case, a solar collector (Fig. 5) consisting of an absorber plate attached to piping inside an insulated glazed box is used (Fig. 6). The tank must be higher than the collector, and the water circulates into the tank from the collector when the collector water is heated by the Sun. A one-way valve prevents backflow at night. The tank is either under pressure or can gravity feed to a faucet at a lower elevation.

An integrated water source and hot water system could combine solar water pumping from a well to a tank on the roof to provide water, together with a nearby breadbox type solar hot water heater that is connected to the water tank. A pipe from breadbox tank enters the roof, allowing gravity-fed

Fig. 4. Integral or Batch Solar Water Heater. (Credit W. Makofske)

Fig. 5. Thermosyphoning Solar Water Heater. *Note*: The Storage Tank Must be Above the Collector. (Credit W. Makofske)

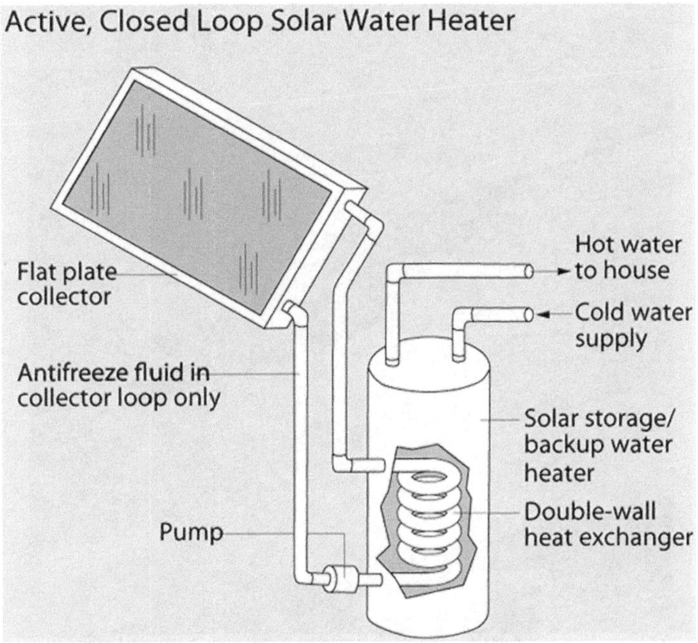

Fig. 6. Components of a Solar Active Collector. *Source*: U.S. Department of Energy, 2012b.

hot water to flow to the occupants. Large tanks of water generally will not freeze, even in freezing climates. By minimizing and insulating the length of the pipe from the water tank to the solar breadbox collector, and then having the solar hot water go directly into the house, there is little chance of water in the connecting pipes freezing.

In mountainous regions or colder areas where freezing may be more of a problem, active collector systems run by a small PV system (about 20 watts) can move heated antifreeze directly from a solar collector into a heat exchanger which then transfers the indirectly heated water to a storage tank in the house by thermosyphoning. Figs. 7, 8, and 9 show two 4 ft by 8 ft collectors, and an 18 watt PV system which runs a small DC pump to move antifreeze to a heat exchanger, and then transfers hot water into an 80 gallon storage tank in the house. The system (Thermo Dynamics Ltd., 2012) runs whenever the Sun is out, increases its pumping speed at higher solar intensity and eliminates the need for any other electrical inputs. Such systems are in use in the United States and elsewhere. Solar water heating

Fig. 7. Solar Hot Water Collectors. *Note*: These Two Collectors Are Powered by an 18-Watt PV Array Visible on the Far Right Side of the Bottom Collector. (Credit W. Makofske)

Fig. 8. DC Solar Hot Water Pump. *Note*: The Pump Circulates Antifreeze and Is Powered by 18-Watt PV Panel on Roof. (Credit W. Makofske)

Fig. 9. Heat Exchanger and Solar Pump. *Note*: The Pump Is in the White Box in Foreground Connected to an 80 Gallon Solar Storage Tank. (Credit W. Makofske)

can be a very simple technology, and there are numerous designs and variations to the basic technology (Build It Solar, 2012).

As with other solar systems, applying energy efficiency allows the system to be smaller and therefore cheaper. The pipes from the collectors to the heat exchanger are typically well insulated. In addition, the storage tank and surrounding pipes should also be well insulated. Very efficient showerheads and hot water energy conservation measures will allow the hot water to last several days. People who use solar hot water can and often do modify their consumption to match the available hot water; this eliminates the need for backup hot water throughout most or even all of the year. This would be relatively easy in a climate with 300 sunny days per year.

CONCLUSIONS

Various direct-solar small-scale applications for electricity production, solar heating of dwellings, daylighting, passive cooling, and solar hot water have been briefly discussed. Although widely applicable to most settings, these technologies and approaches are particularly suitable to rural decentralized populations. In most of these applications, the life-cycle costs will be reasonable and technical complexity will not be overwhelming. Initially, it is likely that government programs will be needed to deal with development issues, potential first cost financing, and technical issues. As noted in the introduction, such decentralized solar programs have been expanding in many countries and have for the most part been successful. However, any program needs to be tailored to the existing culture, and its particular needs. This will be true for countries such as Uzbekistan and Turkmenistan.

REFERENCES

Arizona Solar Center. (2012a). *Passive solar heating & cooling manual, Part 2 of 4*. Retrieved from http://www.azsolarcenter.org/tech-science/solar-architecture/passive-solar-design-manual/passive-solar-design-manual-heating.html

Arizona Solar Center. (2012b). Passive solar heating & cooling manual, Part 3 of 4: Natural cooling. Retrieved from http://www.azsolarcenter.org/tech-science/solar-architecture/passive-solar-design-manual/passive-solar-design-manual-cooling.html

Arzikulov, E. (Chapter 25, This volume). Getting the salt out: Innovative solar technologies for rural clean water. Chapter 25, this volume.

Build It Solar. (2012). *The renewable energy site for do-it-yourselfers (2012) solar water heating*. Retrieved from http://www.builditsolar.com/Projects/WaterHeating/water_heating.htm

Butti, K., & Perlin, J. (1980). *A golden thread: 2500 years of solar architecture and technology* (289 pp.). Palo Alto, CA: Cheshire Books.

California Energy Commission Consumer Energy Center. (2012a). Large solar energy projects. Retrieved from http://www.energy.ca.gov/siting/solar/

California Energy Commission Consumer Energy Center. (2012b). *Passive solar design – Attached greenhouses*. Retrieved from http://www.consumerenergycenter.org/home/construction/solardesign/greenhouses.html

Florida Solar Energy Center, Consumer. (2012). *System types*. Retrieved from http://www.fsec.ucf.edu/en/consumer/solar_hot_water/homes/system_types.htm

Jungfeng, L., Signeng, W., Miaji, Z., & Linguan, M. (2007). *China solar PV report*. Retrieved from http://www.wwfchina.org/english/downloads/ClimateChange/china-pv-report-en.pdf

National Renewable Energy Laboratory. (2012a). *Solar photovoltaic technology basics*. Retrieved from http://www.nrel.gov/learning/re_photovoltaics.html

National Renewable Energy Laboratory. (2012b). *Passive solar heating and daylighting for homeowners*. Retrieved from http://www.nrel.gov/learning/ho_passive_solar_day.html

New Mexico Solar Energy Association. (2012). *New Mexico Solar Energy Association*. Retrieved from www.nmsea.org/

Nygaard, I. (2009). *The compatibility of rural electrification and promotion of low-carbon technologies in developing countries – the case of solar PV for sub-Saharan Africa*. Retrieved from http://www.eeinstitute.org/european-review-of-energy-market/EREM_8_Article_Nygaard.pdf

Thermo Dynamics Ltd. (2012). *Solar heating*. Retrieved from www.thermo-dynamics.com/

Torcelini, P., & Pless, S. (2004). *Trombe walls in low-energy buildings: Practical experiences.* Retrieved from http://www.nrel.gov/docs/fy04osti/36277.pdf

U.S. Department of Energy. (2012a). SunShot initiative. Retrieved from http://www1.eere.energy.gov/solar/sunshot/

U.S. Department of Energy. (2012b). Energy efficiency & renewable energy, energy savers (2012) solar water heaters. Retrieved from http://www.energysavers.gov/your_home/water_heating/index.cfm/mytopic = 12850/

Wikipedia. (2012a). *List of solar thermal power stations.* Retrieved from http://en.wikipedia.org/wiki/List_of_solar_thermal_power_stations

Wikipedia. (2012b). *Photovoltaics.* Retrieved from http://en.wikipedia.org/wiki/Photovoltaics

Wikipedia. (2012c). *Trombe wall.* Retrieved from http://en.wikipedia.org/wiki/Trombe_wall.

Williams, N. (2010). *Solar solutions for the developing world.* Retrieved from http://www.thesolutionsjournal.com/node/548

CHAPTER 25

GETTING THE SALT OUT: INNOVATIVE SOLAR TECHNOLOGIES FOR RURAL CLEAN WATER

Eshkuvat Arzikulov and Qamariddin N. Srojev

ABSTRACT

Uzbekistan easily gets 300 days of sunshine a year, and is a natural choice for development of solar technologies and their implementation. Central Asia is working together to solve critical water problems including contamination and dry wells. This chapter addresses the need to find solutions for rural populations living in the shadow of the Aral Sea disaster, with poor quality drinking water and limited infrastructure. With assistance from generous government funding, the Institute for Solar Energy is developing new point-based solar desalinization and pumping technologies. Preliminary designs are promising and government support for raw materials mining and wafer production are growing. Domestic R&D and production ensures cost-effective dissemination options. The Institute houses both research and project management for all solar technologies and expects to field test prototypes within 2–3 years. Solar energy is a growing sector in the economy of Uzbekistan.

Disaster by Design: The Aral Sea and its Lessons for Sustainability
Research in Social Problems and Public Policy, Volume 20, 373–383

ISSN: 0196-1152/doi:10.1108/S0196-1152(2012)0000020037

INTRODUCTION

Most of the electricity in Uzbekistan is produced in power plants that run on natural gas. Currently, the country has 37 gas-fired and 9 hydropower plants with a capacity of more than 11 million kW and an annual production ability of more than 55 billion kW-h per year. Eighty-seven percent of installed capacity is accounted for by natural gas-fired power plants, and 13% by hydropower plants. The largest gas-fired power plants in Uzbekistan are the Syrdarya (3 million kW), Novo-Angren (2.1 million kW), Tashkent (1.86 million kW), and Navoiy (1.25 million kW), and the hydraulic plants are Charvak (620.5 MW), Hodzhikent (165 MW), Farkhad (126 MW), and Gazalkent (120 MW).

In the Soviet period the Central Asian Power System consisted of two divisions. One part was located in Northern Kazakhstan, and was directly linked to the electricity grid of Russia. The second division, covering much of Central Asia, included the power grids of southern Kazakhstan, Uzbekistan, Turkmenistan, Tajikistan, and Kyrgyzstan. The control center for managing the regional power system for Central Asia was in Tashkent. After the collapse of the Soviet Union, Uzbekistan inherited the status of regional energy coordinator, and Uzbekistan continues to provide electricity transmission services to all the republics of the region through its network. Uzbekistan accounts for about 40% of the total generating capacities of the United Energy System of Central Asia (Sokolov, 2000), which includes 83 power plants with a total capacity of 25,000 MW. Despite the fact that the post-Soviet exchange in electricity fell by 80% according to some reports, the electricity in many areas of Kyrgyzstan and Tajikistan is still critically dependent on the energy supplies from Uzbekistan (Antonov, 2000).

New ideas are also beginning to circulate. The report by President Islam Karimov at the meeting of the Cabinet of Ministers of the Republic of Uzbekistan dedicated to the socio-economic development in 2010 and priorities for 2011, emphasized the need for the widespread adoption of energy-saving technologies in all areas of economic activities.

The important message from this meeting is to increase the use of alternative energy sources. One reason is that increasing energy demand in the world contributes to an increased demand for alternative energy sources. Indeed, in recent years the use of alternative energy sources has increased significantly. If this sector continues to develop at this rapid pace, it will take over most of the common energy market. This is good news for those who want to implement solar energy technologies.

Solar energy holds a special place among alternative energy sources. According to energy experts, the surface of the globe absorbs 174 PW in solar energy each day. And one of the most urgent tasks for global society, with our expanding energy demands, is a reasonable use of this bounty of the Sun.

Undoubtedly, positive results can be achieved in Uzbekistan when we take into account international studies and working solar installations. Australia has built the world's largest solar power plant with a capacity of 1000 MW, while the Spanish government has allocated 1.45 billion US dollars to build a solar power plant in Seville, with a design capacity of 20 MW. Algeria has installed solar panels on land equal to 45 football fields in the Sahara desert. German engineers have designed a ship that runs on solar power. Five hundred square meters of solar panels were installed on this innovative ship, christened the *Solar Planet*.

Nearly 3000 MW of solar energy will go online in Asian countries by 2012 when more large-scale projects are realized, projects for which the Asian Development Bank (ADB) allocated 2.25 billion US dollars in funding (Asian Development Bank, 2012).

ENERGY CAPACITY AND POTENTIAL IN UZBEKISTAN

In Uzbekistan, where there are 300 sunny days a year, solar energy is not only convenient but an obvious technology to develop. The annual total potential solar energy in Uzbekistan is equivalent to 50.97 billion tons of oil.

Solar energy is promising in terms of both ecology and economy. Solar installations are particularly advantageous for providing electricity and heat to the remote territories far away from the existing centralized electricity and heating systems. Rather than investing in heavy old style infrastructure, solar energy offers point-based solutions in western Uzbekistan. For example, desalination of water in the steppes, the drying of agricultural products (such as fruits), the heating, ventilation, and cooling of greenhouses and buildings, and their provision with hot water, as well as the heating of kilns to make bricks and tiles – these are all promising areas of practical applications for solar energy.

In 2010, Uzbekistan began implementing projects that run on clean renewable energy sources. In particular, two hydro-meteorological substations in Sariassiy region of Surkhandarya area are continuously fed by solar photoelectric power panels. Solar photoelectric panels have been installed in the surgical department of the Central Hospital of Kyzyltepa district, and at a nursing home in Chuya village in the Navoiy region. These installations provide continuous electricity to the buildings, and more hospitals are starting to be converted to solar power. In 2011, solar power stations were installed in the regional children's rheumatologic sanatorium of Yakkabag district, at the surgical ward of the Central Hospital of Kamashi district in Kashkadarya region, at the infectious diseases hospital in Kayrokli village in the Samarkand region, and also at the central maternity hospital complex of Kokand city. In other words, solar energy is being operationalized in rural areas across Uzbekistan where medical services are furthest from the main power grids.

This is hopefully only the beginning of how efficient use of alternative energy sources, like our abundant solar energy, will serve the economic development of our country and improve people's health and welfare.

The Asian Development Bank is providing financial assistance in Uzbekistan to establish the Institute of Solar Energy and support technology expansion by 2013.[1] The institute will combine the functions of science and technology research with project design and management. The institute's mission will include the creation of specific projects in the field of solar energy applications. ADB will encourage commercial lending and private investment in the solar energy projects and help finance the development of solar energy in Uzbekistan.

ADB sees the value of Uzbekistan's solar potential within the scope of its plans to increase solar power generation in Asia by up to 3000 MW by the end of 2013, at an initial investment of 2.3 billion USD.[2]

The timing of this foreign direct investment could not be better. According to the Center for Economic Research of Uzbekistan, if the domestic consumption of resources and trends continue at their current pace, natural gas and coal in Uzbekistan will only last for the next twenty to thirty years, while oil reserves are already almost depleted. Rapid conversion is necessary to change Uzbekistan's power supply distribution. At present, the primary energy resource needs are met up to 97% by oil and gas, 2.3% by coal, and 0.7% by hydropower. The share of renewables in Uzbekistan does not yet exceed 1%.

At the same time, the potential for renewable energy in Uzbekistan is about 51 billion tons of oil equivalent with a technical capacity of 17.98 billion tons of oil equivalent. In particular, the potential for solar energy in Uzbekistan, where there are 300 sunny days a year, is calculated at 50.97 billion tons of oil equivalent.

Taking advantage of the abundant solar energy hitting Uzbekistan will reflect a major shift in energy policy and production for our country. Until now, the basis for electricity generation has been natural gas and petroleum sources. Natural gas alone provides 80% of the total generated domestic electricity. The nation has an extensive network of local electrical and high-voltage power lines with a total length of 231,300 km, and available voltage of 0.4–500 kV (Uzbekenergo, 2012). In order to ensure a constant supply of electricity for the economy and the country's population while also creating favorable conditions for a stable and efficient development of higher electricity capacity, a new program has been approved. A governmental limited company (GLC) called Uzbekenergo was formed by decree in 2006 to facilitate the reorganization and privatization of power providers, while at the same time ensuring uninterrupted provision of electric power for economic and domestic purposes, and to create favorable conditions for stable and effective further development of the electric power industry.[3]

SOLAR POWER FOR PEOPLE

Solar energy is not only important because it is abundant and easy to harness in our hot and desertified country, but also because it is an important strategy for growing our economy and building a higher capacity energy grid without relying on more oil imports. Solar energy also promises very important benefits to our people, especially those living in the underdeveloped rural areas. These benefits include more and better access to electricity. Additionally solar energy also stands to completely change the provisioning of fresh drinking water to rural areas. This problem affects not only domestic populations but also our neighbors in the greater Aral Sea region. Providing the Aral Sea population with pure drinking water is one of the major problems facing the governments of Uzbekistan, Kazakhstan, and Turkmenistan; this issue has been discussed many times at international level meetings. Solving this problem will greatly enhance the quality of life for millions of people in Central Asia.

One approach that worked in the past was to successfully attract foreign direct investment. For starters, the government completed a major program to begin to improve drinking water in the Aral Sea region. It was part of a broader package of projects funded by several international institutions with support from the international community to address social and environmental problems of the Aral Sea (UNESCO, 2000). Two countries provided financial support through their national centers for scientific research: France (CNRG) and Germany (DFG). Uzbekistan also secured financing for projects worth 103.2 million US dollars in loans, funded by the International Bank for Reconstruction and Development, the Kuwait Fund for Arab Economic Development (KFAED), and the German Development

Bank. The main focus for these projects has been the improvement of existing water systems and the creation of new networks of water provision, both of which require considerable effort and expense.

Despite these investments, the environmental situation in the area of the Aral Sea has deteriorated so much to date that the crisis has expanded beyond the Central Asian region and has acquired a global dimension. It is worth noting the long-term consequences already happening today to the health and gene pool of future generations living in the disaster zone and the local climate change impacts have implications for the whole world. Tragically, until now the problem of providing pure drinking water to and within the Aral Sea disaster zone remains unsolved.

The global economic downturn has changed the funding scenario for future projects. In today's economically challenging situation, the execution of expensive projects of the kind mentioned above is less likely. Fortunately, it is reasonable and possible to begin solving these problems with a minimum expenditure of funds. With the inauguration of the Institute of Solar Energy in Tashkent, it will be possible to research, develop, and disseminate solar solutions as we expand our capacity (UzDaily, 2012). From an economic and technological point of view, in our opinion it is effective to develop two technologies: first, the off grid solar desalination of salt water and second, to use solar energy to pump the underground artesian waters that are the least damaged by human impact. Both of these solar conversion technologies can be considered reliable sources of clean water. In the remainder of this chapter, we summarize the concepts and benefits of these technologies for the Aral Sea region.

TWO POINT-BASED SOLAR SOLUTIONS

We already know that the energy of the Sun can be dynamically applied in two ways that can help alleviate the crisis in western Uzbekistan: for the desalination of salt water and to ensure the lifting of artesian water, both by what are loosely called solar water systems.

What we are proposing is not entirely new. At this time there are many variants of solar stills in the world already (Bayramov 1964, 1967; Delyannis, 2003). In areas with salinized water supply in Uzbekistan, we propose to employ water stills of two types of solar radiation concentrators: the first concentrator uses a large-diameter optical lens obtained by the sol-gel method, the second solar concentrator is a mirror type (Figs. 1 and 2).

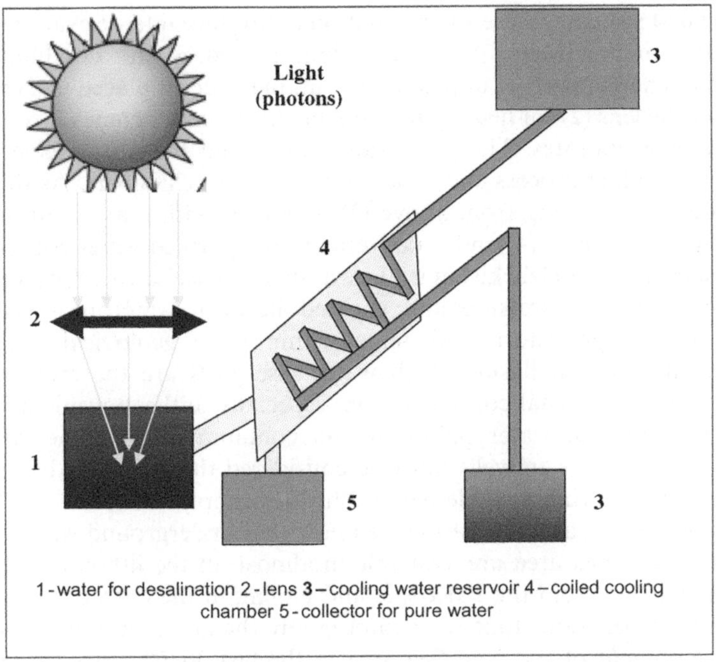

Fig. 1. Solar Still to Desalinize Drinking Water. *Source:* Authors.

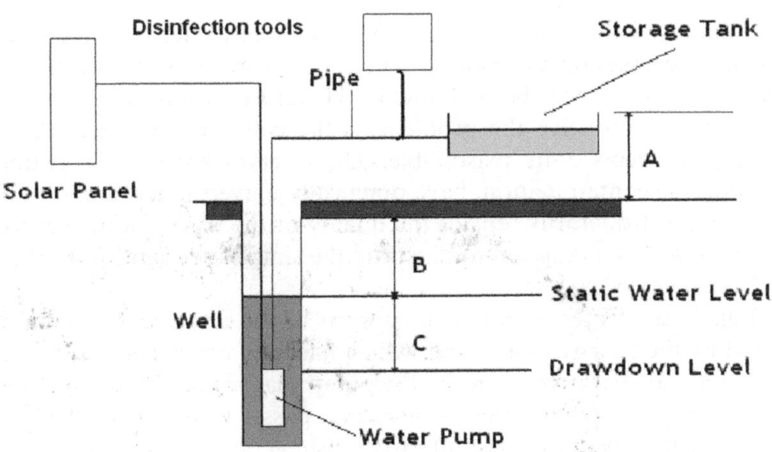

Fig. 2. Simple Water Pumping System. *Source:* Authors.

In the first system, we use solar radiation to produce intense evaporation of salt water and subsequent condensation, providing the desalination of saltwater. The salinized or unpalatable drinking water is placed in a chamber (1) under the lens (2) for heating by the Sun. As the water comes to a boil, the pure water evaporates. The pure water vapor and the salts are separated during the boiling process and collected in different chambers. As the water condenses with cooling from above (3), it travels with gravity through the coils of the condenser (4) and is delivered to the purified water collector (5).

Artesian water in Uzbekistan is like artesian water elsewhere, water which is under hydraulic pressure and encased between waterproof layers. It mainly lies in prehistoric sediments within large geological structures, forming the artesian basins. Such water reservoirs are the most reliably protected from external contamination, especially anthropogenic activities. Given the crisis of water pollution and contamination in the Aral Sea region, drilling artesian wells must be considered the most reliable and the best way of achieving a stable new freshwater supply.

Through testing and evaluation we know that underground water resources in the Aral Sea area are available in almost all the lithological genetic bedrock starting from the Paleozoic and ending in the current Quaternary sediments. At the same time we cannot ignore the fact that intensive economic development of the Aral Sea area in the last 30–60 years caused very negative consequences for the environment in general, including the underground hydrosphere. The greatest impact has been measured on shallow groundwater, which can contaminate the artisan water if not pumped correctly.

Fortunately, we can be sure that the negative influence of modern, agro-industrial processes on the water quality of pressurized water in the Upper Cretaceous layers has been limited. Therefore, using this source of freshwater to help solve the problem of the contaminated Aral Sea area water supply seems quite reasonable. The reserves are less susceptible to anthropogenic contamination, have practically universal distribution underground, and satisfactorily replace the quality of the water supply sources of Aral Sea. Fig. 2 is a diagram to illustrate the simplest design for an artesian water retrieval system.

In Fig. 2, the water pump is submerged in the artisan well water. It is powered by the nearby solar panel, which has been developed to provide all of the electrical energy required by the pump year round. The water is lifted to the surface through a series of pipes and passes through the disinfection system. The pure water passes out of the disinfection device and is collected in a storage tank for use as needed.

As we develop this solar pumping technology, we have to pay attention to one particular problem that has occurred with previous extractive technologies. According to some experts involved in the study of the artesian reservoirs, the salinity of groundwater, which may come in contact with the artesian water, could lead to degradation of the freshwater deposits. Unfortunately, this concern is quite legitimate in our case, since there is poor management infrastructure for the existing artesian wells. For example, if there is a daily off-take of pure drinking water, it is dumped into the same collector-drainage network as regular diverted Aral Sea water and finds further use for irrigation. It should be directly separated into a higher grade primary drinking water cistern. These will need to be installed separately when our pumps are put into service.

RESEARCH AND DEVELOPMENT INITIATIVES AT THE INSTITUTE

The opening of the institute is just the initial step for us as we begin to develop our own solar industry. We are currently a small team concentrated at institutes and universities, but our efforts are set to expand as we add projects, and will include academics, small businesses, and factories. Currently we already involve the physics, materials sciences, and technology departments, and are poised to expand collaborations. This will happen with the involvement and support from local governments inside Uzbekistan.

Given the influx of financial resources in the last year and the direct governmental support we have, our time frame for the implementation stage is 2–3 years. We have parabolic mirrors and we are beginning work on solar collectors. The size of our scientific research projects expands every year, and will be coordinated by the institute. We are presently developing the capacity to produce our own silicon wafers. An Uzbek-Korean joint venture factory is under construction in Navoiy for obtaining bulk silicon. Tashkent will be the production hub for polycrystalline silicon and monocrystalline silicon, so that we can begin manufacturing silicon wafers for solar panels domestically.

As discussed above, a critical concern for us is to develop and implement the technologies to alleviate the freshwater crisis for the population in western Uzbekistan. We have both short- and long-term projects directly working on increasing supply of fresh, safe, clean drinking water across the Aral Sea region in Karakalpakstan. In the short-term, we are not mandated to improve or address irrigation water quality. Our anticipated short-term outcome is securing a basic freshwater supply for the entire rural population across the republic. We know that the initial costs will be higher, but expect the cost of drinking water to decrease with time. This is a real consideration because of widespread poverty in the area, one reason why freshwater supply has not been guaranteed for more than 50% of the population in the past. From the government's perspective, after short-term needs are met, the long-term goal is to ensure supply of freshwater to schools, hospitals, and other state institutions. Our role in this will not change, as we are the central institute and knowledge hub for solar development.

CONCLUSIONS

We have reviewed two promising technologies under development for our country, both capable of ensuring the daily minimum requirement of fresh water for everyone. These are the portable solar water desalination infrastructure and off-grid solar water systems to pump and extract artesian water. Moreover, these two technologies can be applied in practice to provide the population of the Aral Sea disaster zone with drinking water, as a dramatic improvement over the health compromising conditions there today. Lastly, both systems can be developed and implemented cost-efficiently through domestic capacity at our Institute for Solar Energy. Last but not the least, these technologies help protect the environment.

NOTES

1. See a public announcement of this new project at http://www.uzdaily.com/articles-id-17451.htm.
2. Asian Development Bank. Retrieved from http://www.adb.org/news/adb-helps-uzbekistan-establish-solar-energy-institute.
3. Uzbekenergo has a trilingual presence on the Internet, in Uzbek, Russian, and English, at www.uzbekenergo.uz. This suggests its forthright efforts toward transparency, public awareness, and openness to outside investment.

REFERENCES

Antonov, V. N. (2000), *Water resources of Uzbekistan, their use in modern conditions and in the future*. Scientific papers: *Water resources, the problem of the Aral Sea and the environment*. Tashkent University Press.

Asian Development Bank. (2012, February 16). *ADB helps Uzbekistan establish Solar Energy Institute.* Retrieved from http://www.adb.org/news/adb-helps-uzbekistan-establish-solar-energy-institute

Bayramov, R. (1964). *Comparative tests for the "hotbed" type solar desalinization device* (in Russian). Izvestiya Academii Nauk Turkmenistana, Nr. 1. Ashgabat, Turkmenistan.

Bayramov, R. (1967). *Modern methods water desalination* (in Russian), Ashgabat, Turkmenistan.

Delyannis, E. (2003). Historic background of desalination and renewable energies. *Solar Energy*, *75*(5), 357–366. Retrieved from http://dx.doi.org/10.1016/j.solener.2003.08.002

Sokolov, V. I. (2000). *Definition of boundaries of water collection basins, of trans-bordering, local and mixed types of surface water resources in the basin of the Aral Sea and quantitative assessment.* Collection Scientific Works of SRC ICWC, Issue 2. Tashkent (pp. 35–53).

UNESCO. (2000). *Water related vision for the Aral Sea basin for the year 2025.* Division of Water Sciences, Paris. Retrieved from http://unesdoc.unesco.org/images/0012/001262/126259mo.pdf

UzDaily. (2012, February 16). International Solar Energy Institute opened in Uzbekistan. Retrieved from http://www.uzdaily.com/articles-id-17451.htm

Uzbekenergo. (2012). *About the company.* Retrieved from http://www.uzbekenergo.uz/eng/about_us/

CHAPTER 26

RENEWABLE ENERGY AS A KEY FACTOR FOR SUSTAINABLE DEVELOPMENT IN UZBEKISTAN

Ashraf Khodjaev

Sustainable development is the imperative of the 21st century.
—Ban Ki-Moon (2011a), UN Secretary General

ABSTRACT

Uzbekistan currently relies almost entirely on oil and gas for its energy resources. However, it has massive potential for solar energy, with short-run viability particularly for adoption of solar hot water and solar thermal. Wind power potential is also strong. In this chapter policies are discussed which will promote the rapid adoption of renewable energy in Uzbekistan.

INTRODUCTION

The arguments for wide development of renewable energy in Uzbekistan are gaining strength (Nasyrov, Vasikov, Juul-Kristensen, Zavialova, & Pozychanyuk, 2007; Saidova, Salihov, Kabulova, Eliso, & Shadybaev, 2011).

Disaster by Design: The Aral Sea and its Lessons for Sustainability
Research in Social Problems and Public Policy, Volume 20, 385–389

ISSN: 0196-1152/doi:10.1108/S0196-1152(2012)0000020038

Chief among them are increasing energy consumption in combination with rapid economic and population growth, the depletion of some of Uzbekistan's major oil fields, increasing export obligations for the supply of natural gas abroad, and our Central Asian legacy of environmental degradation. All of these factors are emerging against the backdrop of adverse climate change impacts.

Recent global disasters also point the way toward safer choices found with renewable energy. Such disasters include the oil spill in the Gulf of Mexico in 2010 and the impact of the earthquake and tsunami to the nuclear power plant "Fukushima" in Japan in 2011. In many countries, government policy has met with public and private sector support for global and national strategies to accelerate the introduction of renewable energy technologies and expand their respective markets. As numerous states declared during the recent Rio + 20 Conference, such efforts are crucial for the transformation of energy systems and to give impetus to the future green economy.

In Uzbekistan, there is a clear understanding that, despite the existence of rich hydrocarbon reserves in the country, these sources are not renewable and are limited in nature. According to up-to-date forecasts, proven reserves of natural gas, oil, and coal in Uzbekistan, given the current level of consumption, will last for the next three to four decades.

According to experts of the Center for Economic Research (CER, Uzbekistan), in the period 2015–2020, an acute shortage of liquid hydrocarbons will be temporarily mitigated by new production of synthetic liquid fuels from natural gas. However, after 2020, due to possible complete exhaustion of existing oil and gas fields, the shortage of liquid hydrocarbon will return. After 2030, it will pose a threat to the energy and overall economic security in Uzbekistan. Thus, the total liquid hydrocarbon resources that must be replaced by 2030 with alternative energy sources may exceed the equivalent of 4.5 million tons of oil equivalent.

EXISTING CAPACITY

Currently, the share of renewable energy in the fuel-energy balance of Uzbekistan is negligible. The national economy is hydrocarbon oriented. The share of oil and gas in the structure of primary energy resources is about 97%, coal 2.3%, and hydropower 0.7%. However, due to geographical location and climatic conditions, the country has considerable potential for renewable energy, particularly solar and wind. It is estimated that the potential for renewable energy in Uzbekistan is about 5100 million tons of oil equivalent, and the technical potential is estimated at 182 million tons.

It is important to stress, that a significant portion – about 97% – of the abovementioned potential belongs to *solar energy*. Data obtained during long-term observations at a network of meteorological stations in Uzbekistan shows that the number of "sun days" for the various regions of the republic varies from 2410 to 3090 hours per year, fluctuating during the day by season, and with duration of 11 hours in summer time and 4 hours in winter. The highest gross potential of solar energy is found in Karakalpakstan, Navoiy, and Bukhara regions of Uzbekistan.

Bearing in mind the existing level of technological development, the appropriate approach for capturing the potential of solar energy in Uzbekistan is with solar collectors designed to provide hot water supply and heating in rural areas. At present the population in Uzbekistan for domestic needs consumes more than 12 billion cubic meters of natural gas per year. Taking into consideration that the ratio of replacement by solar collectors in this sector that can be technically accomplished is around 50% per year, CER calculates that the introduction of solar collectors for widespread domestic use may displace as much as 6.3 billion cubic meters per year (or about 5.17 million tons of oil equivalent). The estimated cost of saved natural gas will be around $1.4 billion a year (at a cost of $220 per 1,000 cubic meters). A lump sum investment of some $8 billion will be required for solar collector installation on all houses in rural areas, with an expected payback of 6 years.

There is less incentive for the rapid adoption of photovoltaic solar energy for electricity supply due to high investment cost. Mass construction of solar power stations will require huge investments and this kind of power supply will be comparatively expensive, exceeding the cost of thermal power by a factor of between 3 and 6. Nevertheless, it is attractive to consider use of solar power stations to augment existing thermal power stations in areas of high demand and good solar gain. CER finds it feasible to build enough independent solar power stations and combined energy units by 2020 to replace 15% of gas consumption (about 1.8 billion cubic meters). The costs would be offset by profits from exports of natural gas estimated at $270 million a year.

Uzbekistan also has potential for development of *wind energy*. Given the geographical location and some adverse atmospheric surface layer conditions, wind energy in the country is seasonal in nature. The specific wind flow capacity for Uzbekistan averages about 84 W/m^2. The technical potential of wind energy is estimated at 0.43 million tons of oil equivalent per year. The most optimal corridors for windmill installation are near three windy cities in the Tashkent region, Ahangaran, Angren, and Bekabad. Wind will serve as an optimal source of alternative clean energy to supply local farms in remote areas.

At the same time, considering the capacity of the agricultural sector of Uzbekistan's economy, the possibility of using *biomass* (agricultural waste) is another way to provide renewable energy for country's needs.

ENCOURAGING RENEWABLE ENERGY DEVELOPMENT

In order to stimulate the accelerated development of renewable energy, taking into consideration the existing potential and realities in Uzbekistan, the following guidelines can be offered.

First, to create the necessary legal framework that governs the process of large-scale introduction of renewable energy. In particular, it is proposed to adopt the "Law on Renewable Energy" which will designate a government body to take the lead on renewable energy by coordinating work, introducing renewable energy programs, creating incentives, encouraging investors, and others measures.

Second, to develop and adopt a national long-term program for the implementation of renewable energy with clear goals, phases of realization, funding sources, responsible agencies, expected results, etc. This program should become a key component of a broader strategy for sustainable development and ensure transition to a "green economy" based on the priorities of Rio + 20. In particular, this would include the "Sustainable Energy for All" initiative launched by the United Nations Secretary General in 2011 (Ki Moon, 2011b). One of the main objectives of this initiative is doubling the share of renewable energy in the global energy mix.

Third, to create a special Fund for Renewable Energy funded by profits from the export of natural gas and products of gas processing, payments by large consumers of electricity and fuel, and by grants from international financial institutions and donor countries. The fund, in turn, will finance implementation of projects for greater use of renewable energy, research and development, capacity building, and training in accordance with the national program.

CONCLUSION

Sustainable development meets the needs of the present without compromising the ability of future generations to meet their own needs. Implementing a renewable energy strategy in Uzbekistan and throughout Central Asia will have concrete and practical results, reaching a contribution to the energy balance of 20–30% by 2030. This program of alternative renewable energy adoption will help to ensure sustainable development in Uzbekistan, strengthen its energy security, gradually eliminate the dependence on imported liquid hydrocarbons, significantly reduce harm to the environment, optimize the energy balance, create a new "green" branch in the economy, create jobs, and ensure sustainable and long-term use of existing hydrocarbon resources.

For Uzbekistan and the whole region of Central Asia, the development of solar power is particularly important as an alternative to the construction of huge dams and hydropower plants in the region. There is a regional abundance of solar energy while water resources are limited. In a region where water is the key factor for food security and the source of life for the majority of the population, solar energy is a more appropriate, safe, and promising alternative than that represented by large hydropower.

REFERENCES

Ki-moon, B. (2011a). *Sustainable energy for all.* A Vision Statement by Secretary General of the United Nations, pp. 3–5. United Nations, New York.

Ki-moon, B. (2011b). *Promotion of new and renewable sources of energy.* Report of the UN Secretary General (A/66/306), pp. 3, 4. United Nations, New York. Available at http://www.un.org/esa/dsd/resources/res_pdfs/ga-66/SG%20report_Promotion_new_renewable_energy.pdf

Nasyrov, T., Vasikov, A., Juul-Kristensen, B., Zavialova, L., & Pozychanyuk, P. (2007). *Perspectives for development of renewable energy in Uzbekistan.* UNDP Uzbekistan, pp. 61–66.

Saidova, G. K., Salihov, T. P., Kabulova, K., Elisov, A., & Shadybaev, T. (2011). *Alternative sources of energy: Opportunities of use in Uzbekistan.* Analytical Report of the Center for Economic Research, pp. 25–39. *Development Focus*, 2011, 13, Center for Economic Research, Tashkent, Uzbekistan. Available at http://www.cer.uz/upload/iblock/216/one%20pages_13_eng.pdf

CHAPTER 27

PERMACULTURE RESTORATION OF THE ARAL SEA WATERSHED

Andrew Jones

ABSTRACT

Permaculture, as a design science supporting "permanent culture," has developed experience globally in ecologically regenerative design. The Aral Sea watershed provides a unique opportunity to apply a relevant design process aiming at restoration of ecological function. Permacultural practice follows a core of ethics and principles and a basic design process based on such factors as stakeholder analysis, water harvesting principles, and the scale of permanence. The focus of permacultural interventions is on water, forests, and perennial vegetation, balanced with human needs. This chapter highlights key elements of the design process and summarizes the process of permaculture rehabilitation. Given success in similar situations working at a smaller-scale suggests a potential approach for regeneration in the Aral Sea Basin.

INTRODUCTION

Permaculture is a word co-developed in the late 1970s by David Holmgren and Bill Mollison referring to a design science for supporting a "permanent culture" (Holmgren, n.d.). The goal is to create holistically functional systems

Disaster by Design: The Aral Sea and its Lessons for Sustainability
Research in Social Problems and Public Policy, Volume 20, 391–409

ISSN: 0196-1152/doi:10.1108/S0196-1152(2012)0000020039

that successfully mimic natural ecosystems in diversity, productivity, and stability. To achieve this goal, the practice of permaculture is premised on a series of ethics and design principles appropriately applied to site-specific conditions that promote the use of supportive techniques, strategies, and patterning. Drawing on diverse land management traditions from cultures that found balance with nature, permaculture seeks to honor the traditional wisdom gained from observing earth patterns in the past, as well as the advances in modern science that allow us to analyze complex and dynamic systems (Mollison, 1988).

In the 34 years since Mollison and Holmgren's (1978) groundbreaking text proposed this approach, permacultural practice has been disseminated across the globe and rapidly evolved. As it bears fruit, it offers evidence that our positive future lies in design and cooperation with each other, as well as with nature. One aspect of this can be seen in the expanding number of permaculture projects and practitioners now working globally (Worldwide Permaculture Projects, n.d.).

The Aral Sea disaster related literature, as has been documented in the chapters above, primarily proposes technological and large-scale solutions. The beauty of permaculture is its reliance on principles that work from the smallest "garden" scale on upward. It offers the ability to approach the huge Aral region from the perspective of appropriate scale. Perhaps large-scale solutions to a huge disaster may prove more costly, impactful, and risky than an approach based on many smaller but carefully integrated projects tailored to specific factors and having considerably less cost, impact, and risk. That approach invites experimentation and learning from failures that prove helpful rather than fatal to the overall project. This chapter aims to lay out an appropriate general approach to rehabilitation in the Aral Sea region to see the opportunity within the disaster. How might we approach such a situation from the perspective of a "permaculture" restoration? What are we intending to restore? What are appropriate steps to begin the journey? Before we begin a more detailed analysis, it is useful to review the key ethics and principles of permaculture around which the other processes will be oriented.

PERMACULTURE ETHICS AND PRINCIPLES

The ethical foundation of permaculture was laid out in the now classic work by Mollison (1988). His approach is holistic and inclusive, encompassing:

- *Earth care* – care of the Earth and all its ecological functions and systems

- *People care* – care of the human inhabitants of the Earth
- *Fair share* – return of surplus energy to the above two functions.

These are the ethical foundations upon which the rest of our design process will rest and to which it must conform.

A core set of 12 permaculture principles help to guide permaculture design thinking (Holmgren, 2002):

1. *Observe and interact* – Observation of nature is the key to understanding a system and appropriate solutions or interventions.
2. *Catch and store energy* – We live in an open system, fuelled by the sun. Nature stores this energy in numerous ways, but at different rates and in different seasons. We can design to capture peak productivity.
3. *Obtain a yield* – A yield is a useful resource surplus to the needs of the local system, and thus available for additional life-enhancing services. Yields are best attained through conservative resource use, and should be sustainable over time. True cost accounting is used to measure yield, so upstream and downstream impacts are accounted for.
4. *Apply self-regulation and accept feedback* – Self-regulation applies to limiting consumption beyond our basic needs, particularly of resources that are nonrenewable. Feedback is provided in nature by the overall quality of the system and her individual components. We can, for example, monitor qualitative and quantitative indicators for water quality and volume over time to understand the impact of our actions on our target area.
5. *Use and value renewable resources and services* – Renewable resource use avoids the pitfalls of exhausting nonrenewables and encourages a conservation society mindset. Our ecological footprint shrinks dramatically as we apply this principle.
6. *Produce no waste* – There is no waste in nature, everything is cycled into another system. As we replicate such cradle to cradle design, we too can turn our straight-line systems that end in landfills into interactive, infinite loop systems where materials continuously cycle from one useful form to another, the final resting place being organic materials to form soil.
7. *Design from patterns to details* – Natural patterning contains complex design intelligence. Ensuring harmonious general patterning on a broad design scale is the first step to producing more detailed design that is appropriate in scale and relationships between elements.
8. *Integrate rather than segregate* – Integrative design allows for the benefit of connections between elements to be realized. It is the number of connections between elements which ultimately determines the resilience

of a system, which – like a spider's web – can continue to function even when a majority of such connections are broken.

9. *Use small and slow solutions* – Our problems often arise in part due to scale and due to our impatience to "see results." As we begin small, we can gain the benefit of experience, and feedback to tweak our system as we extend it.
10. *Use and value diversity* – Nature inherently produces and values diversity within itself. Mature natural systems generally support large diversity on all levels, which helps to create more stable systems.
11. *Use edges and value the marginal* – Edges exist between different landforms, and represent additional resource gathering opportunities, for example, the edge of water and land supports mangroves and marshes. As we recognize the value of edges, we can use them creatively in our design to enhance our potential yield.
12. *Creatively use and respond to change* – Change is a constant. When we actively read the situation in front of us, we can quickly adapt our processes in response to change. This approach keeps us focused, flexible, and adaptable. We avoid the pitfalls of over-structured hierarchies of decision-making, and rigidity in our systems and consciousness (Design Principles, n.d.).

OVERALL DESIGN PROCESS FOR AN ARAL SEA RESTORATION

We might approach this process, like any good design process (Jacke & Toensmeier, 2005a, 2005b), with a series of activities progressing across three stages: site assessment, conceptual and detailed design, and implementation. The following is an outline of each of the three stages. It does not take into account any political or economic decision-making made in Uzbekistan to date, or the existing structures, but rather starts from a clean slate. We approach restoration from the point of view that takes in the shrunken sea, the bloated soils, the compromised river flows, and works forward from there.

Site Assessment

Careful process works wonders. In our site assessment we consider from a variety of angles the vision and goals, cultural and ecological contexts, and the scale of permanence, among other attributes.

Vision and Goals

While the vision needs to drive the design, sub-goals help to specify different areas of focus. In order to do this, we need to specify the boundaries for the design. In the case of the Aral Sea, we might propose the Aral Sea watershed as the design boundary for this process.

Our vision is an ecological functioning watershed for the Aral Sea region. Our intention is to restore ecological function in terms of water, soils, forests, and the human communities within the region. Specific goals could be further developed and detailed. It is unclear what the final outcome might be in terms of restoring the Aral Sea as a body of water. However, we know that by restoring ecological function, the needs of the local inhabitants can be better served, and in the long-term, the true wealth of the region will increase according to fundamental markers of soil biology, water quality, air quality, tree biomass, and available biological resources and services. Ultimately, when these ecological systems are functional again, it may be possible to rebuild the original water management system. This assumes the cessation of current human engineering practices involving inappropriate water removal at unsustainable extraction levels (Granit et al., 2010).

Cultural Context and Stakeholder Analysis

What is the cultural context in which the process is to operate? What are the practical ramifications of permaculture in terms of appropriate strategies for decision-making and management processes or even financial systems within the intended human system?

There is always a need to understand this context, and on the scale we are considering, to undertake a stakeholder analysis to understand how to engage with different stakeholders. Basically, any time we are looking to introduce change into a human system, we are affecting the balance of who "wins" and who "loses" economically and to what extent. By understanding the current context and how that might change, we are better able to explain the benefits of permaculture to stakeholders to engage more of them in the process, and re-engage any people who were previously marginalized. It is possible to conceive of systems via design where everybody "wins."

This is particularly important for transnational projects, which in the case of the Aral Sea implies activities in eight countries (Ackmansoy & McKinney, 1997).

Ecological Context and Site Use History

While we do not consider political and economic history in our preparation for implementing permaculture approaches, we do perform a rigorous sector analysis on ecological history and biological characteristics of our designated site. In this case we would be looking at the watershed and the sub-watershed of the Aral Sea.

Sector analysis essentially works as a process to map the "wild" energies affecting our design, e.g., seasonal wind, flooding, and solar patterns, as well as any major human-based energy such as major pollution or contamination from industry, agriculture or other sources.

Zone analysis maps the human energy being expended from a single point location and looks at how this energy pattern might be shrunk to make it more efficient. Depending upon the scale of the project, this might be a house for a family-scale design or a town or village for the human communities of the Aral Sea watershed. For the scale considered, inputs and outputs are analyzed in an "opportunity analysis" to determine how to shrink ecological footprints and acquire and generate basic resources closer to where the people are. This can be achieved, for example, by incorporating food production within the towns and villages, by maintaining coppice woodlots for firewood, and mixed agroforestry in surrounding belts to produce needed food, fiber, and fuel materials. This has the added benefits of promoting enhanced air quality, and stabilized microclimates. Ultimately, such a design process integrated at an individual family level would bring the process back to a level of individual responsibility and personal application that is understandable and actionable for most people (Mollison, 1988). Is this process amenable in Uzbekistan? Given the universal applications of permaculture since it originally spread worldwide from Australia, the methods has proven adaptable and acceptable in a great variety of contexts and will succeed here as well.

This final aspect of the site assessment can benefit from a useful tool called the *scale of permanence*. This tool helps to orient our data collection so as not to miss any key information for our site. In the case of the Aral Sea watershed, it would be necessary to break up watersheds into sub-watershed units to facilitate the data gathering process. A more thorough overview of the *scale of permanence* can be found in the next section.

CONCEPTUAL AND DETAILED DESIGN

As we gather the data, we begin integrating the conceptual design phase. There is no hard line of demarcation between these processes, and it can happen that the overall process flips between them. The concepts of water and perennial agriculture are key elements of conceptual design that will need to be addressed in light of their clear importance to the success of Aral restoration.

Rigorous design is carried out at a level of detail that includes fully quantified engineering, planting lists etc. It is the basis upon which detailed budgets can be developed and projects managed. On a project scale of this size, it would be necessary to break up the process into sub-catchments and projects within sub-catchments. However, the overall vision and integrity through appropriate patterning must be maintained. The projects must create results that interlock again when implementation is complete.

The Scale of Permanence

Permaculture uses a range of different tools to help guide design decisions or processes. One of these is called the scale of permanence, which was developed originally by P.A. Yeomans (Yeomans, 1981), further refined by Bill Mollison (Mollison, 1979) as well as David Jacke (Jacke & Toensmeier, 2005a, 2005b). Scale of permanence lays out a framework to make sure that fundamental considerations are not missed when too many elements are considered as we strive for a complete understanding as our starting point at a given site. With this approach, we first identify the least changeable elements of a site or landscape. We begin with these fundamental permanent features and consider the most readily changeable factors last. If any error were to be tolerated, it should be with the most changeable and thus most flexible features. We will examine 11 elements of place that will be considered in this review.[1]

Climate

Climate is the first factor and, despite current usage of the term, is a relatively fixed characteristic of place. Broad climate parameters include tropical, temperate, and continental zones. Uzbekistan has a predominantly continental climate and the appropriate patterning, strategies, and tools employed are consequently also different from what we would apply in a temperate or tropical zone. In Uzbekistan, we must pay particular attention to drylands where annual precipitation is less than the evaporation. In the case of the Aral Sea region, we are dealing with natural fluctuations in climate extremes between summer and winter, due in part to distance from any major ocean, as well as the effects of the watershed evaporating and overall altitude.

Landform

In using the watershed to define our area of focus, we are taking into account the landform, which defines the watershed boundaries and characteristics. A bioregion is a useful concept here, being a region of ecological function defined by specific natural boundaries such as a watershed or by culturally defined characteristics. Thus, we have the Aral Sea bioregion as our defined focus, and we include the ecological and social fabric and all interactions within the bioregion.

Water

Water, its flow, quality and seasonal availability, is a key feature. Water is the lifeblood of the planet, and its flows can be likened to energy. As the universal solvent, it carries dissolved nutrients with gravity on its path. It connects the top of the watershed with the bottom, and is the linking element for both ecological and human communities on this path. We may not be able to predict the final outcome of restoring functioning hydrology to the Aral Sea bioregion, but it is obviously one of the fundamental aspects to be addressed.

When we address water in our project efforts, we can usefully refer to the work of Lancaster and others who have summarized water harvesting principles in line with a permaculture approach to landscape restoration and management. Our fundamental strategy in this case is to slow water down in its inevitable gravitational movement from the top to bottom catchments by causing it to linger and infiltrate, supporting ecological processes and restoration on its path. We will return to Lancaster's eight water harvesting principles later.

Access and Circulation

This element is among those defined and created by human activity. It mainly consists of how people move resources, goods, and services into and out of the bioregion. It thus considers the how: reviewing the roads, transportation networks, and distribution channels. The value here is to see it for what it is, whether it is well apportioned, insufficient, seasonally hazardous, appropriate, in short to determine its current characteristics and to integrate its optimization into our project design.

Vegetation and Wildlife

With this element we get at the underlying substance of the bioregion, and we understand all existing strengths and any gaps created by human manipulations of the landforms. We begin from the premise that vegetation and wildlife in the region are our potential collaborators in restoring proper function. Looking at the web of life for the Aral Sea region we ask, is there a fully functioning natural ecology in terms of traditional predators? What is the food web? What are the traditional vegetation characteristics in different zones of the deltas, the deserts and so on? Once we understand some of these aspects, we can look at how to support restorative and reparative function by promoting the process of succession. The vegetation and wildlife analysis very much proceeds in layers. Our analysis of vegetation inevitably ends up in the soil food web, to understand how we can support biologically active soils. As the preceding chapters have shown, the soils are severely compromised today, but we can still look for the underlying health within the soils and begin from there.

Microclimates

Microclimates form when we create spatial niches in the temperature and humidity profile. A pond creates a microclimate, as does a patch of mulched earth. Nature is filled with microclimates, which tend to get lost when we reduce the ecosystem function; just consider the irregular patchiness of a forest compared with a monocrop cotton field. Although forests are the great stabilizers, they also provide diverse niches for a multitude of different life forms. Microclimates can be manipulated in favor of different outcomes on the small and large scale. Microclimates are our allies in restoring and rehabilitating overall bioregions and fostering self-sufficient ecological integrity in biomes.

Buildings and Infrastructure

Buildings and infrastructure need to be appropriate to climate, and the more local materials they can utilize, the more likely they are to be ecologically adapted to place. As our human communities use appropriate ecological design and construction, life becomes more comfortable with wiser resource use, with less work needed and more efficiency gained. Concrete boxes in desert climates for example tend to be uncomfortable at both hot and cold extremes, while cob or mud-brick materials tend to offer high thermal properties and better year-round comfort. These are traditional building materials in the area and readily available.

Zones of Use

As with buildings and infrastructure, zones of use investigate the broad and highly variable range of human activities carried out in different parts of our site. These are inherently adaptable as they are based on day-to-day decisions and choices. The zones of use can be analyzed somewhat objectively to understand if they are serving the broader goals. These may be influenced by political and economic agendas, but in a permaculture analysis should be as free of assumptions as possible. For example, objectively asking the question "Does high input, industrialized and toxic cotton irrigation agriculture make sense any longer within the Aral Sea region?" can be a catalyst for dynamic discussion and new frameworks of agricultural activity. Another useful question could be, "Are our resource use patterns congruent with the maintenance of the quality of life base, and of meeting fundamental needs?" Simply analyzing our zones of use often helps to identify new opportunities and might call into question embedded system assumptions.

Soil Fertility

Soil can be built and enhanced relatively quickly, but it takes active biology to achieve this. The cyclical activity of the plants and trees that grow, and the soil life web contribute to soil enhancement on any given piece of land. We can stack the functionality of each element on our land, not only with selected trees and plants, but also through the endemic bacteria and fungi. Stacking here refers to the multiple services our elements perform, such as providing nutrients, shade, mulch or aesthetic value. Missing components that would complete a healthy zone can be collected from areas of higher existing ecological function and re-introduced to other areas as they are restored. It can be surprising to learn how many biological soil enhancement and restoration techniques exist. These have been developed especially to support food production by those with limited resource bases. In other words, restoration can be achieved without heavy government subsidies, particularly where these tend to support the high input, industrial scale agriculture that cause or contribute to the problem.

Aesthetics

Aesthetics is more than a matter of beauty in the eye of the beholder. There is a form following function rule of nature which is graceful and elegant in its expression. Working with it rather than against it is one of our guiding design principles. Recent human aesthetics has been more a matter of taste and fashion, and often worked to the detriment of a more holistic, natural process. Permaculture practitioners embrace the myriad forms of nature following as they do rules of patterning, scale, and the fractal geometry of the natural world. Aesthetic solutions are important, for they add pleasure to life. When we attune our eyes to the patterns of nature, non-beauty tips us off that we have a wrong solution or design.

Legal Issues

All legal matters are a purely human instrument for governance and often shape the flow of many processes in a society. Though on the established scale of permanence this element often ranks higher, Uzbekistan is reshaping itself as an independent nation, so there is reason to hope for change. We end here to reflect on the big picture of how permaculture might be introduced in Uzbekistan. There is room for consideration of new opportunities, and writing new laws for better outcomes. The big questions that come to mind here consider our second element, water: Who is able to extract water from a watershed flow? How much, and when? When do societal desires based on economics or other matrices trump ecological flows? Here, historic decisions about flows of water and all resources within the Aral Sea watershed do need to be accounted for and reviewed. Moving forward from the current reality in the Aral Sea Basin, which aspects of existing legal structures are in support of watershed restoration? If none or few, then they need to be modified.

Restoration Strategies Around Water

In thinking about the Aral Sea, we propose to work on a watershed and bioregional basis to plan the restoration process. As stated earlier, water integrates fundamentally with landform to create these ecological boundaries such that watersheds provide a fundamental level of division. Moreover, the system has a clear ecological connection from top to bottom.

Though a watershed in nature can be vast and complex, Lancaster (2006a, 2006b) boiled permaculture supportive water management down to eight principles:

- Begin with long, thoughtful observation.
- Start at the top – or highpoint – of your watershed and work your way down.
- Start small and simple.
- Spread and infiltrate the flow of water.
- Always plan for an overflow route, and manage that overflow water as a resource.
- Maximize living and organic groundcover.
- Maximize beneficial relationships and efficiency by "stacking functions."
- Continually reassess your system: the "feedback loop."

APPLYING PERMACULTURE CONCEPTS IN THE ARAL SEA BASIN

Above we have reviewed the fundamentals of permaculture with a brief overview of the guiding principles that we apply to gaining a holistic, site-centric analysis of land, water, and society. Here we discuss the how of the possible implementation phases for permaculture in Uzbekistan.

Where to begin is always a key question. It would be helpful to analyze the watershed information in relation to ecological function and process, and identify zones of high strategic value in terms of such things as groundwater recharge, existing natural forests etc. It makes sense to begin working in locations that can maximize impact with existing net positives in energy and resources available to support our overall strategy. As a general rule, we want to begin at the top of the watershed to foster success in some of our key water harvesting strategies.

However, we might consider taking a different approach in Uzbekistan, given the acute nature of some of the ecological problems downstream. In the case of the Aral Sea Basin, we currently face a large area of former seabed, which is now a source of toxic dust, affecting the broader region during sandstorms (Granit et al., 2010). This represents a major ecological hotspot worthy of immediate attention, so we will require a strategy to identify key ecological hotspots throughout the watershed for preservation or remediation. In the case of the former seabed, the most effective solution would be to stabilize the sediment by covering it with water again. This is, in turn, reliant on enhancing water use efficiency through the watershed in order to restore ecological flows of the two major rivers. A net result will be the dilution of the now concentrated seabed toxins through the rising volume of water.

In addition to restoration of water volume to cover areas of dry seabed, it will be valuable to explore bioremediation solutions featuring both aquatic plants and mycelium. Active soil ecology, whether aerobic or anaerobic, has the capacity to bind toxic elements into inert forms. Mycelium has unique capacities to bioremediate toxic substances, including radioactive elements, in ways which are only beginning to be understood, but hold promise (Stamets, 2005). In the absence of restorative water flows being achieved, the next best option would be to explore contour plantings of salt tolerant species, beginning from the 1960s shoreline, and working downslope on a contour model.

Arguably less crisis-driven than the containment of the toxic dust is consideration of the agricultural production model. It would also be useful to identify the major agricultural and land use systems in operation so that these could be targeted for specific redesign and enhancement. This would initially be done as a proof of concept, to learn what is already working well, and show people what is possible to enhance. Current high-input industrial agricultural models would benefit from redesign to save soils from eventual salinization and loss of productive land, given that such systems have already demonstrated salinization issues (Granit et al., 2010).

These are a few of the obvious, specific and critical pieces that need to be addressed in moving toward sustainability in the highly degraded and unhealthy Aral Sea Basin. There are other components we can add from a general portfolio of restoration strategies which are already applied with success elsewhere. We turn now to water, forests and soil as anchoring features of our permaculture approach.

OF WATER, FORESTS AND SOIL

In the continental climate and dryland environment of the Aral Sea Basin, it makes sense to focus on water harvesting techniques first. Our aim is to slow the flow and infiltration of water. "Keyline design" again based on the original ideas of Yeomans (1981) has become synonymous with water harvesting in the permaculture approach.

Our water harvesting strategies are hands-on manipulations of the land in order to achieve desired alterations in water flow and water infiltration. These include earthworks for rainwater harvesting and infiltration, such as contour swales, check dams, micro-dams, and contour mindful tree planting. Our main goal is to store water in the soil. We do this to support micro-scale biological processes and perennial plant systems that provide long-term stability. Eventually, the interim, patchy or mixed plantings are replaced by a forest, or by a natural grassland root web, returning to the original stability of perennial, soil-building systems.

Our reforestation strategies are intended to create diverse agroforestry systems. In a dryland environment, key concerns are always reducing evaporation, shading soil, and building organic matter. The opposite is often the case as drought and desiccation lead to desertification. However, where reforestation outcomes have been achieved in the past, several things generally occur: rainwater infiltration leads to lower salt concentrations, and the living soil biomass locks some salt molecules into inert, large organic molecules. Organic matter content in the soil increases. Fungal biomass increases, improving the soil's capacity to break down larger, more complex molecules, such as lignins found in wood, and persistent organic pollutants.

In short, a positive feedback loop is started where overall soil life increases as measured by density of bacteria, fungi, and beneficial nematodes, as well as soil macro-arthropods (Proact Network, 2008).

Specific strategies to achieve net positive results include the use of fast growing hardy pioneer legume trees within broader agroforestry strategies. These pioneers are planted in high numbers initially, and used as nurse trees for longer lived and perhaps more delicate trees. The legume trees fix nitrogen, provide shade and wind protection (affecting microclimate) and ultimately can be used as part of a "chop and drop" or coppice mulch system to enhance the soil quality through the increase of organic matter, and promotion of fungal biomass.

Species selection will depend on climate and soil characteristics; however, the beauty of this strategy is that there are always available species with the appropriate characteristics for a given project's needs. A good starting point for selecting plants that will thrive under a given set of conditions (full sun, partial sun, west facing wall, near a pond etc.) is an online database for the temperate zone, *Plants for a Future*.[2] Some of these plants have a wide enough range that they can still be used in a more arid continental climate. The composition of the forest zones to be restored and expanded will depend on their location. It is wise to begin with the principle mentioned earlier, to restore the upper watershed and the steep slopes (over 20 degrees) to native forest first. This maintains a zone of "wilderness function" at the top of the watershed, which supports the work on the more heavily polluted, manipulated, and difficult downstream locations (Mollison, 1988).

The importance of forests and perennial agriculture (using non-annual species to provide for a more stable vegetation regime) can be summarized in the mind-map below (Fig. 1), an excellent summary provided by Toensmeier (Toensmeier, 2012).

The figure portrays the sum total of all the benefits we know to be provided by perennial agriculture. Reviewing our two key points so far, the importance of restoring forests and optimizing water filtration, we hit upon a significant benefit that will accrue in this region just by achieving these two. It should replace some of the precipitation cycle lost with the Aral Sea's inherent water cycle regulatory functions.

We know that once we travel inland from an ocean approximately 100 km, nearly 100% of the rainfall is generated by evaporation and evapotranspiration from trees and vegetation (Mollison, 1988) – except where a large inland body of water or a mountain range is part of the cycle. The Aral Sea Basin is in a landlocked area, far away from an influencing major ocean. One of the key benefits of investing in permaculture in the case

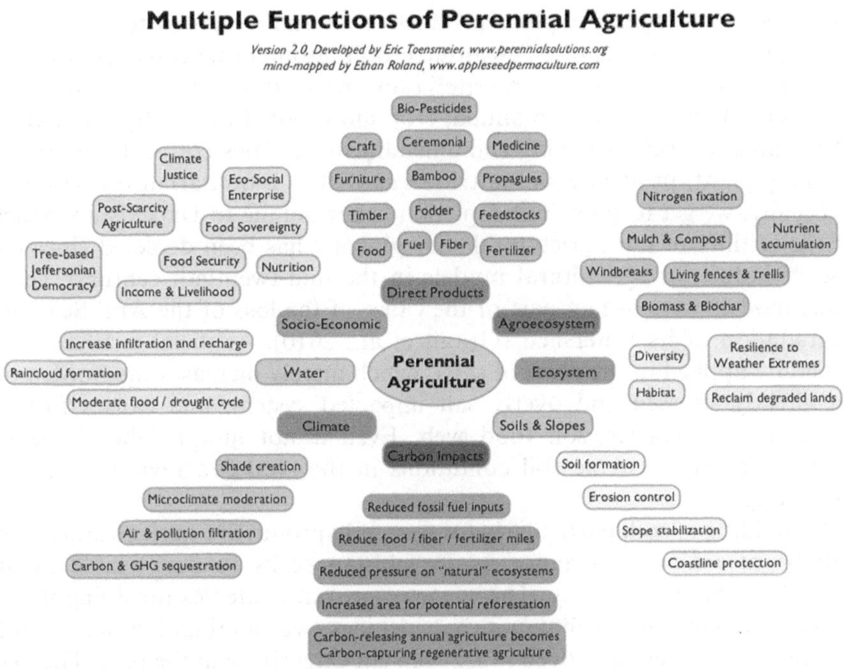

Fig. 1. Benefits of Perennial Agriculture (Toensmeier, 2012). Used by permission of the author.

of the Aral Sea disaster would be to increase rainfall. Therefore, the generation of precipitation through the primary activity of reforestation is of major importance to the general ecological restoration in the Aral Sea Basin.

There are multiple additional benefits of reforestation. As Mollison has pointed out, forests are one of the great climate stabilizers, helping to level out the extremes of temperature, wind etc. which may otherwise be present. Thus forests reduce vulnerability to weather extremes. Perennial agriculture, which may act like a forest, also provides a host of additional direct and indirect benefits.

Thus, our major strategy in parallel to harvesting and managing our water is to re-establish forests and a perennial agricultural regime. In doing so, we are able to exponentially increase the productivity of the landscape in a manner that supports the needs of both human and nonhuman populations. It is thus not the primary focus of this work, but rather more like a subsequent strategy to explore productive alternative crops to cotton. To name just one example of a substitute crop that might be considered with a new appreciation for selection criteria and restoration priorities, industrial hemp could be used. It is an annual crop more suited to the dryland nature of the area, as well as having soil amendment qualities cotton lacks. It is a stacking plant, providing soil nutrition as well as a utilitarian textile fiber.

Finally, we get to soils. Soil depletion is not unique to Uzbekistan. Much of the fertility of our agricultural soils globally has been depleted since the rise of industrial agricultural models in the mid-twentieth century. These same models were in fact part of the cause of the loss of the Aral Sea, and degradation of its watershed (Granit et al., 2010). Artificial fertilizers are generally applied in the form of salts, which merely increases the burden on an already stressed and overly salt impacted system, and causes further harm to the depleted soil food web. Even if not unique, the challenge of restoring the extreme soil conditions in the Aral Sea region is no less rewarding.

Rebuilding of soil, particularly for currently productive agricultural areas, will be of major importance as Uzbekistan seeks to stay viable as an agriculture-based economy. The range of proven strategies for doing this is beyond merely using techniques, but includes two dominant systems, both of which likely were practiced in Uzbekistan extensively in the past. The first is *holistic management*, which includes appropriate inclusion of grazing animals in a rotational system. This mimics natural grazing patterns that aerate soil through hoof action and fertilize naturally with manure. This integrated system promotes land restoration, stabilization, and health of soils (Savory & Butterfield, 1998). The second we call *soil food web enhancement* and uses compost, vermicompost, green manures, compost tea, biofertilizers, and mulching.

PEOPLE DRIVE THE DESIGN AND THE FUTURE THEY SEE

In the Aral Sea Basin watershed restoration, there is one mitigating factor that we need to integrate with our strategy and when determining our initial focus. There are people in our watershed. Many of the not insubstantial number of people are experiencing direct challenges related to the loss of the Aral Sea. They are bearing the brunt of the diminished ecological function, increased pollution and other degradation in the Aral Sea Basin. Thus, we need to consider any additional impacts that will occur to the population as we proceed with our restoration efforts. Based on our social ethics, it would also be necessary to work outward from population centers, incorporating a design process that is twofold. We achieve stakeholder involvement as described earlier in this chapter, and we maximize community resilience through improved local food, shelter, and water access through the choices we make throughout the design and implementation stages.

A design process applied to towns and villages as an early step provides opportunities to identify and develop the human resources to support the regeneration on the broader scale. Human population centers can form restoration hubs (see "access" above), where materials and resources are developed. These are locations of high levels of human ingenuity and human capital – human "forests" so to speak. Elsewhere in the world we have already seen that they become incubators of memes, ideas, and processes that can impact the areas around them in terms of the human systems. Ultimately, Uzbekistan can stand to benefit from these localized cultural incubators from whence new ideas might take seed.

Thus, towns and villages can be seen as highly strategic by such actions as catching water at the top of the watershed, building soil and planting perennial forest systems. Whereas human civilization achieved long-term survival in the past by passing on old knowledge, this is the opportunity to create a regenerative focused culture in the Aral Sea watershed. Fostering this new "culture" will help determine the success or failure of any restoration attempt.

A lot goes into picking the right design processes before anyone is ready to write the plant list or dig the first holes. Appropriately focused design processes, combined with a broad stakeholder engagement, along the lines of the transition town process (Hopkins & Heinberg, 2008) would be an excellent starting point. Parallel to this, specific materials and curriculum could be developed for schools in order to begin mainstreaming the key concepts, linked to hands-on restoration and food production in school gardens and public spaces in towns and cities within the region. These grassroots efforts help build momentum and address pockets of activity that enhance quality of life but are outside the scope and thus missed by large-scale Aral Sea restoration efforts.

CONCLUSION

The broad restoration strategies proposed in this chapter have been proved numerous times on the small scale. In this sense, there is nothing new being proposed, other than the scale, which would be a new frontier for this work. The restoration of the Aral Sea watershed provides a unique opportunity, given the extent of the crisis, the deterioration of the water supply system and the negative feedback loops that are taking hold. Uzbekistan is in a time of transition as it stabilizes its place among independent nations, and the need to address the diminished water flow is not going to go away.

Permaculture is a timely solution as the need for restorative work is apparent. Uzbekistan could choose to continue down the worn pathway of outmoded paradigms from the early part of the last century. But let us hope its leaders see the tipping point and choose to invest resources and ingenuity into restoring the Aral Sea Basin. As a downstream country they already know the dangers of attempting to parcel access to rivers. Seeing the Aral Sea Basin for the bioregion that it is, and investing time, money and human energy into rehabilitating the region, if not the Aral Sea, will do much more than stabilize the ecology, but that result is the necessary starting point to plan for.

NOTES

1. Yeomans originally defined eight in his work.
2. See references for complete URL.

REFERENCES

Ackmansoy, S., & McKinney, D. (1997). *Spatial hydrology of the Aral Sea using Geographic Information Systems (GIS)*, Center for Research in Water Resources, University of Texas. Retrieved from http://www.ce.utexas.edu/prof/maidment/grad/akmansoy/aral/content/pro1e.htm

Design Principles. (n.d.) *Permaculture principles.* Retrieved from http://www.permacultur eprinciples.com/principles.php. Accessed on July 5, 2012.

Granit, J., Jägerskog, A., Löfgren, R., Bullock, A., Gooijer, G., Pettigrew, S., & Linström, A. (2010). *Water intelligence report Central Asia: Baseline report.* UNDP Water Governance Facility at the Stockholm International Water Institute (SIWI), Paper No. 15.

Holmgren, D. (n.d.). *Wikipedia.* Retrieved from http://en.wikipedia.org/wiki/David_Holmgren. Accessed on July 1, 2012.

Holmgren, D. (2002). *Permaculture: Principles and pathways beyond sustainability.* Hepburn, Victoria: Holmgren Design Services.

Hopkins, R., & Heinberg, R. (2008). *The transition handbook: From oil dependency to local resilience.* Devon, UK: Green Books.

Jacke, D., & Toensmeier, E. (2005a). *Edible forest gardens, Vol. 1: Ecological vision and theory for temperate climate permaculture*. Vermont, USA: Chelsea Green.

Jacke, D., & Toensmeier, E. (2005b). *Edible forest gardens, Vol. 2: Ecological design and practice for temperate climate permaculture*. Vermont, USA: Chelsea Green.

Lancaster, B. (2006a). *Rainwater harvesting for drylands, Vol. 1: Guiding principles to welcome rain into your life and landscape*. Arizona, USA: Rainsource Press.

Lancaster, B. (2006b). *Rainwater harvesting for drylands and beyond, (Vol. 2: Water-harvesting earthworks)*. Arizona, USA: Rainsource Press.

Mollison, B. (1979). *Permaculture two: Practical design for town and country in permanent agriculture*. Tasmania, Australia: Tagari Publications.

Mollison, B. (1988). *Permaculture: A designer's manual*. Tasmania, Australia: Tagari Publications.

Mollison, B., & Holmgren, D. (1978). *Permaculture one: A perennial agriculture for human settlements*. Tasmania, Australia: Tagari Publications.

ProAct Network. (2008). *The role of environmental management and eco-engineering in disaster risk reduction and climate change adaptation*. Case Study, Jordan Valley Permaculture Project, Jordan, pp. 98–112. Retrieved from www.proactnetwork.org

Savory, A., & Butterfield, J. (1998). *Holistic management: A new framework for decision making* (2nd ed.). Washington, DC: Island Press.

Stamets, P. (2005). *Mycelium running: How mushrooms can help save the world*. Berkeley, California: Ten Speed Press.

Toensmeier, E. (2012). *Multiple functions of perennial agriculture*. *The Overstory*, Agroforestry ejournal, *247*, June 17.

Worldwide Permaculture Projects. (n.d.). *Worldwide permaculture network*. Retrieved from http://www.permacultureglobal.com/projects. Accessed on July 1, 2012.

Yeomans, P. A. (1981). *Water for every farm/using the keyline plan*. Manly, N.S.W, Australia: Second Back Row Press.

SECTION SIX
LESSONS OF THE ARAL SEA DISASTER: IMPLICATIONS FOR SOCIAL LEARNING

The Aral Sea Disaster was a failure of social learning. Rather than evidencing a society that learned from its experiences and modified its behavior to create outcomes that could be sustained over time, there was a self-destructive element in the Soviet approach that, in this case, proved amazingly effective. Once an idea was advanced from the top, a whole array of social mechanisms were deployed to take the concept at its literal or face value and make it into an unassailable truth, even when many of the actors fully understood the futility, foolishness or destructiveness of the venture. Moreover, it was the nature of the system to employ a rigid system of accountability even for a lax system of feasibility. People lost careers, homes and lives for failure to meet quotas. More often, numbers were doctored. Blame was shifted to the least powerful while the true weakest links were bolstered. Corruption further distorted a system that was based on distortion. Everything in the system was a distortion but for the one crucial ability to deliver the required action, regardless of the outcome. The western concept of gridlock, where a project faces innumerable hurdles, applies here only to the correctives, mitigations and solutions, but not to principle action.

What are the most important lessons of the Aral Sea disaster? How do we prevent replications of such errors and promote, instead, a shift toward world sustainability? After all, our task is not only to think about how to untangle messes already created, but to additionally evade new and bigger ones. The scope of the Aral Sea disaster is so huge. How does it help us anticipate the even greater scale of global climate change? Or does it help us think about climate change and other global problem sets in a different way allowing us to finally meet such challenges?

As Michael Edelstein documents in Chapter 28, "Aral Sea Demise as a Dry Run for Climate Change: From Cumulative to Cascading Impacts," there are important similarities between the Aral Sea disaster and the current crisis of global climate change. Critical points were missed where correctives might have been introduced to stop or slow or reverse the Aral's demise before it was too late. Although the scale, causes and details of the two disasters are hardly identical, people around the world, much like the people of the Aral region, watch helplessly as the climate disaster is allowed to deepen while, everywhere, human behavior continues to exacerbate the conditions of its advance. Where is the global *mahalla* when we need it, the community voice that can change course? The chapter summarizes key elements of the two disasters, in terms of cause, consequence and controllability. Building upon Glantz' famous notion of "creeping environmental disaster," Edelstein adds nuances from his own work for understanding how a disaster of such magnitude was allowed to occur. Connecting the multiple interacting points of the Aral disaster, what Glantz calls hotspots, Edelstein speaks of cumulative and even "cascading" disaster. The role of social learning in preventing and managing cascading disaster is then set forth as necessary for achieving a sustainable society.

Our last word has been saved for Mickey-Aka.[1] Mickey Glantz has been one of the key observers, most prolific and thoughtful writers and important actors over the course of the Aral Sea disaster. His involvement, always forward looking, stretches back to points when solutions might have been applied. Recall Alaya's homage to Glantz's brilliant effort to seek World Heritage status for the remaining sea in a last ditch effort to stop its continued demise. Glantz has indeed been able to approach the Aral situation in the rare integrative manner that is absent from most social thought and social learning. In conversations with Glantz, one detects the bitterness of one who knows that the disaster might have been reversed and good outcomes salvaged if only crucial moments of action were not missed. Having often visited the entire region and worked with the various nations, Glantz has earned the status of a senior researcher and keen observer and commentator. But he is respected, equally, as if an elder statesman. In Chapter 29, "Highlands-to-Sea Cooperation in the Aral Sea Basin: Linking or Sinking?," he weaves these roles together to offer a concept that derives from deep knowledge, wisdom and relationship. Here Glantz advances ideas that resonate with Kamalov's bioregional thinking in Chapter 21. For Glantz, as for Kamalov, the river basin is a whole region that cannot be broken, yet boundaries do just this. However, for Glantz, it is not just the downstream nation(s) that are at the mercy of upstream. Rather there is a mutual interdependence that keeps upstream nations on the hook, as well. It is not as if their success at the expense of downstream can be sustained for long. It is only by realizing the inherent unity and interdependence of the entire bioregion that the river can be managed for common good. Conflict is not a solution. Rather, the choice is for cooperation across the watershed or collective failure. Linking or sinking, says Glantz, are the only two options out there. And we know which one he favors. The challenge to the region is to make the right choice, one that requires recognition of the commonality of interest for which this chapter makes a clarion call.

Michael R. Edelstein
Editor

NOTE

1. Aka is an Uzbek suffix indicating respect for a male elder or wise person.

CHAPTER 28

ARAL SEA DEMISE AS A DRY RUN FOR CLIMATE CHANGE: FROM CUMULATIVE TO CASCADING IMPACTS☆

Michael R. Edelstein

ABSTRACT

This chapter explores the lessons of the Aral Sea disaster for social learning and for achieving future sustainability in the Aral region. Evident is what might be called "the law of delayed response" which states that social response tends to lag behind deteriorating ecological conditions. The bounding of the Aral Sea problems at various scales of consideration from local to global is explored, with a particular emphasis on the importance of a bioregional approach. Not only is the Aral disaster a global problem, part of the overall climate crisis, but it is also a much replicated disaster, sharing a global but decentralized impact. Proposals address the need for sustainable outcomes in the Aral region.

☆I am greatly indebted to Astrid Cerny, PhD for her thoughtful and critical comments on drafts of this chapter.

Disaster by Design: The Aral Sea and its Lessons for Sustainability
Research in Social Problems and Public Policy, Volume 20, 415–441

ISSN: 0196-1152/doi:10.1108/S0196-1152(2012)0000020041

INTRODUCTION

This chapter discusses how the Aral Sea disaster, perhaps the worst single ecological disaster yet to befall the Earth, represents an important dry run for climate change and for other sustainability crises. The disaster underscores a failure of social learning of multiple dimensions. On one hand, the disaster is a microcosm reflecting global processes underway everywhere, on the other it is a macrocosm, a model whose characteristics are increasingly reflected in other events worldwide. When it comes to global climate change, we are all Karakalpaks now.

SLOW DEVELOPING DISASTERS

In my nearly 40 years of work on toxic disasters, I have distinguished between high intensity, short duration acute events and low intensity, long duration chronic events. Chronic events incubate during an extended period of low level and silent activity, perhaps releasing small increments of harm continually until they accumulate sufficiently or change into a more lethal form to be recognized as a hazard. Formal revelation of the calamity and public disclosure may be stalled or may not occur at all. Consider the US case of Love Canal, where contaminants were unknown except to the polluter until accidentally discovered in the process of another investigation. Once made public, the incident became one of the most celebrated controversies in the later 1970s and early 1980s. Then there is the question of how and when the hazard is addressed. The response may be as slow in developing as the chronic condition itself, maybe even slower. Relocation of the local population from the inner rings around Love Canal came only after creative and risky and very public citizen action. Relocation may be necessary, as at Love Canal. In the case of long-term contamination, there may be no cleanup feasible beyond encapsulation, treatment, and long-term abandonment (Edelstein, 2004).

Such chronic events can be easily contrasted with acute disaster, where significant effect, discovery, and public awareness are often encapsulated in one precipitous event. Chernobyl was a major acute event that the Soviets attempted to pretend had not occurred, delaying public disclosure at the most crucial juncture. Eventual relocation of the population and abandonment of a wide area were required.

Another acute disaster is connected to my own life story. I was born and raised in Johnstown, Pennsylvania, the U.S. city historically most associated with natural disaster. There were lots of chronic environmental issues evident there, but none became the basis of community identification the way that the Great Johnstown Flood did. That flood, actually several of them, shaped the psychology of Johnstown. Chronic disaster was of minimal threat in contrast to our collective memory of the horrors of one bleak day in 1889 when 2,000 people were swept away and drowned. A second major flood hit Johnstown in the 1950s just after my birth. I grew up staring at the high water marks left on city hall, reminiscent of the lines parents draw on the wall as their children grow except that these high water marks exceeded even the height of a giant. When the third major flood occurred in 1977, it was the first impetus for my 35 year career working with disaster.

By any metric of acute vs. chronic, the Aral Sea disaster occurred by whimper rather than force. It was manifested not in the rush of waters, but by their ebb and then diminishing tide. It was characterized not by disaster events, but by their absence, by non-events! Rather than an acute event compacted into hours, minutes, and seconds, it stretched more than a generation. There is an important disconnect between human experience of the disaster and its ecological rate of transformation. The Aral disaster was experienced in increments below the threshold for daily or immediate observation, although the fishermen plying its seas must have sensed the "sea change" early on. But they had no way to know if the changes they saw were part of a pattern or part of a fluctuation. Glantz (1999) terms such events "creeping environmental disasters" and suggests that the slow evolution and imperceptible development were key reasons for the long period of inaction that allowed the Aral Sea disaster to advance to so critical a stage before it was recognized. The imperceptible changes were cumulative and eventually reached the point of being visible and recognized as a sign of disaster. Such incremental steps tend to be ignored by government because at any given point other problems are likely to seem more urgent. Unaddressed, these cumulative changes emerge as a system threatening disaster.

It is interesting that the Aral disaster was perceptively a chronic event even if at an ecological scale it was an acute event. When the magnitude of the disaster is contrasted to its interval of occurrence, it happened at lightning speed. At one juncture, Mickey Glantz made a "flip book" out of yearly aerial photographs of the Aral Sea to show the rapidity of its decline. One could thumb through it quickly to make the huge Aral disappear right before one's eyes, an animator's trick to make a chronic event seem acute.

THE DISJUNCTURE BETWEEN RAPID ENVIRONMENTAL DECLINE AND SLOW HUMAN RESPONSE

The failure of social learning in the case of the Aral Sea of course reflects its being a disaster by design. As discussed previously, in Chapter 9, the entire Soviet system implemented policy absent feedback mechanisms that might have afforded the opportunity to take corrective actions. Here I want to examine a facet of this failure of social learning, the lag evident in the Aral disaster. Inspiring an emergent "law of delayed response," we find that the evolution of the ecological disaster caused by human activity advanced at a much more rapid pace than did the social response. We seem to always be learning about our errors after it is too late to correct them. When we make errors at an Aral scale, or even more modestly, this lag represents a failure of social learning that must be generically addressed if we are to address lessons of the Aral demise.

To better examine this question of delay in responding to a slow moving event, Fig. 1 was created to explore the details of the disjuncture between when action is required and when it happens.[1] To the left, the figure matches hierarchies of escalating ecological threat with the corresponding social

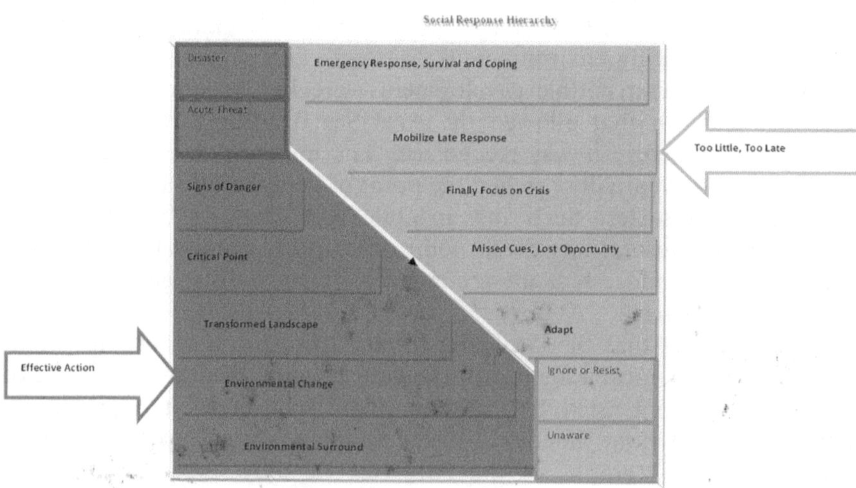

Fig. 1. Disjuncture Between When Action Is Needed and When it Occurs. *Source*: Author.

response for a generic situation, shown to the right (in application, the model would be tailored to specific situations). While the ecological change hierarchy starts off at the bottom at its widest point, showing its broadest range of options, and narrows at the top as options disappear, the social response hierarchy starts off at its most narrow point, reflecting minimal attention paid, and widens as issues gradually start to catch attention and eventually demand action. The point is that full attention is paid only after it is too late for effectual action.

Note that Fig. 1 can be tailored to examine a specific case. What would it look like, for example, for our Aral Sea disaster? To the left, under the hierarchy of ecological changes, we find at the bottom the baseline condition of the Aral Sea as it is systematically altered during Soviet times by water diversion and changes to its ecology. As we move up the left side, environmental crisis intensifies as the system moves toward a critical juncture in its ability to be self-regulating and self sustaining. At this point, signs of ecological distress and water loss begin to multiply. As systems went out of control, problems were apparent. Catch was down, fish died, people were sick, and soil was less fertile. Local people increasingly became focused on survival and coping with the challenges of abnormal daily life. Then positive feedback loops kicked in. With vegetation gone, sand was freed to erode. The sea began to visibly retreat, leaving areas to dry and then blow, adding toxic dusts to the air. The decline of the sea and its vegetation allowed temperatures to soar. Higher temperatures increased water evaporation and salinization and harmed plants. The sea shrank faster; the salt content rose higher; flora and fauna were under greater stress. At some point, the sea could not be sustained and its systems died back, making it even less viable. The overuse of fertilizers and pesticides to compensate for poor soil and poisoned water caused the soil to further decline and fouled the water.

Now if we were watching the right column, I assume that fisherman and other locals noticed anomalies as the landscape became transformed and entered a critical juncture, but were either afraid to speak up, felt helpless, or were unheeded. As I noted earlier, they may have seen what they thought was a periodic cycle or temporary problem, perhaps previously encountered, and mistakenly assumed that conditions would reverse. Presumably, scientists also noticed negative change, but at least some of them viewed the demise of the sea positively. Widespread, global attention and mobilization did not occur over the deteriorating environment until the signs of danger had appeared and the disaster was already moving into an acute phase and then escalated into full blown disaster. At that point, the sea would be lost unless some big, dramatic and costly fix was undertaken. Whatever was done was much too little and much too late. When no serious effort to save the sea materialized, it vanished, and along with it the ecological, physical, social, and economic health of the region.

If further Aral scale disasters are to be avoided, the action arrow on Fig. 1 needs to be moved so that effective action is matched to the time of environmental change and transformation: fixing problems before they get out of hand, restoring ecosystems so that they are self-regulating and self-sustaining and can generate surplus resources, and developing habits and organizational systems for monitoring and corrective action. In short, systems for social learning are required so that it becomes possible to know to act in a timely fashion.

WHAT ARE THE MEANINGFUL BOUNDARIES AROUND THE ARAL SEA DISASTER?

In my work on toxic disaster, I have always been fascinated by the ways that events are bounded by responders and victims and in the mind of observers. As I ponder the significance of the Aral Sea disaster, I begin by examining this perplexing yet significant question. The issue of boundaries is important in many ways. Boundaries may, with some degree of accuracy demarcate a phenomenon that can be described, such as "belongs within a drainage area" or "downwind in an airshed" or "affected by drought" or "inside the plume of contamination." Boundaries may have legal significance or may delineate where certain actions begin and end. Assistance, aid or testing may be made available within a given boundary but not just outside at its margins. Such boundaries are descriptive and useful, so long as they are accurate.

But boundaries are also perceptual, tied to our expectations for what is found within and without. They indicate the space within which certain characteristics or features or people belong, where we are looking for or expecting something – or not surprised when it occurs. In contrast, outside those boundaries we pay little attention because we do not expect to find such occurrences. As humans, we draw boundaries in an effort to understand and in order to limit the scope of our attention to what we can easily grasp. As such, bounding is a form of heuristic device, a form of mental shortcut that allows us to cut through complexity to find what we are looking for. Of course, such heuristics also introduce error into our understanding along with whatever clarity they bring (Kahneman, 2011;

Kahneman, Slovic, & Tversky, 1982). In the end, such boundaries represent the contexts within which phenomenon are understood. They also illustrate the ways that they are not comprehended – the blind spots in our thinking.

Boundaries also play an important role in determining whether one is outside or inside the territory of threat. The psychology of how environmental threats are perceived rests on a remarkably simple axis. If the threat is personal, its importance is magnified. If the threat is not personal, then it may or may not hold some objective interest, but it is not cause for alarm (Edelstein, 2004). We can empathize with others caught in some plight without having to view ourselves as similarly in danger. In fact, it is the fact that the victims and danger are far enough away that maintains a clear boundary between the victims' and observers' realities.

It was interesting to find that none of the Tashkent university students I addressed in 2012 had ever been in far away Karakalpakstan. For them, the Aral Sea story was largely new information, an abstract lesson, and they were able to understand the plight of the peoples of the Aral region without themselves feeling threatened. Given this emotional as well as physical distance, their response was similar to that of my New Jersey students contemplating the same disaster. Both were clearly far away and out danger. The Aral disaster was an abstract, not personal issue. Of course, this distance and abstraction interferes with paying attention and learning from errors. We are more attentive and amenable to new learning when we feel threat.

My immediate interest in boundaries is to view the Aral Sea disaster sequentially as a local event, a regional event, a global event, or as bounded in still other ways. These are all different contexts for social learning. To feed this learning, I have created Fig. 2, a "naan chart." It illustrates that, just as there are many ways to divide the traditional Uzbek bread, there are different logical ways to bound the Aral Sea disaster. It is traditional to rip the loaf all the way across so that one can eat the thinnest and tastiest middle first. The Aral Sea is located conveniently in the low point at the center. We will work our way outward from there.

Viewing the Aral Sea Disaster as a Local Disaster

The Aral Sea is the site of a local, albeit massive, disaster directly affecting the western part of Uzbekistan and Kazakhstan. It is an ecological disaster, but it is every bit as much a social, health, and economic one as well, as discussed in Section 4. A huge sea, vast wetlands, river deltas, forests, productive fields, wells, villages, and people, all have been sacrificed. The

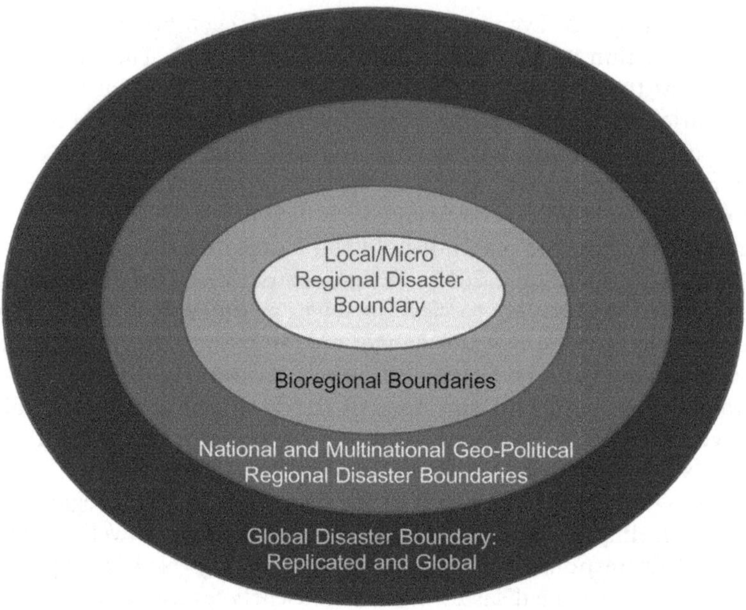

Fig. 2. Some Ways of Bounding the Aral Sea Disaster. *Source*: Author.

biotoxic island of Vozrezhdeniye is invisibly merged into the Aral Kum or Aral desert, as discussed in Chapter 4. A salt sea nearly too saline for its fish, almost fully desiccated; farm fields too salinized for their crops; falling soil productivity suggesting that over time, the more you plant, the less you grow (as discussed in Section 3). To reference a biblical passage that would be understood by most, to look back on this region is to risk being turned into a pillar of salt. The challenge truly, then, is how to look forward.

But maybe the key issue for the Aral Sea as a dry run for climate change is the loss of habitability for a region that was historically a place of verdant forests and wetland systems, blue waters, healthy productive ecosystems, and viable livelihoods. The Aral region was a textbook sustainable environment in many ways until intensive chemical farming and military weapons development was forced on the region. This was before the dazzling, blue gem of the region, the Aral Sea, was taken away.

Here in lies a further challenge, for with the sea gone and the desert spreading, the people of the local region are also scattering to the winds. The communities that made up the Aral region represent the social capital for future restoration activities. And they have borne the brunt of the disaster.

One cannot expect them to endure indefinitely before some turnaround is undertaken. They will go the way of the fish if no action is taken.

Viewing the Aral Sea as a Bioregional Disaster

Central Asia is defined by its two great rivers. The Amu Darya runs 1,578 miles from its origins in Afghanistan's Pamir Mountains, flowing through Tajikistan, Turkmenistan, and Uzbekistan before its original endpoint in the former Aral Sea. The Syr Darya begins in Kyrgyzstan and flows through Tajikistan and Kazakhstan, weaving in and out of Uzbekistan before its 1,370 mile long run ends now in the Little Aral Sea (Glantz, Rubinstein, & Zonn 1993).

Thus, Central Asia is essentially a bioregion, not merely a geopolitical region. Here the watershed drains to a landlocked sea. Such drainages imply that water and whatever the water carries with it moves downstream, drawn by gravity toward their final sink. Watersheds define the relationship of upstream and downstream. Upstream controls the flow of water. Downstream receives wastes and nutrients. And what connects them is the sheer power of the water.

Of course, the Aral Sea was the destination, or sink, for this watershed. But since historical times, there has been a further dimension to this watershed that has been much expanded upon in the past century, with devastating effect. This is the necklace of oases that is supported by the flowing waters of the Amu Darya in both Uzbekistan and Turkmenistan. Very early settlers learned to cultivate the rich alluvial soils and find the natural springs, then to support population clusters along the river's principal and tributary offshoots. Knowledge and appreciation for the provisions and limitation of the bioregion are what made the Silk Road possible. Today, desiccation, dry channels, contaminated springs, and drinking water supplies are just some of the encroachments on this bioregion.

The watershed is just one bioregion defined by the Aral Sea disaster. A second of importance is the airshed. Winds blowing off the Aral Kum blow east toward China and sometimes west over the Mediterranean and into the Middle East and Europe. Toxic dusts are thus distributed widely.

Related to the blowing winds is a third bioregion spreading rapidly in the region, namely desert. The region is dry historically with areas of sizable desert. But living desert has enough flora to minimize the blowing of the substrate. The Aral Kum, as a new desert, blows intensively. Mickey Glantz somewhere referred to it as the biggest succession experiment in the world. It remains to be seen whether plants will colonize and stabilize the barren landscape naturally or if that task awaits human reclamation efforts. What is clear is that drought and other microclimatic changes, combined with the blowing toxic dusts, allow for a different type of succession. Rather than rugged plants moving in and multiplying to create a ground cover, instead, with desertification, it is the desert that spreads.

Viewing the Aral Sea as a National Disaster

As Flavia Alaya astutely observed, in Chapter 20, Uzbekistan has raised the devastation of the Aral region at every opportunity since independence. And, at the same time, it has carefully evaded having its national identity captured by devastation and disaster. My observation during a visit to Tashkent in winter 2012 was that the Aral disaster was far from people's minds and from the national agenda. Uzbekistan has a fast growing and substantial population to educate and to employ. It has an active sustainable development agenda. It has little time to ponder its far western Autonomous Region. But the issue of water is never far from the minds of its leaders and policy makers. And the fate of its rivers is literally the fate of the nation. While the other country surrounding the Aral Sea, Kazakhstan, was able to attract international funding to restore the Little Aral Sea, this project entailed cutting off Syr Darya water from the larger Aral Sea.[2] As an act of triage, to save a part but sacrifice the whole, it was perhaps justifiable. But, from an Uzbek perspective, it was a losing proposition. Sometimes it almost seems that Uzbekistan is waiting for its own turn at international funding to build a big project to finally save the Aral Sea. I think it has a long wait ahead. My concern is that Uzbekistan explores funding for other alternatives for reclamation without waiting for a mega-project.

Viewing the Aral Sea as a Disaster of the Geopolitical Region

Central Asia is an interesting canvas for our assessment because the upstream and downstream issues are relatively clear, interconnecting all the nations of the region. The history of the region as conscripted territories and marginalized peoples of the Soviet Union is held in common. And the lessons of the Aral Sea disaster, although not impacting all equally, is impossible to ignore within the region, even if possible to forget on a global scale. IFAS, the International Fund for Saving the Aral Sea, is jointly run by the Central Asian nations. It has produced many reports but not moved very far toward resolving the Aral disaster or creating a framework for future cooperation.

A number of converging sources emphasize the significance of geopolitics as the near term of greatest importance to the region.[3] At the 2012 Earth Summit in Rio, it was interesting to see Central Asian nation sessions that advanced concepts seeking some kind of common ground. Kazakhstan's Green Bridge Initiative (Ministry of the Environment, 2012), a framework for regional cooperation on green infrastructure, is a good first step at advancing concepts for regional cooperation.

A glacier protection fund initiative from Tajikistan surfaced at that country's session. It was an interesting expression of the growing cooperation between mountain regions of the world who see themselves as keepers of the glaciers and who see common vulnerabilities to climate change among mountain nations. It is obvious why mountain countries want glacier funds. But, why would a downstream country participate? Glacier preservation is in Uzbekistan's long-term interests, however. As a downstream country, Uzbekistan's interests are tied to the integral health of the entire system. This kind of bioregional cooperation is an important step toward finding ways to talk about other obviously contentious but vital issues, such as the Rogun dam. Uzbek support for the Tajikistan glacier fund is a step toward winning their support for a whole system perspective where mountain and valley interests, upstream and down, are integrated.

Viewing the Aral Sea as a Global Disaster

This brings us to the global level of disaster. The Aral Sea became an international cause de celeb, a poster child for what we have done to the Earth. Its international standing has helped to attract international aid and attention. And during perestroika, it was hoped that international pressure and science would help create momentum to save the sea. The issue continues to generate reasonable levels of publicity and coverage, and has attracted some ongoing international assistance.

The Global Climate Meta-Disaster

The Aral Sea disaster is also part of the global climate meta-disaster. Global warming and other climate changes are well documented and have even been the subject of a well deserved Nobel Prize shared by Al Gore, who single-handedly placed climate on the issue map over many years of hard work, with the IPCC, the Intergovernmental Panel on Climate Change, which has organized the largest meta-study ever undertaken by science to confirm the validity of the issue. Since their Nobel Prize, climate change has steadily gained traction as an issue, and the literature on it is now substantial. The United Nations has advanced a series of climate negotiations, beginning in Kyoto, aimed at creating an international agreement for reducing greenhouse gas emissions and, therefore, minimizing the full extent of damage done. Clear tipping points exist for climate, and there is general agreement that the global climate must be stabilized with no more than a 350 ppm contribution of CO_2 to the atmosphere (Hansen, 2009). Despite this attention to the urgency of the topic, a cadre of well-connected issue deniers with media access has blocked global action on climate change.

That's not all. The current generation watches while the disaster is allowed to intensify for Small Island Nations – in their case facing inundation – as it did for the Aral Sea desiccation. The prospects of global sea level rise spell doom, and the large mass of coastal world populations may face catastrophe and relocation. Mountain nations are trying to emulate the Small Island Nations' call for attention, portraying themselves as islands in the sky. As the keepers of the glaciers, they have an important role to play. Climate refugees are becoming a countable demographic, straining the political boundaries of the world, shifting populations to as yet unaffected areas, often rapidly growing urban areas, and tragically affecting the lives of millions. Consider Astrid Cerny's discussion of Aral Region out-migration in this regard (Chapter 14, this volume); although, at least for Kazaks, the prospects of being refugees in Kazakhstan is buffered by that government's efforts to bring its diaspora home. In most other cases, climate refugees do not find a similar warm welcome (Kolmannskog, 2008; Kolmannskog, 2009; McLeman, 2011).

Given the theme of dry run for climate change, some additional exploration of the global nature of the Aral disaster is in order. In particular, I want to say a few things about a new category for considering global events that I term the replicant or replicated disaster.

The Replicant Global Disaster

The understanding of global scale environmental disaster requires a recognition for what I term replicants, or replicated regional disasters. I have long commented in my work that the problem of local or local to regional disasters is magnified worldwide when the same type of disaster is repeated in place after place, region after region, and continent after continent. These "replicants" often have amazing similarities in cause and consequences and their potential for controllability may also share many similarities. In a sense, they are the same disaster repeated in a new place with new victims, possibly on a different continent. The details differ, but the core dynamic is virtually a clone.

Global scale phenomena are, by their nature, hard to address due to complexity, lack of control, and perception of distance. But causes of global scale threats come from some place, and often from the replicants. It is these repeated patterns that we need to address.

Before its desiccation, the Aral Sea was considered to be the fourth largest inland sea, ranking behind its neighbor to the west, the Caspian Sea, by far the largest such water body; Lake Superior, largest of the American Great Lakes; and Lake Victoria, once Africa's largest lake. Recent perusals of the many lists on the Internet of the world's largest inland water bodies found the Aral no longer even listed as a sea on most of them.

When one starts examining lists of large inland water bodies, it is only natural to wonder whether others are desiccating. In this sense, even as a localized event, the Aral Sea disaster has global implications when its key elements are replicated – together or singly – across the globe. In the words of Mickey Glantz (2004), "the future for some places already exists elsewhere on the globe." And in fact, they do. In Chapter 8, Horowitz reviews several lakes in California with strong parallels to the Aral disaster. There, the implications of human activity showed an interesting twist.

Elsewhere, human activity around lakes is showing a familiar pattern. In Africa, Lake Chad is rapidly desiccating, having fallen from its status as Africa's sixth largest lake and lost about one twentieth of its 1963 surface area as of 2004. Shared by Chad, Cameroon, Nigeria, and Niger, Lake Chad has been impacted by diversion of about half its inflow to irrigation. A shallow lake like the Aral and now neighbor to the Sahara desert, it has been very susceptible to hotter and drier weather, severe regional drought, and other facets of climate change. Intensively siphoned off in an area of rapid population growth, there is now conflict around its much shrunken perimeter between residents, herdsmen, farmers, and fishermen, all adversely affected. As with the Aral, a basin commission was formed to attempt to stop destruction of Lake Chad, and diversion of rivers into the lake is seen as a magic solution by some (Glantz, 2004, 2009).

Then there is Iran's Lake Urmia, up until recently the largest salt lake in the Middle East. With 13 tributaries, the lake draws upon diversified sources. However, as many as 39 dams have been built or are planned to be built across these tributaries in order to divert water to irrigation. The shift from subsistence to intensive agriculture that the irrigation waters allowed is similar to the scenario found in Africa at Lake Chad and also, broadly, with the Aral region. Other factors also impacted the lake, including erection of a bridge, whose structure now divides it. A huge local population of 70 million lives within 300 miles of the lake. Health problems familiar in Karakalpakstan are being found here. As sandstorms and dry winds begin to carry desiccated salts and sands across the region, their dispersion airshed may at times overlap windblown dusts from the Aral Kum, reaching Iraq and Turkey and the Caspian countries. The Urmia desiccation is almost contemporaneous with the Aral's, beginning here in the 1970s but not becoming publicized until the 1990s.

A large outmigration ensued, with people particularly moving to the regional city of Tabriz. The array of associated social issues only came to light in 2000. A campaign, "Go to Lake Urmia and Fill it up with our Tears" culminated in 2009 with a demonstration where people each emptied a bottle of water, symbolizing their tears, into the lake. A subsequent rally in Tabriz in 2011 involved thousands of demonstrators protesting government inaction to save the lake. On both occasions, police confronted protesters, leading to arrests and imprisonment, some for long terms; two protesters died. The press was prohibited from printing any news about Lake Urmia. Subsequent large demonstrations in 2012 found hundreds detained. A new tactic of holding protests outside of Iran was then begun. Diversions of other water sources to lake Urmia are being discussed, but the federal government wants the local region to bear the costs of any actions (Ayboga & Ilhan, 2012).

Other bodies of water could be similarly described here, including Lake Lop Nur in Xinjiang, China, now largely dried up, the shrinking Dead Sea, and threats to the Nile and Ganges Rivers (see Nihoul, Zavialov, & Micklin, 2004; Zavialov, 2005). Pictures of rusting fishing fleets sitting on deserts, formally sea bottoms, are becoming hard to identify. There are too many candidates.

A similar replicant analysis could be made based upon locations suffering other common elements from the long list of problems associated with the Aral Sea disaster: desertification, toxic sand storms, water logged soils, unecological agriculture and massive pesticide poisoning, irrigation that is inefficient and unsustainable, and farming for export not for subsistence. There is a long list of elements of the Aral disaster that are replicated across the globe. Impacts are in theory local, but at the scale of replication, they are massive.

There is a considerable social learning agenda associated with how we bound issues extending from the Aral Sea disaster. Boundaries help both to define what we see and what is limited from our consideration. For example, evident in how the Aral disaster occurred is the traditional absence of bioregional approaches reflected in map drawing that assumed nature did not matter. And, too often, it has not. Table 1 illustrates how each of the ways of bounding the Aral disaster generates opportunities for new social learning.

In short, there is an invitation here for a literal explosion of new social learning initiatives capable of invigorating participation, creativity, and

Table 1. Opportunities for New Modes of Social Learning Over Different Boundaries.

Scale of Boundary	Some New Opportunities for Social Learning
Local	Mahalla Permacultural initiative Community focus on restoring habitability Application of community-scale renewable energy Potable water and sanitation initiatives Community public health Regional monitoring and advocacy
National and geopolitical	New Aral initiative to create restorative actions at an appropriate scale Silk Road initiative to promote integration of regional economies Green Bridges to promote regional sustainable development Environmental impact statements Water rights
Bioregional	Joint benchmarks setting Monitoring Independent stewards Highlands to sea thinking Win-win framework created
Global	Policy focus Central Asian nations meet and beat Kyoto IPCC meta-analysis of climate Global leadership
Replicant	Networking with peer victims Cross-linking comparable cases Going after the overarching causes

Source: Author.

future collaborative know-how. In a society preparing for a democratic future in transition from a totalitarian past, these approaches are important proving grounds for broad participation in forging a sustainable future.

Uzbekistan is undertaking a massive expansion of its higher educational system. As I indicated in the chapter I prepared for the conference inaugurating this initiative (Edelstein, 2012), the direction that this educational system proceeds will have a lot to do with whether Uzbekistan prepares itself for a sustainable future. Conventional higher education does not provide opportunities for learning to approach problems from a sustainable mindset. Thus, another type of boundary that we confront in examining the Aral disaster is the division of the disaster between topics such as climate, agriculture, water, chemical contamination, energy, economy, society, and so on, as if they were totally separable fields addressable in the absence of other considerations. Yet, these problems are all connected and people thinking about them need to think *across* not just *within* topical areas of concern. All aspects of sustainability (and the real world) are interdisciplinary by nature. Likewise, there is an overpreparation of technical specialists and an underpreparation of broad generalists.

In sum, our naan has served us well. Each ring reflects a way of bounding our issues of concern. But in reality, a cross-cutting approach allows us to move from the most local to the most global scale, capturing multiple ways of bounding the disaster knowing that we are unwise to ignore any scale if we want to fully understand how to sustainably address the restoration of the region even if we never see restoration of the Aral Sea itself. Each speaks to a different group of stakeholders, potential actors, and takers of responsibility for addressing some portion of the plethora of issues remaining to be addressed. And it reminds us that the Aral Sea disaster belongs simultaneously in each and all of these boundaries and across all of them. It is all one integrated naan.

CUMULATIVE AND CASCADING DISASTER

Cumulative and cascading disasters are multiple, interlocked events, each potent on its own, but in combination, potentially lethal. Not only does one disaster push the next to its tipping point, but interacting in complex ways, their cumulative expression creates a cascade of disasters. In our discussions of the multiple Aral Sea disasters, we see just such a cascade. And in looking beyond the Aral to the broader question of climate change, cumulative and cascading catastrophe may be as important to expect as silent and creeping disaster.[4]

The concept of wicked problems (Rittel & Webber, 1974) is often used to explain the inability of planners to address complex and unpredictable events. In my view, this brilliant analysis has become a frequent excuse for a failure to set up the appropriate social learning mechanisms. Despite their complexity and the unpredictability of the details of the disaster, cumulative and cascading disasters are events that can to a great degree be anticipated. In fact, once the mechanisms for monitoring and modeling are put in place, a regional study center could do a reasonable job of managing such disasters. The principle follows Fig. 1, namely if one monitors changes to baseline ecological conditions as they become transformed and move toward critical levels, one can act in a timely fashion. The real unpredictability occurs when one allows conditions to proceed to levels of acute decline and disaster without acting.

To further explore the dynamics of cumulative and cascading disaster as it relates to the Aral Sea bioregion, I enlisted the assistance of Tobias Siegfried, a physicist and climate scientist at HydroSolutions in Switzerland[5] who works on emerging areas of sustainability that relate to security. He has done a lot of thinking about these issues in Central Asia. And his thoughts are not only interesting for their content but for modeling an important social learning approach. The quotes in the following paragraphs are all from Siegfried.

Within an interactive system, not all of the cumulative environmental stressors have the same weight. Some are "drivers" or "game changers." Climate is one of the "big drivers" in Central Asia. Population pressure is the other. However, "climate change ... should always [be] put ... in context with ... other stressors" such as "religious extremism, economic deprivation because of soil degradation, and large inter-annual and intra-annual variability of the availability of water resources."

Significant climate change is projected for Central Asia (see Siegfried et al., 2011; corroborating Miller et al. (Chapter 5) and Hasanov (Chapter 12)). There are three major climate effects worth watching in the region. The first and most significant involves climate's influence on the movement of water through the bioregional system.

> ... the runoff regimes of these largely glacier and snow-melt driven rivers, the Syr Darya and Amu Darya, will change. The seasonality will change. So the peak runoff will come earlier in the season. Then in the downstream, especially in the unregulated catchment, there will be less water available for irrigation.

Second, climate change may indirectly cause landslide and flooding problems.

> Third, what can be important implications of vanishing glaciers is the destabilization of slopes because of permafrost no longer gluing the loose pebbles together. So we have increasing landslide problems. Also, the retreating glaciers leave behind this unstable terrain, called moraines, behind which you can have some meltwater accumulating. And if these kind of lakes spill over for whatever reason, than you can have devastating flooding in the immediate catchment.

The flooding and landslide issue is one of the dangers Ambassador Askarov warned about in Chapter 6. Siegfried, however, does not believe that glacier melt has the "game changer" status often ascribed to it.

> Often you read these days that, "Oh, the glaciers are gone. There will be one big desert." That's just not going to happen .... This melting is going to happen over the next 50 to 100 years. There are still going to be glaciers, they are so high up there in the Central Asian ranges. But, even today the glacier contribution to runoff is only on the order of 2–3% total runoff ....

Overall, climate impacts will not occur rapidly. Rather its influence is drawn out over a longer period within which another environmental stressor may become the "game changer in the region." Climate change impacts may serve as significant intensifiers of other dynamics even if they are not themselves the drivers. In particular, it may serve to exacerbate social unrest.

> It is my humble opinion that the region is facing even greater problems than climate change. The game changing issues that I foresee are more of the political-economic challenges which can really uproot these current economic systems, as they are. Because, think of it, what you witness now, especially in the downstream, the Aral Sea is polluted and poisoned and it is a toxic dump and there is a lot of salinity accumulating. Over time, the problems have slowly migrated upstream.[6] Now they've already reached very close to the Fergana valley, where you see a lot of soil salinization. And with negative impacts on agriculture, this effects welfare and this can further destabilize the societies there. And they're already fragile, because some population segments, especially the young ones, do not see themselves really in the position of being adequately empowered. So doubt will fill their dreams and wishes. So it can all sort of lead a little bit to a tipping point in which climate change, and its impact on water, might play a role.

But the biggest short-term threat to the region is geopolitical. And it arises less from climate than from the unequal distribution of water in the region.

> Central Asia is probably one of the most fascinating water-stressed regions on the planet with regard to the distribution of water over time and space. It's definitely a region that will see significant challenges coming up. These challenges are not necessarily due to

climate change. I'm not saying that climate is not important, but we have to be open also to political dynamics being much more important. Inter-state relationships will determine whether this region faces a peaceful or conflicting future.

Siegfried views the key game changers as the very large scale dam projects that were the focus of Ambassador Askarov's concern in Chapter 6.

...And, as we see already, now, that the discourse in Central Asia is all about development of new hydropower resources .... The mountain republics, Kyrgyzstan and Tajikistan are now trying to develop to fullest potential their vast hydro-power resources in Central Asia. Kambarata 1 and 2 are the two significant dams in the Syr Darya in Kyrgyzstan. Tajikistan is developing the Rogun dam. It is a big, significant dam on the Vakhsh tributary to the Amu Darya that would turn the Vakhsh cascade into a multi-reservoir cascade .... These are game changers. These are large dams in the Syr Darya and Amu Darya. And whether they are being built or being closed, then there is a real potential for conflict.

While Ambassador Askarov presented Uzbekistan's downstream perspective on the Rogun dam, Siegfried explained the upstream view of both projects. The Kambarata dams lie in Kyrgyzstan on the Syr Darya. Substantial hydropower existed there until the 1970s, when the Soviets closed two dams located just above Uzbekistan, promising the two Kambarata dams as replacements to provide both hydropower and water storage for downstream irrigation during the summer. When the USSR became broke and then folded, the Kambarata twins never were constructed. Kyrgyzstan wants to restore its lost hydropower capacity. But, while within the USSR, these questions occurred within one nation, now the dams will place control over downstream water in one country with a different nation. A similar problem occurs with Kyrgyzstan's neighbor to the south.

Look at Tajikistan, which is currently a poor country. Yet, it has the 8th largest hydropower potential of any country in the world. So it has a totally legitimate reason to develop these resources. But the problem is that the development of these resources upstream makes the downstream perceive a potential threat on their water supply.

And therein lays the potential for conflict, one nation's assets come potentially at the expense of the next.

There is an old Uzbek saying that the five fingers of your hand are all different.[7] It is a colloquialism used to express the inherent inequality in a situation. The bioregional geopolitics of Central Asia is driven by upstream to downstream inequality. The water flows from above. But it brings risk as well as benefit. And its power is the power of the mountains. But its ability to give life is expressed in the fertile fields below. For the five nations of the region, the conditions are indeed different.

Within the bioregion, the problems end up in Uzbekistan. Ambassador Askarov's intense concern over the Rogun dam in Chapter 6 illustrates Siegfried's concerns. They are echoed as well in Chapter 29 by Mickey Glantz. Picking a thread found in Yusup Kamalov's discussion in Chapter 21, Glantz proposes a cooperative concept to govern the entire watershed from Highlands to Sea. Siegfried's analysis gives great urgency to these ideas. In the immediate term, it is the political climate that must change.

I should interject an issue that Siegfried and others have mentioned to me but which has not been greatly stressed in print. Afghanistan is the origin point of the Amu Darya River. If the nation ever emerges from its long period of conflict long enough to function as a peaceful nation, it can rightfully take its not insubstantial cut of the flow of the river off the top for its own agriculture, population, and industry. This future action is always in consideration by those worried about the long-term distribution of the river's waters. Turkmenistan has its guaranteed cut through the Karakum Canal; Kazakhstan has access to the Syr Darya and other water sources. Kyrgyzstan and Tajikistan are upstream countries. Uzbekistan, as the downstream recipient of whatever is left over, has half the entire region's population and an economy currently dependent on irrigated agriculture. A future where the great rivers not only fail to reach the Aral Sea, but are greatly depleted before they ever reach Uzbekistan, is not farfetched to imagine. In our travel across Uzbekistan, we witnessed both the Syr Darya and the Amu Darya at historic low flows for the late spring and early summer. The potential for drought, already intensely experienced in the region in recent years, and all of the uncertainties noted above, makes the potential for cumulative cascading disaster all the more dire.

For our purposes, Siegfried has helped to establish some of the cumulative ingredients that may contribute to cascading impacts. Not all are of equal importance. They each may act on different timelines, asserting more authority at various points. All interact. But at a given moment, one has to pay attention to the stressors that are driving the overall current impact and serving as potential game changers. Thus, the question becomes, what geopolitical world will future glacial melt find as it runs down from the highlands toward the sea and will any of the meltwater reach this endpoint?

SUSTAINABLE OPTIONS

The Aral Sea disaster is a failure of social learning. Rather than evidencing a society that learned from its experiments and modified its behavior to create outcomes that could be sustained over the long term, it is the model of a society that extracts value from place (people and environment) and then leaves them behind as waste. The Soviet Union is long gone. It is high time to decide if Uzbekistan and its neighbors can break from the Soviet model and develop the capacity to effect qualitatively different and better solutions.

As we employ new ways of thinking to restore a sustainable future for the Aral region, we need to apply our thinking globally to the issues of climate, food security, water security, desertification and salinization, health deterioration due to exposure to toxins, and all the other elements of the Aral story. We must stop replicating the Aral disaster elsewhere, adapt active strategies of restoration to cut off and reverse whole system degradation before critical points are missed, and, perhaps most urgently, we must adopt an approach that does not lead us blindly down an Aral path, using new sustainable and attainable endpoints, as well as the means for social learning required to achieve them.

The first place to look for social and cultural capital for long-term sustainability in an ancient society is at the living historical record.

Search for Locally Rooted Solutions

When we examine a case study demonstrating a colossal failure to act in a sustainable manner of the magnitude of the Aral Sea disaster, it is interesting to reflect on the fact that Central Asia is a 4,000 plus years long experiment in sustainable living in an arid environment. Farmers used crop rotation and allowed fields to lay fallow to restore them. Although irrigation was taken from the rivers, it occurred at a scale that the rivers were not depleted.[8] Populations of substantial size were supported. Some communities overused their resources or salinized their soil and failed,[9] littering the desert with their ruins,[10] the shift to unsustainable agriculture is a substantially modern phenomenon, one that risks failure at a scale that prior practice did not.

Sardoba *Sustainability*

The Silk Road was a loose pathway of connectors between oases in the deserts of Central Asia that allowed camel caravans to cross the region connecting various trade routes (Golden, 2011; Liu, 2010). Stops were made at oases. Each oasis contained a permanent spring fed or dug well water source and might have been the basis for a permanent community as well as supporting nomads. The path between oases might be long and a shelter was offered at *caravansarai* (walled campgrounds). These provided some modicum of shelter from sandstorms and attack so that camel caravans could find rest and shelter. A highly developed water storage technology, the *sardoba*, was often located nearby the *caravansarai*. These in-ground stone cisterns stored water so as to keep it cool and minimize evaporation. Gates helped prevent animals from falling in the water. Tolstov (2005, p. 294) defines *sardoba* as "a reservoir for water fed by rain and snow waters gathering in *takyrs* (low areas) from where the water was taken into the reservoirs." A *sardoba* was often located close to the *caravansarai*, although not so close as to be controlled by groups camping there. Rather it was set apart as a kind of commons where all might gain entry. An 1881 account of an expedition in Aral region came across a *sardoba* in the Kizil Kum desert (Morgan, 1881).

> Sardoba Chil-gumbez is a fair type of the conical buildings erected to store snow-water for use in summer. It stands in a small hollow and is carefully built of baked bricks. The steps leading down to the water have all been broken and the descent to it is now very steep. In order to prevent horses, sheep, and camels from falling into it, a low mud wall has been built across the entrance and no cattle are ever admitted inside the enclosure that its waters may not be polluted. We found this excellent at the end of August. Every winter Illebai Turkomans, nomadizing in these parts, fill the cistern to the brim with snow, which thaws and remains pure throughout the summer, an unusual circumstance in open stagnant pools.

Sardoba still dot Uzbekistan. Our joint American-Uzbek team explored one near Navoiy on our way west (see Fig. 3). We cannot know what the surrounding landscape was like back in the heyday of caravans, but today it is orange to ochre fine sandy soil as far as the eye can see. The heat is intense and dry, shimmering in the distance. The glare is stark; the shade, provided only by the *sardoba,* is a distinctive contrast. Arriving at the pace of an air-conditioned minibus on a paved highway, instead of in a weary caravan plodding across the sand, it was nevertheless refreshing and cool inside, a marvelous surprise of human ingenuity and a gift from nature.

Scaling a Sustainable Future for the Aral Region

What is the meaning of sustainable development for Uzbekistan? Does it imply utilizing "green methods" to engage in modern land transforming

Fig. 3. *Sardoba* Water Storage Building. *Source*: Photo by the author.

development and resource extraction? Or might there be another sensibility in *sardoba* sustainability, a different path to follow?

The challenge now is to invent the future. Does the future entail bringing back the lost sea? Or are there other sustainable alternatives? It has been my own position that the path forward for Uzbekistan is not well served by thinking at a large scale, hoping for Siberian river diversions and the like. These are not likely and cannot achieve sustainable outcomes. Rather, buttressed here by Andrew Jones (Chapter 27) and by Michael Wilson (Chapter 10; see also the rest of Section 3), it has been my idea that an ever expanding network of small-scale projects can stabilize the Aral desert substratum, by addressing the toxic wind storm problem, and eventually helping the desert bloom. These generate local livelihoods and sustainable economies along the way. Over time, such remediation activities may allow for restoration of waterways and parts of the Aral Sea, eventually to be interconnected into larger lakes, gradually restoring the fisheries, micro-climate, vegetation, and wildlife of the region.

A key component will be a major shift toward sustainable agricultural practices and permaculture, establishment of water use efficiencies, and capture of water for lake filling. Such small projects offer the potential for experimentation, for failure to be confined to small scale and for learning from what goes right and what does wrong. It is the kind of project scale that could involve the social capital of the Aral region proactively, employing locals and providing hope for a better future in place. The best historic practices can be combined with the best new practices, merging indigenous with global learning. The toxic legacy issues and health issues cannot be erased from the slate and must be addressed, as well. If dam scale funding were put to these tasks, rapid progress could be made.

These ideas of course defy the logic of white gold – of the endless fields of thirsty cotton grown for export. The white gold perspective is deeply ingrained from the top to the bottom of the decision-making and decision reinforcing structure in Uzbekistan. And, yet, in our new relationships with professors of agriculture, several found in this volume, and with officials, alike, there appears to be a readiness for new ideas and approaches. Building on the American cooperative extension model, already present in some form, new ideas can be disseminated quickly and supported. Perhaps the *mahalla*, introduced by Gulyamova in Chapter 22, can be employed as a tool for social networking and participatory decision-making.

My ideas also contradict the concept of scale that governed Soviet thinking and left a lasting imprint in Uzbekistan. Small scale is important for sustainable planning for reasons I outlined above, experimentation, flexibility, ability to modify for local conditions, and minimizing impact. Small scale also allows for modest costs and thus greater feasibility of implementation. Notions of water payment need to be considered along with other elements of Kamalov's proposals (Chapter 21). The potential integration of renewable energy, permaculture, and ecological agriculture, along with elements of water security, can go a long way toward restoration of local self-sufficiency. The reintroduction, as needed, of private garden and orchard plots needs to be encouraged in every way. Uzbekistan already has a strong direct market economy. This needs to be protected and expanded.

In the end, the potential exists to recreate the Silk Road as a connective concept within Central Asia aimed at unifying economies and collaborative inner connections. The concept of the Silk Road, during its shift to the cotton road, became divisive rather than connective. In line with the ideas floated by Kazakhstan's delegates at Rio + 20 earlier this year, the nations of the region need to be looking at common sustainable economic models for mutual benefit. In line with the Tajik proposal described above, Central Asians need to be building bioregional support into their relationships. The rise of the new Silk Road can be based upon this and other similar ideas, creating a common economic and resource zone that will allow a region with strong assets but major problems to advance jointly, thus evading the worst outcomes of geopolitical conflict. The new Silk Road initiative can help push the region ahead on factors of mutual benefit and sustainable advance.

CONCLUSION: LOST MEMORY OF A LOST SEA

In a recent conversation, Mickey Glantz, the climate expert who advocated for the Aral Sea from the time of the earliest recognition of its doom through the end, lamented:

> When my generation is gone, people like Yusup Kamalov and myself, no one will even remember that the sea was there. Young people in the region grew up with the Aral Kum [the Aral desert], not the Aral Sea.

As someone who did not visit the region until 2011 and then, of course, found no sea, I could appreciate the fact that the Aral Sea today is a memory held by those who knew and loved it and saw it slip away. Can those having never witnessed the Aral Sea feel the same way? Can they work as fervently to bring it back, or at least see a restoration of the Aral region?

In February 2011, I took the new high speed train from Tashkent to Samarkand. By luck, I heard a snatch of conversation about the Aral Sea from the row in front and soon found myself engaged in conversation with a remarkable young Karakalpak high school student, Aziza Kaipbergenova, granddaughter of perestroika era Karakalpak leader Tulepbergen Kaipbergenov, and student activist in the Uzbek student organization Kamelot. Aziza explained to me her project to translate her grandfather's works into English so that more would understand what was lost with the destruction of the Aral Sea. Even though she never saw the sea, I came away assured that she knows it in the ways Mickey feared were lost. Perhaps the older generation can rest more easily that the Aral is not forgotten. Aziza offered a ray of hope.

NOTES

1. The figure is inspired by Glantz' Hotspot Pyramid, seen, for example, in Glantz (2007).

2. By all accounts, the Little Aral Sea is thriving. However, at the Rio + 20 Earth Summit in Rio in June 2012, I asked the Kazakhstan government panel if the Aral Sea disaster was over for their people. The response to my question was interesting. The responding official indicated that the Aral Sea disaster was not over and that the Little Aral success did not erase all the problems to be addressed. While this volume has focused on Uzbekistan, there are clearly many impacts to be explored in its

northern neighbor. The official indicated that the failure to find closure on the Aral issue was one factor motivating the Kazakh government to issue a global proposal at Rio for regional cooperation through what they have called the Green Bridge initiative, which seeks to create a green economy for the region (Ministry of Environmental Protection, Kazakhstan, 2012).

3. Including Glantz (Chapter 29), Askarov (Chapter 21), Kamalov (Chapter 6), and Swiss Climate Scientist and Security Expert Tobias Siegfried, cited at length later in this chapter.

4. The triple Japanese disasters of 2011 are an extreme example and, of course, one caused the next.

5. Siegfried's comments come from an interview I conducted with him for the October 2011 Conference on Exchanging the Lessons of the Aral Sea disaster held at Ramapo College of New Jersey. The interview is available on the conference page of our project website www.aralseadisaster.org.

6. Note this counter-intuitive observation of downstream impacts migrating upstream. Siegfried is referring to the occurrence of salinization and waterlogged soils and pesticide contamination arising progressively further and further from the Aral Sea as the same irrigation-intensive cultivation is practiced further afield to make up for falling soil productivity further south.

7. "Besh barmoq bir (xil) emas" (the five fingers [of a hand] are not the same).

8. Although they were redirected. Recall the discussion in Chapter 9 of the Khorezmians diverting the Amu Darya to the Aral Sea from its prior Caspian endpoint.

9. A society can be sustainable in many of its practices but not in others. The latter become Achilles heels undermining the entire social enterprise.

10. Others failed because of attack during one of the many invasions of the region. This, in its own right, demonstrates a key point. Unstable political environments are also unsustainable social and physical environments.

REFERENCES

Ayboga, E., & Ilhan, A. (2012). *Iran's dam policy and the case of Lake Urmia*. July 12 posting on ekopotamya, the web of Tigris and Euphrates. Retrieved from http://ekopotamya.net/index.php/2012/07/irans-dam-policy-and-the-case-of-the-lake-urmia/. Accessed on July 14, 2012.

Edelstein, M. R. (2004). *Contaminated communities: Coping with residential toxic exposure*. Boulder, CO: Westview Press.

Edelstein, M. R. (2012). Educating for sustainability. *Presentation to the international conference on upbringing of an educated and intellectually advanced generation as the most important condition of sustainable development and modernization of the country*. Tashkent, Uzbekistan, February 17.

Glantz, M. H. (Ed.). (1999). *Creeping environmental problems and sustainable development in the Aral Sea basin*. New York, NY: Cambridge University Press.

Glantz, M. H. (2004). *Lake Chad and the Aral Sea: A sad tale of two lakes. Fragile ecologies*. Retrieved from http://www.fragilecologies.com/sep09_04.html. Accessed on August 2011.

Glantz, M. H. (2007). Aral Sea basin: A sea dies, a sea also rises. *Ambio, 36*(4), 323–327.
Glantz, M., Rubinstein, A., & Zonn, I. (1993). Tragedy in the Aral Sea Basin: Looking back to look ahead? *Global Environmental Change*, (June), 174–198.
Golden, P. (2011). *Central Asia in world's history*. New York, NY: Oxford University Press.
Hansen, J. (2009). *Storms of my grandchildren: The truth about the climate catastrophe and our last chance to save humanity*. New York: Bloomsbury.
Kahneman, D. (2011). *Thinking fast and slow*. New York, NY: Farrar, Straus and Giroux.
Kahneman, D., Slovic, P., & Tversky, A. (1982). *Judgment under uncertainty: Heuristics and biases*. New York, NY: Cambridge University Press.
Kolmannskog, V. (2008, April). *Future floods of refugees: A comment on climate change: Conflict and forced migration*. Oslo, Norway: Norwegian Refugee Council.
Kolmannskog, V. (2009). *Climate change, disaster, displacement and migration: Initial evidence from Africa. New issues in refugee research*. Research Paper No. 180. Norwegian Refugee Council, Oslo, Norway.
Liu, X. (2010). *The silk road in word history*. New York, NY: Oxford University Press.
McLeman, R. A. (2011). Settlement abandonment in the context of global environmental changes. *Global Environmental Change, 21S*, 108–120.
Ministry of Environmental Protection, Kazakhstan. (2012, June). *Republic of Kazakhstan: A global proposal Rio + 20*.
Morgan, E. D. (1881). Steppe routes from Karshi to the Amu Daria. Proceeding of the Royal Geographical Society and Monthly Record of Geography. *New Monthly Series, 3*(12), 723–731. Retrieved from http://www.jstor.org/stable/1800735. Accessed on July 1, 2012.
Nihoul, J. C. J., Zavialov, P., & Micklin, P. (Eds.) (2004). Dying and dead seas: Climatic versus anthropic causes. Nato Science Series iv. Earth and Environmental Sciences, Vol. 36. *Proceedings of the Nato advanced research workshop on dying and dead seas – Climatic versus anthropic causes, liege Belgium*. May 7–10, 2003 Norwell, MA: Kluwer Academic.
Noury, V. (2009). A dying lake. *New African, April*, 21–24.
Rittel, H., & Webber, M. (1974). Dilemmas in a general theory of planning. *Policy Sciences* (4), 155–169.
Siegfried, T., Bernauer, B., Guiennet, R., Sellars, S., Robertson, A. W., Mankin, J., … Yakovlev, A. (2011). *Will climate change exacerbate water stress in Central Asia? Climatic change*, doi:10.1007/s10584-011-0253-z. Retrieved from http://water.columbia.edu/files/2011/11/syrDarya_ClimateImpacts_Siegfried.pdf. Accessed on July 12, 2011.
Tolstov, S. P. (2005). *Following the tracks of ancient Khorezmian civilization*. Tashkent, UZ: UNESCO.
Zavialov, P. (2005). The Aral crisis in global perspective. In P. Zavialov (Ed.), *Physical oceanography of the dying Aral Sea* (pp. 115–128). New York: Springer Praxis Books. doi:10.1007/b138791

CHAPTER 29

HIGHLANDS-TO-SEA COOPERATION IN THE ARAL SEA BASIN: LINKING OR SINKING?

Michael H. Glantz

ABSTRACT

Given the failure to save the Aral Sea, new ways of thinking are needed to avert future disasters unleashed by climate change or other events. The threat to glaciers calls attention to the people of mountain regions who are the stewards to these "water towers in the sky." Mountain partnerships are forming across the globe to help build public recognition and support for the unique vulnerabilities and assets of such regions to rapid climate change. A parallel Mountain Coalition would reach down the mountain to form relationships between upstream and downstream countries. Such linkages are a viable approach to pursue common cause and avoid conflict while helping to address and limit negative climate consequences.

INTRODUCTION

Most discussions about the Aral Sea Basin focus on the Aral Sea and the surrounding lowlands where a large portion of the population of the Central

Disaster by Design: The Aral Sea and its Lessons for Sustainability
Research in Social Problems and Public Policy, Volume 20, 443–451

ISSN: 0196-1152/doi:10.1108/S0196-1152(2012)0000020042

Asian states of the former Soviet Union lives. That is likely because of the decimation of the sea following a Politburo decision in the 1950s to almost double the amount of desert soils to be put under irrigated cotton production. The water would come from the two major rivers – the Amu Darya and Syr Darya – that were flowing freely and in great volume into the Aral Seas, Big and Small. Today, the direct and indirect consequences of that flawed economic decision are now visible to the world. Such abuse of the environment always creates more disorder in the long run than the short-term benefits they garner. The task then becomes one of how to sustain livelihoods in the water scarce, environmentally degraded circum-Aral region and to restore the region to some level of reasonable habitability.

But in a world of interacting systems, it is never that simple, and the Aral restoration is no exception. The decimation of the Aral Sea captured attention *worldwide* as early as 25 years ago earning the sea the dubious label of being the world's worst ecological disaster of the twentieth century. Making a bad regional environmental situation worse, today there is mounting concern about climate change and, more specifically, about its impacts on the sources of the water flowing in the Amu Darya and Syr Darya, water resources flowing from the glaciers and snow cover in the basin's surrounding mountains. The threats to the Aral region and other points downstream may no longer come just from water diversions for irrigating agricultural fields. Now, one is forced to look upstream for trouble resulting from upstream diversions, potential glacier lake outbursts (GLOFS) and changes in seasonal stream flow. With the acceleration of glacier melt, downstream countries and peoples must pay considerable attention to the water affairs of upstream states.

THE MOUNTAINS AS GLOBAL COMMONS

Like the more densely populated lowlands, mountain resources (land and water) are subjected to human impacts as well as to natural changes. Both drivers of change are occurring simultaneously and intersecting in obvious as well as not so obvious ways and at different rates of change. A result of increasing levels of fossil fuel burning and CO_2 emissions at the local to national level worldwide has had a cumulative effect on the atmosphere, causing a warming of the global climate regime.

In a world dominated by belief in private property or enclosed by a state's political boundaries, it is easy to forget that the atmosphere is a "global commons" like the high seas, outer space, and Antarctica. Like these commons, mountains ranges, even though they may be located within specific states, are not so politically bound.

The point is that in some important ways, mountain regions can also be viewed as "global commons." After all, mountains are, in essence, water towers that provide an increasingly scarce life-sustaining water supply to downstream nations, regions, and local settlements. As such, the international community and especially governments in the lowlands have a responsibility and a need to keep mountain ecosystems as well as mountain people healthy.

The point is that the regional impacts of global warming on mountain systems worldwide affect not only mountain ecosystems and those people who live in the mountains but also those who live in the highlands and in the lowlands through which a mountain's river waters are expected to flow. Hence, mountains are of worldwide concern and that concern will only heighten as the climate warms, as regional water balances change or degrade, as regional and local climate regimes shift, and as water insecurity as an international issue comes to rival present-day concerns about food insecurity.

The recognition of the importance of mountains is not new. The mountain regions of Asia – the Himalayas, the Hindu Kush, and the Pamirs – have in the past been referred to as "Asia's Water Towers." A subsection of these mountain ranges (the Hindu Kush, Karakoram, and the Pamirs) have been referred to as the "Roof of the World [*Bam-i-Dunya*]." These words not only reflect the majesty of these places, they also capture the life sustaining roles that mountains play. The upstream-downstream unifier, in every way, is the river course itself, as it wends its way to the lowlands. Take as an example what authors Mock and O'Neil (2006) wrote about the Wakhan Corridor as the headwaters of Central Asia's rivers (37°N, 73°E):

> Wakhan has two distinct parts. In Lower Wakhan, the Oxus River, or Amu Darya as it is known in Persian, flows along the valley floor, past villages inhabited by the hospitable Wakhi people. Five thousand metres of vertical relief commands the southern horizon where snow-capped Hindu Kush peaks tower over the valley. The Amu Darya or 'mother river' forms much of Afghanistan's northern border with Tajikistan and Uzbekistan as it courses more than 2400 km across Central Asia to the Aral Sea. Beyond Lower Wakhan rises the Afghan Pamir where … [we can see] rolling grasslands, at more than 3500 m, the domain of the last remaining Kyrgyz nomads.

This description is a romantic one that represents how people tend to see the mountains in general: Beautiful vistas, idyllic settings, snowcapped peaks, dripping glaciers, pastoral people, small happy settlements, a chain of mountain ranges blessing several countries and cultures in the greater

Central Asian region all the way west to the Caucasus. The view was meant to speak of nature and therefore did not present the details of struggles and difficulties faced by mountain people and mountain countries. Yet, the chain of interconnections that flows with the river down the mountainside begins not only with the glaciers but with the lives of mountain people as well. The goods and services provided by mountain ecosystems and inhabitants (their caretakers) benefit not only mountain communities just below the glaciers but also the ecosystems of the highlands and middle reaches of rivers as well as the heavily populated settlements downstream often in neighboring countries. In recognizing this downstream dependency on the mountain regions, it is often overlooked that the relatively sparse populations in the Pamir to the Himalayan Hindu Kush mountains are in essence the "guardians" of these regional "water towers" and as such are the stewards of the stored waters in the melting glaciers.

The peaks of Central Asia share with mountainous regions worldwide at least some similar set of impacts associated with a changing climate: melting glaciers, changes in seasonal characteristics of stream flow, changes in the "normal" (e.g., expected) flow of the seasons along with biological, ecological, and social changes that could foreseeably occur under a warmer global climate regime. The major mountain ranges of Greater Asia are collectively the sources of water that flow from highlands to lowlands to well over 2 billion people. Most of those people dependent on water supplies from mountain regions live in the lower elevations where water is used for agriculture, irrigation, raising livestock, and domestic, municipal, and industrial consumption.

From a natural science perspective, Greater Asian Mountains are important for several reasons. For example, the spatial extent of the snow cover of the Eurasian landmass has long been determined, as a result of its interannual and decadal scale changes in albedo, to influence not only the behavior (robustness) of the Indian monsoon but also the behavior of the atmosphere on a global scale. Comparisons of photos of changes in glaciers from decades ago to the present (their shrinkage) in Greater Asia and elsewhere underscores to the public and to policy-makers that the climate has been warming at the global scale. Such comparisons serve to highlight the increasing rates of ice and snowmelt that are presently underway. These rates appear to be accelerating. With regard to global warming, mountains are not just of interest because they provide photo opportunities for scientists to stand in front of as they seek to highlight climate change. They must be seen in their own right as the highly vulnerable and dwindling sources of fresh water resources along watercourses from the highlands to the lowlands to the sea.

In the absence of a clear threshold of change, policy-makers must be cautious. Here, a concept from the legal profession called "foreseeability" is quite valuable. Foreseeability can be viewed as a qualitative expression of probability: even in the absence of quantitative assessments a person can make responsible and correct decisions by a careful use of existing information and observation. The reality is that most policy-makers make their decisions every day without full information.

ECOSYSTEM GOODS AND SERVICES FOR HUMAN WELL-BEING

In light of the decimation of the Aral Sea it becomes extremely important to reconsider a stated goal of the United Nations' Millennium Ecosystem Assessment issued in 2005: "ecosystems goods and services for human well-being." This goal statement most likely reflected the sentiment of the 1987 Brundtland Commission Report – "Our Common Future" (World Commission, 1987). The Brundtland report sought to encourage industry's "buy in" to caring about environmental health, fostering the belief that exploitation of ecosystems can be profitable without having to sacrifice the well-being of the ecosystems. This reasoning, that ecosystems must provide goods and services to societies, narrowly interpreted as "pay your way or else…," could be used to suggest that if an ecosystem is perceived to have little economic value, then it is expendable (e.g., an ecosystem: "use it or lose it"). There is implied here a one-way relationship whereby all of the care and benefits are granted to humans. And that, even uncared for, these resources will keep flowing. This is a short-sighted, unrealistic perspective.

Turning the Millennium Ecosystem Assessment phrase around to read as "Human goods and services for ecosystems well-being" is, therefore, important. It recognizes an essential fact. Humans need (e.g., depend on) ecosystems; ecosystems do not need (e.g., depend on) humans. Taking the original statement by itself – without reversing the emphasis to one of "human goods and services" – leads to a distorted (anthropocentric) view of how societies must interact with ecosystems, especially in a changing climate situation. Some characteristics of ecosystems just do not lend themselves to a quantification of value in terms of costs and benefits. The Aral Sea was viewed by Soviet leaders as a useless body of water evaporating in a desert, and of much less importance than cotton export. The outcome speaks for itself. How then do we reimagine our relationships with nature so as to not replicate similar avoidable disasters over and over?

H2O: AN INITIATIVE WHOSE TIME HAS COME

The H2O – "Highlands to Oceans" (or "Seas," in this case) – initiative is a concept that emerged in my thinking to improve cooperation and avoid foreseeable conflicts between highlanders and lowlanders, who in fact need each other's resources such as water downstream in summer and energy upstream in winter. It is designed explicitly to connect the people and policies of the highlands and lowlands using stream flow as the common shared resource. The flow of water from the highest elevations starts with seasonal glacial melt. People all along the mountain watercourses rely on that water for food production and water resources, as the glacier's waters wend their way through lower elevations toward the open sea or toward inland depressions.

As an initiative, H2O makes explicit the concrete linkages needed between highland and lowland governments and people so that all can thrive through peace and cooperation at least when it comes to sharing life-giving water resources. H2O highlights as well as strengthens the view that the mountains are the planet's water towers (not just its reservoirs!) and, equally as important, that mountain people are the guardians of those water towers. The H2O initiative also reinforces the need for generating awareness and interest in as well as support for the sustainable development activities that take place upstream. For its part, those upstream must be aware of and sensitive to the dependence of downstream populations on the reliability of receiving their water resources in a timely way to sustain their livelihoods.

Greater Central Asia provides a perfect example of why an H2O initiative is vital. As Fig. 1 indicates, there is a startling divide between upstream and downstream water users. For long-term peace and water security to succeed in the region, a method of collaborative diplomacy is required to bring a balance to conflict-laden interests.

Under so-called "normal" climate conditions of the past, relations between highlands and lowlands were a mix of competition and cooperation. Scientists now agree that for the past few decades that global warming as a result of the burning of fossil fuels and tropical deforestation has been underway, and the warming continues. As the climate warms in Central Asia, the H2O initiative can lead to mutual support between highland and lowland governments and people as they search for ways to cope with a changing climate and its impacts on their societies and ecosystems. This would be a step toward "disaster avoidance diplomacy."

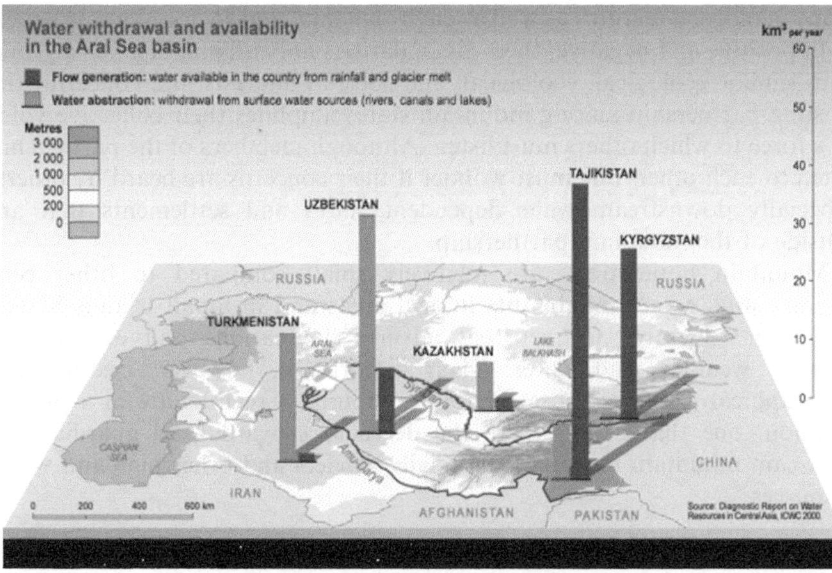

Fig. 1. Water Withdrawal and Availability in the Aral Sea Basin. *Source*: Philippe Rekacewicz, UNEP/GRID-Arendal. http://www.grida.no/graphicslib/detail/water-withdrawal-and-availability-in-aral-sea-basin_85d9

INSIDE THE "*MOUNTAIN BOX*"; OUTSIDE THE "MOUNTAIN COALITION"

The human mind is capable of packaging things differently and it matters what we include in the box and what we leave out. Sometimes not all the parts needed to reassemble a working model are left inside. I propose two ways of packaging common interests, the first enabling the second.

The Mountain Box

My reference here to a "mountain box" is meant to draw attention to the existence of a UN FAO-sponsored mountain partnership among countries with mountains within or across their borders. Such countries have in the past felt neglected in global climate change related negotiations and have supported the partnership. Mountain countries around the globe share many geological features as well as environmental and socioeconomic issues, both positive and negative, their dissimilarities notwithstanding. As a group with similar geological, ecological, and societal interests and concerns, the existing partnership among mountain states amplifies their collective voice as a force to which others must listen. Although members of the partnership listen to each other, one must wonder if their concerns are heard by others, especially downstream water-dependent states and settlements that are outside of the mountain partnership.

Mountain populations are relatively small compared to other eco-regions, and, as are inhabitants in arid areas and in Small Islands States, they are vulnerable to recurrent adverse fluctuations and extremes in climate, weather, and water. Their importance, however, lies in their geographical position: those inhabiting mountain regions are in a pivotal location, one that makes them the de facto stewards and guardians of upstream mountain resources, especially glaciers and other highland water resources.

The Mountain Coalition

A mountain coalition does not yet exist. However, the development of a mountain coalition would help to draw attention to the value of forming a different kind of partnership, a coalition of states with mountain-related interests and concerns such as the equitable sharing of water resources between upstream and the heavily populated downstream states. The strengths derived from a coalition would complement the strengths of the existing (and still growing) mountain partnership. While the partnership represents an "inside strategy" to develop external support for mountain issues, a coalition represents an "outside strategy" to link those outside the mountain box with those inside, specifically downstream countries.

Thus, an H2O initiative seeks to make concrete the linkages between highland and lowland governments and people. It highlights as well as strengthens the view that the mountains are the planet's water towers (not just its reservoirs!) and that mountain people are the guardians of those water towers. Because water as well as pollution flows downhill, there is a need to assure that mountain states and people receive from the international community the levels of support that they need to protect those resources for their own future needs as well as for the needs of downstream states and people. The H2O initiative provides a mechanism for seeking such support.

LOOKING BACK TO LOOK AHEAD

The Aral Sea has been lost. There were opportunities in the 1980s and 1990s when the attention of the world was focused on the shrinking sea and the plight of circum-Aral inhabitants. The time was right for arresting the decline and saving the Aral at half its size of the 1940s. That moment was lost: there was a way but there was no political will to save the sea. The newly independent Central Asian states and Kazakhstan (as they were then called) were coping with the challenges of building new nations. Their leaders were often at political loggerheads and no agreement on a sea-saving strategy ever emerged for the Big Aral. To its credit, Kazakhstan chose to save the Little Aral, fed by the Syr Darya. That smaller body of water was saved by a barrier and has since returned to a viable, productive livelihood-supporting body of water. That opportunity may never come for the Big Aral, which is now referred to as the Aral Kum, the Aral desert, sandwiched between the region's two deserts – the Kara Kum and the Kyzyl Kum. We now need to focus on the next looming water-related pressures: climate change and the likely increase in diversions of Amu Darya water by the government of Afghanistan, a highland country that occupies about 16% of the territory of the Aral Basin. And we cannot afford to compound the errors of the Aral – first, to create the disaster and then to miss the chance to act in a timely way. An initiative such as H2O is a way to go.

> "Water is Life" is a saying in Central Asia. Mountain governments and people in Central Asia are the "Guardians of the region's Water Towers."

REFERENCES

Millennium Ecosystem Assessment. (2005). *Ecosystems and human well-being: Synthesis* (Retrieved from: http://www.millenniumassessment.org/documents/document.356.aspx.pdf). Washington, DC: Island Press.

Mock, J., & O'Neil, K. (2006). *The roof of the world. The lonely planet guide to the middle of nowhere.* Footscray, Australia: Lonely Planet Publications.

World Commission on Environment and Development. (1987). *Our common future.* New York: Oxford University Press.

ABOUT THE AUTHORS

Bakhtiyor Abdullayev is the leader of the Samarkand region branch of eco movement, a green political party in Uzbekistan and professor of Soil Science at Samarkand Agricultural Institute. Dr. Abdullaev earned his Ph.D. in Agriculture in 1990. He held the position of vice dean for academics for several years. He actively participates in the ecological political movement at the regional and national levels.

Mirza Nomman Ahmed is a PhD fellow at the Agricultural Policy and Market Research Institute, Justus Liebig University Giessen, Germany. He does project coordination for the Climate change Network for Central Asia (CLINCA). CLINCA is a cooperative project dedicated to the training of young scientists from Central Asia. In content the project handles socio-economic and environmental consequences of climate change in this region. The research focus is good governance, transition total factor productivity, agriculture and climate change.

Farhod Ahrorov is associate professor in the Agricultural Economics and Management Department of Samarkand Agricultural Institute. He defended his Ph.D. research (completed at Samarkand Agricultural Institute) on the topic of tobacco production and economy before the scientific council at Samarkand Institute of Economics and Service. In 2008–2009 he held a postdoctoral research fellowship in the Human Ecology Department of Free University Brussels, Belgium. His current research theme is "Sustainable development of rural areas of southeastern regions of Uzbekistan with special emphasis on economic and environmental issues." He has published 14 research articles and 18 conference papers. He is coauthor of the textbook, in Uzbek, on "agrarian policy and agricultural markets." He is actively involved with international research and educational projects.

Flavia Alaya is emerita professor of cultural history at Ramapo College, and a scholar, biographer, memoirist and cultural activist with a special interest in the evolution of historic cities. In the 1990s she chaired the City of Paterson (New Jersey) Historic Preservation Commission and was advisor to the National Trust for Historic Preservation, as well as a consultant in the process that later achieved National Park status for Paterson's Great Falls

National Landmark Industrial District. A Columbia University Ph.D., she has received Fulbright, Guggenheim, and Dodge Foundation awards over the course of her career. In 1997, as a Kress Foundation fellow, she took part in an international seminar in Urban and Territorial Conservation at ICCROM/UNESCO in Rome, exploring preservation issues in World Heritage perspective with cultural representatives from around the globe. Active in her local community, she recently helped found the Center for Historic American Building Arts (CHABA), a community partner for urban preservation that emphasizes environmental justice and regional sustainability.

Laylo L. Almatova, MD, graduated from the Samarkand State Institute of Medicine. She has been director of the Samarkand branch of the Uzbekistan Health and Medical Statistics Institute since 2005.

Eshkuvat U. Arzikulov is chairman and head of solid state physics in the Physics Department at Samarkand State University, Republic of Uzbekistan. He is the head scientist for the new scientific laboratory researching microelectronic materials synthesis. This groundbreaking laboratory was organized and funded in 2011 by decree of the President of Uzbekistan. Arzikulov's primary work is in semiconductor physics, nanotechnology, solar energy, and solar water systems. He is a two-time winner of ISTEDOD Foundation award grants with which he traveled to Russia and China (1999, 2006).

Rakhman Ulashevich Arzikulov is a medical doctor and graduate of Samarkand State Institute of Medicine (1964). His main research interest is the fundamentals of a healthy lifestyle, its relationship with philosophy and the spirituality of human beings. He is the author of several books, including the textbook *Healthy lifestyle* in 2005 (in Uzbek).

Murad Askarov has been the Permanent Representative of the Republic of Uzbekistan to the United Nations in New York since February 2009. He has over a decade of experience as a diplomat. Prior to arriving in the United States, he held a senior foreign policy analysis position in the Ministry of Foreign Affairs in Tashkent, after serving as first and second secretary in the same department. Askarov served as attaché and head of chancellery at the Embassy of Uzbekistan to the United States in Washington, DC. He holds a Ph.D. in political science and a master's degree in international relations from the University of World Economy and Diplomacy in Tashkent, Uzbekistan.

Gulmira Boboeva currently holds the position of Senior Lecturer in the "Water Supply and Water Resources Protection and Rational Use" Department at Samarkand State Architectural and Civil Engineering Institute (SSACEI). She graduated from Kiev National University of Construction and Architecture in the field of water resources management. Ms. Boboeva holds a master's degree in civil engineering. In addition to her native Uzbek, she speaks fluent Russian and Persian.

Astrid Cerny is a human and environmental geographer who has a passion for dry land environments and pastoralists. She researches ecological and political adaptation strategies for pastoralists, with a particular interest in Central Asia. Additional work in progress is dedicated to sustainability issues of food security, adaptations to climate change and sociocultural behavior change. Her dissertation research as a Fulbright grant recipient considered the socioecological conditions for Kazak pastoralists in western China. Prior to earning her Ph.D. in geography, she completed an MA in international studies, both at the University of Washington. She has held teaching and consulting positions in China, Germany, the Czech Republic, and Finland. Astrid currently teaches world sustainability at Ramapo College. She provides qualitative research and consulting services to international development organizations for effective project design that promotes social sustainability.

Utkur Djanibekov, a development economist, is currently junior researcher at the Center for Development Research (ZEF), University in Bonn, Germany. He conducts his research in the framework of the project "Opportunities for climate change mitigation and adaptation through afforestation of degraded lands in Central Asia" supported by the Robert Bosch Foundation. Utkur's research interests include ecological and agricultural economics, farm risk management, economics of land degradation, carbon markets, and virtual water analysis. He has participated in several international conferences on environmental management issues, as well as published articles related to that topic.

Gavkhar Salaevna Durdieva is an honors graduate of the Samarkand State Architectural and Civil Engineering Institute. She subsequently achieved academic rank status in the technical sciences. Her major research effort since 1998 has been on the subject: "Structural strengthening of architectural monuments of Khorezm" at the Mamun Academy in Khorezm. She carries out complex scientific tasks to develop methods of physical and biological protection for architectural monuments in Khorezm. A major career achievement for Durdieva was to chair the conference "Reconstruction of architectural monuments of Khorezm and their use" in 2003. The conference led to publication of the conference proceedings and a patent. An accomplished expert in her field, Durdieva has published over a hundred articles, three monographs and three expert opinions on methodology. She is an experienced specialist for conducting scientific experiments in the Southern Aral region.

Michael R. Edelstein, Ph.D., is an environmental psychologist, a sustainability theorist and practitioner. He heads the Institute for Environmental Studies at Ramapo College of New Jersey, where he has taught for 37 years. Edelstein is on the undergraduate faculty of Environment Studies and the graduate faculty of Sustainability Studies. He has practiced social (and psychosocial) impact assessment since the 1970s. Although concerned with all aspects of sustainability, the legacy issues of environmental contamination and degradation are a particular interest. Edelstein's personal research and writing since the late 1970s has focused on the social and psychological impacts of environmental contamination and undesired environmental change. Beyond articles, chapters and lectures, the work has resulted in a number of books, including *Contaminated Communities: Coping with Residential Toxic Exposure* (Westview, 2004), considered a classic in the field, the coauthored Radon's *Deadly Daughters: Science, Environmental Policy and the Politics of Risk* and the coedited *Cultures of Contamination: Legacies of Pollution in Russia and the United States* (Elsevier, 2007).

Abror Gadaev is professor in the "Water Supply and Water Resources Protection and Rational Use" Department at Samarkand State Architectural and Civil Engineering Institute (SSACEI). He furthermore serves as head of the International Relations Department at SSACEI. He's experienced in groundwater resource management, mainly in the water wells rehabilitation and reconstruction field. His research area is the Aral Sea disaster zone and rational use of groundwater in the Central Asian region, an urgent need in the arid zones of Central Asia. In addition to being involved with international and national activities in water resources management for many years, he has published numerous textbooks and manuals for university students on water-related environmental situations. Currently he is an active member of the International Association of Hydrological Sciences, International Gender and Water Alliance, Uzbekistan International Ecological & Sanitation Foundation "Ecosan" and the Global Water Partnership in Uzbekistan.

Ibragim Ganiev is associate professor in the Department of Agricultural Economics and Management, Samarkand Agricultural University, Samarkand, Uzbekistan. He received his Ph.D. from Samarkand Institute of Economics and Service in economics. The author of many books and articles, he has collaborated in international projects with scientists from the United States, Germany, Sweden, Poland, and Russia serving as coordinator of several research projects supported by international funds and scientific organizations. His work focuses on agrarian policy, food production market, and food security in Central Asia.

Debjani Ghatak, Ph.D., is a postdoctoral associate in the Institute of Marine and Coastal Sciences at Rutgers University. Her research focuses on Arctic climate, changes in northern hemisphere snow, and the responsiveness of snow to atmospheric and other cryospheric variabilities. She includes both observational datasets and climate models in her research to understand cryospheric teleconnection, climate feedback and water and energy budgets.

Michael H. Glantz, Ph.D., studies, writes and lectures about how climate, water and weather affect society and how society affects climate, water and weather, especially how those interactions affect quality of life issues. In October 2008 Glantz and his group, with support from the Rockefeller Foundation, established a new international Consortium for Capacity Building (CCB) within the Institute for Arctic and Alpine Research (INSTAAR) at the University of Colorado. Before this initiative, he was a Senior Scientist at the National Center for Atmospheric Research. Glantz is currently working to bridge the activities of meteorological and hydrological science communities with the social sciences and humanities around the globe to increase awareness among political leaders and the general public about the interactions of climate, water, weather, and society. Much of his recent work has focused on how to cope with a changing climate, especially in developing countries. He has also coordinated joint research in the Central Asian Republics of the ex-USSR ever since 1976. Glantz has convened more that 30 international, multidisciplinary meetings and written and edited more than 30 books in his 40 years of climate-related impacts research.

Lola Gulyamova has a Ph.D. from Moscow State University, and over 35 years of extensive experience gained throughout Uzbekistan and in Russia, Estonia, Belarus and the United States. In 1996–1997 she spent a year as a Fulbright Visiting Scholar in the National Center of Geographic Information and Analysis, at the University of California Santa Barbara. She has a wealth of social expert experience with the Asian Development Bank, USAID, and other entities; for ADB this includes holding the position of Social and Gender Expert for ADB's Surkhandarya Water Supply and Sanitation Project, and as Social Expert for the ADB-funded Developing Strategic Urban Assessments Project. From a wider perspective, Lola has many academic achievements; beginning with several research assignments, she later became Department Chair at the National University of Uzbekistan, and is serving as Chair of the Board of the Fulbright Association of Uzbekistan.

Shavkat Hasanov, Ph.D., is head of the international department and lecturer in the agricultural economics and management unit. A member of EAAE (European Association Agricultural Economists), he serves as local coordinator for TOSCA (Erasmus Mundus) and EPASAT (Tempus) projects. Hasanov has more than twenty research and teaching articles in publication. A major current topic of focus is "Problems Associated with the Use of Scarce Resources in the Agriculture of the Zerafshan Valley."

Howard Horowitz is professor of geography at Ramapo College, where he has taught physical geography and natural resources courses since 1982. He earned his doctorate at the University of Oregon, and was a forestry worker on western US public lands during the 1970s. He was employed as an expert witness for the US Environmental Protection Agency during the cancellation hearings of the herbicide 2,4,-T from 1979 to 1981, and has remained active in opposing the overuse of herbicides on public forestlands and lakes. He has led 25 summer and winter field trips to the American West, including to Mono Lake. He is a poet who has published geographical poems in the form of maps.

Andrew Jones is Australian-born and educated internationally. With a background in applied science (ecology) and international development studies, he has traveled and worked extensively in the United States, southeast Asia, the Pacific, Europe, Middle East, Mexico, and Peru. Since taking his permaculture course in 1999, he has had the opportunity to apply permaculture strategies in postconflict and development projects, as well as private sector contexts and for private clients. Working as an educator since 2003, Andrew has led numerous permaculture courses. He currently works as an "ecopreneur," based in Baja California Sur, Mexico.

Yusup S. Kamalov is chairman of the NGO Union for Defense of the Aral Sea and Amudarya (UDASA) in Nukus, Karakalpakstan. In 2010,

Kamalovwas elected Chairman of the Global Water Partnership (GWP) for the Caucasus and Central Asia. Kamalov has been working on the issue of the Aral Sea since 1977.

Ashraf B. Khodjaev has been the Second Secretary of the Permanent Mission of the Republic of Uzbekistan to the United Nations in New York since April 2010. Between 2006 and 2010 he was a scholar analyst at the Center for Political Studies in Tashkent and was covering the Asian regional sector with a primary focus on regional security, political and economic development, energy. During this period he obtained a Ph.D. in economic science on "Factors of energy security in the economic priorities of Uzbekistan and China." Prior to that he served as attaché at the Ministry of Foreign Affairs. He has been the recipient of an international scholarship which he used to study at Peking University in Chinese studies in 2004–2005.

Ramona Lall, Ph.D., holds a doctorate in Environmental Health Sciences from New York University (2008) and currently works for the New York City Department of Health. She previously served as a postdoctoral associate research scientist at NYU's Department of Environmental Medicine. Her research focus is air pollution exposure assessment and environmental epidemiology. Past research projects include air quality assessment for New York City (including a study of air quality at the World Trade Center), and the study of health effects as a result of short-term exposures to particulate matter (PM) air pollution from different "sources" (e.g., traffic, regional sulfate, residual oil, etc.) in NYC. She teaches graduate courses in Environmental Health Science at NYU. Originally from Calcutta Calcutta, India, she is particularly interested in studying environmental and public health issues in Asia.

John P.A. Lamers from the Netherlands, studied agronomy and plant nutrition at Wageningen University, The Netherlands, and agroeconomics (Ph.D.) at Hohenheim University, Germany. He has more than fifteen years of professional experience in long- and short-term missions in Africa as researcher, government advisor, development worker, and consultant. He has worked with international and national agricultural research institutes, extension systems, NGOs and farmers' organizations. Since 1997, he has been working in countries of the former Soviet Union including Georgia, Azerbaijan, Tajikistan and Uzbekistan. His research interests are financial opportunities of farm forestry, alternative approaches to agricultural service provision, crop production and soil fertility. Between 2001 and 2011 he headed the ZEF/UNESCO project in Uzbekistan as lead coordinator. He returned in January 2012 as senior researcher to the Center for Development Research (ZEF) at Bonn University.

William J. Makofske, Ph.D. in Physics, Rutgers University, and professor emeritus of physics, Ramapo College of New Jersey, has worked in energy efficiency and renewable energy as an energy educator and researcher for over forty years. He was a Visiting Scientist at Argonne National Laboratory, the Building Research Establishment in England, and a Fulbright Scholar in Energy and Environment in Germany. At Ramapo, he initiated and ran the Alternative Energy Center which provided hands-on renewable energy education in passive solar, solar hot water, PV, and wind energy for 25 years. He served on two State of NJ Governor's Energy Task Forces; one on grid-connected renewable energy, and the other on renewable energy policy. His self-designed solar house in Warwick, NY, is near Zero Net Energy. He continues to teach in the undergraduate and graduate programs on energy and climate at Ramapo College as adjunct professor. He also consults and develops curricula to train energy auditors for BPI certification.

James R. Miller, Ph.D., is a professor in the Department of Marine and Coastal Sciences at Rutgers University. His research involves the use of global climate models to examine how climate might change in the future as atmospheric greenhouse gases increase, with an emphasis on changes in the hydrologic cycle, the Arctic climate system, and mountainous regions. A primary focus is on climate feedbacks at high altitudes and high latitudes. Of particular interest are water and energy budgets and the interactions and feedbacks among climate variables such as snow, clouds, atmospheric water vapor, radiation, and atmospheric circulation.

Aziz Murtazaev is a software engineer at Samsung Electronics in the Republic of Korea. He holds a master's degree in Computer Engineering from Ajou University, Korea. Before coming to Korea, he completed his Bachelor's degree in economics at National University of Uzbekistan in Tashkent in 2007 and worked in the computer software field for two years. He was born and lived in Karakalpakstan until 2003. He belongs to the Khoja tribe, a distinct branch of the Karakalpaks whose early ancestors first brought Islam to Central Asia. He speaks Karakalpak, Russian, Uzbek, English, and conversational Korean.

Olim Murtazaev is head and professor of agricultural economics and management unit, Samarkand Agricultural University, Uzbekistan. He is the author of several textbooks within the agricultural economics area. His work focuses on agricultural markets, food policy and food security, and resource use economics. He has significant experience in research and educational projects.

Ilkhomjon Niyazov is currently an undergraduate student in the Agricultural Economics Department at Samarkand Agricultural Institute. His major is Agricultural economics. His field of research is ecology. He is involved in economical efficiency issues of the Aral Sea saving projects. He has made three oral and poster presentations at international conferences in the United States and Sweden and has published his first articles. He participated in the Aral Sea expedition with an Uzbek–American team of researchers. In 2012 he received a grant to study for one academic year in Poland at Warsaw University of Technology. In addition to his native Uzbek, he speaks English, Russian, and Tajik.

Kudrat Nurmetov is junior researcher at the Slovak University of Agriculture in Nitra, Slovakia. He holds an M.Sc. in Water Resource Economics from Tashkent Institute of Irrigation and Melioration. His Ph.D. topic is on "Combining water footprint and value chain analyses for increasing water use efficiency and food security issues: Case study Khorezm region, Uzbekistan." His main research interests are food security, international trade, value chain, and water footprint analyses. He is experienced with international projects on water and land resource management and worked in state organizations of Uzbekistan. He published a number of articles on international trade of agriculture products, water use efficiency, and the food processing industry.

Imtiaz Rangwala is research associate in the Department of Marine and Coastal Sciences at Rutgers University. His research focuses on investigating regional climate change with an emphasis on the high elevation regions such as the Tibetan Plateau and Colorado Rocky Mountains. He employs the use of observations and climate models for understanding climatic feedbacks, water and energy budgets, and impacts of climate change on water resources. He is also interested in elucidating if mountain regions are more sensitive to global climate change and what mechanisms lead to that.

Leslie Raucher, MA, was introduced to GIS and map-making while receiving her BA in geology and environmental geosciences from Lafayette College. At the time of the expedition to Uzbekistan, Leslie was a master's student in Sustainability Studies, with Dr. Michael Edelstein at Ramapo College. Other major mapping endeavors have included delineating watersheds, remote sensing with ground checks, and preliminary water quality data collection for a long-term international collaboration. Another project helped determine land use change for the Bergen County Department of Planning and Economic Development. She continues to work with the New Jersey Higher Education Partnership for Sustainability.

Inna Rudenko graduated from Urgench State University (B.Sc.) and the Tashkent Institute of Irrigation and Melioration (M.Sc.) in Uzbekistan before earning her Ph.D. in economics from the Institute for Environmental Economics and World Trade, Leibniz Universität, Hannover, Germany in 2008. She has long-term working experience with international projects such as the ZEF/UNESCO Khorezm project, German Agro Action, and UNDP, as well as local organizations, universities and NGOs. Her specialization is farm economics, pre-assessment and evaluation, and agro-service provision. Her experience is in transitional economics, agricultural economics, research for development, value chains analysis, water footprint analysis, farm level budget and profit modeling.

Saodat Safarova is currently an undergraduate student at Samarkand Medicine Institute. She is interested in the field of public health and healthy lifestyles in Uzbekistan. She is an honors student and a winner of the 2012 Healthy Lifestyle competition, organized by the Kamolot Youth Movement's Samarkand regional committee of the Republic of the Uzbekistan. In addition to her native Uzbek, she speaks English, Russian, and Tajik.

Qamariddin N. Srojev is completing his Ph.D. in semiconductor physics from Samarkand State University. His primary work is experimental investigation of electro physical and photoelectrical properties of silicon-based materials for application with solar panels.

Mukhtor Turaev, Ph.D., is associate professor and chair of the Ecology Department at Bukhara State University, Uzbekistan. His research fields include issues related to ornithological fauna in desert areas. He has published more than 85 scientific works and has participated in several domestic and international research projects, including projects of Birdlife International.

Michael Wilson began his career in horticulture in 1986 at the New Jersey Botanical Garden at Skylands. In 1989, he became the Horticultural Utility Foreman for the Frelinghuysen Arboretum in Morristown, New Jersey and has recently retired from the Morris County Park Commission. Currently, as adjunct professor at Ramapo College, he is teaching sustainable agriculture and native plant landscapes. Michael is also the proprietor of Michael Wilson Environmental Horticulture; a consultation design firm for sustainable landscapes and habitat restoration. Michael has an A.A.S. in agri-business, BA in environmental studies from Ramapo College and an MA in environmental management, Magna Cum Laude, from Montclair University. He also has attended continuing educational courses to supplement his studies. Michael is active in the North American Rock Garden Society and the Native Plant Society of New Jersey.

Zikrilla Yasakov is a graduate of the Samarkand State Architectural and Civil Engineering Institute, majoring in "Security of life activities" with a minor in environmental protection. He is the author of several scientific articles. His main research interest is rational use and protection of water resources. In 2009 he took the 3rd place in a national Kelajak Ovozi competition with his project "cleaning sewage water with solar energy technology."